Understanding General Chemistry

Understanding General Chemistry

Atef Korchef

CRC Press
Taylor & Francis Group
Boca Raton London New York

CRC Press is an imprint of the
Taylor & Francis Group, an **informa** business

First edition published 2022
by CRC Press
6000 Broken Sound Parkway NW, Suite 300, Boca Raton, FL 33487-2742

and by CRC Press
2 Park Square, Milton Park, Abingdon, Oxon, OX14 4RN

CRC Press is an imprint of Taylor & Francis Group, LLC

Library of Congress CataloginginPublication Data

Names: Korchef, Atef, author.
Title: Understanding general chemistry / Atef Korchef.
Description: First edition. | Boca Raton : Taylor and Francis, 2022. |
Includes bibliographical references and index.
Identifiers: LCCN 2021042249 | ISBN 9781032189406 (hardback) | ISBN
9781032189147 (paperback) | ISBN 9781003257059 (ebook)
Subjects: LCSH: Chemistry.
Classification: LCC QD33.2 .K67 2022 | DDC 540--dc23/eng/20211112
LC record available at https://lccn.loc.gov/2021042249

ISBN: 978-1-032-18940-6 (hbk)
ISBN: 978-1-032-18914-7 (pbk)
ISBN: 978-1-003-25705-9 (ebk)

DOI: 10.1201/9781003257059

Typeset in Warnock Pro
by Deanta Global Publishing Services, Chennai, India

Contents

Preface...xi
About the Author..xiii

Chapter 1 Matter...1
    1.1   Objectives...1
    1.2   Chemistry and Matter...1
    1.3   States of Matter..1
    1.4   Types of Matter..1
            1.4.1   Substances..1
            1.4.2   Mixtures..2
            1.4.3   Methods for Separating Compounds and Mixtures...3
                        1.4.3.1   Separation of Compounds...3
                        1.4.3.2   Separation of Mixtures...3
    1.5   Properties of Matter...4
            1.5.1   Physical Properties..4
            1.5.2   Chemical Properties..5
    1.6   Chemical and Physical Changes...5
            1.6.1   Chemical Change...5
            1.6.2   Physical Change...5
    Check Your Reading...6
    Summary of Chapter 1...6
    Practice on Chapter 1...7
    Answers to Questions...10
    Key Explanations...11

Chapter 2 Measurements...15
    2.1   Objectives...15
    2.2   Measurements..15
    2.3   Examples of Physical Properties and their Units...18
            2.3.1   Volume..18
            2.3.2   Temperature...20
            2.3.3   Density and Specific Gravity..21
    2.4   Uncertainty in Measurement...22
            2.4.1   Exact and Inexact Numbers...22
            2.4.2   Significant Figures...22
            2.4.3   Rules Determining the Significant Figures...23
            2.4.4   Rules for Rounding-off Numbers...23
    2.5   Calculations with Measured Numbers..24
            2.5.1   Addition and Subtraction..24
            2.5.2   Multiplication and Division..25
            2.5.3   Exact Numbers in Calculation...25
            2.5.4   Scientific Notation in Calculation...26
    2.6   Accuracy and Precision...27
    Check Your Reading...27
    Summary of Chapter 2...27
    Practice on Chapter 2...29
    Answers to Questions...36
    Key Answer Explanations...38

Chapter 3   Atoms, Molecules, Ions and Moles..43
    3.1   Objectives...43
    3.2   A Brief History of the Atom 43
    3.3   Atomic Structure 44
    3.4   Symbol of an Element 44
    3.5   Isotopes..45
    3.6   Molecules...45
    3.7   Chemical Formulas...46
    3.8   Ions..47
    3.9   Ionic Compounds 47
    3.10   Moles 48
    Check Your Reading...48
    Summary of Chapter 3 49
    Practice on Chapter 3...50
    Answers to Questions 53
    Key Explanations 54

Chapter 4   Calculations in Chemistry 57
    4.1   Objectives...57
    4.2   Atomic Mass (Atomic Weight)...57
    4.3   Molecular Mass (Formula Mass) 58
    4.4   Molar Mass 59
    4.5   Number of Moles 60
    4.6   Number of Atoms, Molecules, or Ions...60
    4.7   Number of Moles, Number of Atoms and Mass of an Element in a Compound.......................61
    4.8   Law of Definite Proportions..62
    4.9   Mass Percent of an Element in a Compound..62
    4.10   Law of Multiple Proportions..64
    4.11   Determining the Empirical and the Molecular Formulas 65
        4.11.1   From Molar Masses 65
        4.11.2   From the Masses of the Constituent Elements of the Compounds 65
        4.11.3   From the Masses of Products 66
    4.12   Balancing Chemical Equations and Stoichiometry 67
        4.12.1   Law of Conservation of Mass 67
        4.12.2   Balancing Chemical Equations..67
        4.12.3   Stoichiometry..68
    4.13   Solutions..70
        4.13.1   Types of Solutions..70
        4.13.2   Mole Ratio and Mole Percent...70
        4.13.3   Molarity...71
        4.13.4   Molality 71
        4.13.5   Solubility...72
        4.13.6   Dilution 72
    Check Your Reading...73
    Summary of Chapter 4 73
    Practice on Chapter 4...75
    Answers to Questions 80
    Key Explanations 82

Chapter 5   Thermochemistry...93
    5.1   Objectives...93
    5.2   Energy, Heat and Work...93
    5.3   System and Surroundings..93
    5.4   Energy Exchange...94

5.5   Internal Energy...96
5.6   Calorimetry...97
        5.6.1   Principles of Calorimetry...97
        5.6.2   Heat Capacity and Specific Heat...98
5.7   Enthalpy..99
        5.7.1   Definition...99
        5.7.2   Enthalpy Changes During Phase Changes – Latent Heat................................99
        5.7.3   Enthalpy of Reaction...102
        5.7.4   Standard Enthalpy of Formation...103
        5.7.5   Hess's Law..104
5.8   Bond Enthalpy..105
5.9   Lattice Energy and the Born−Haber Cycle...107
Check Your Reading..108
Summary of Chapter 5..108
Practice on Chapter 5...112
Answers to Questions..114
Key Explanations..116

Chapter 6   Introduction to Quantum Theory..123
6.1   Objectives..123
6.2   Electromagnetic Radiation – Light..123
6.3   Wave-Particle Duality of Light..124
6.4   Photoelectric Effect...124
6.5   Line Spectra of the Hydrogen Atom..125
        6.5.1   Bohr's Model..125
        6.5.2   Successes and Limitations of the Bohr Model..127
6.6   De Broglie Hypothesis – the Wave-Particle Duality of Matter.......................................129
6.7   Heisenberg Uncertainty Principle..129
6.8   Introduction to Quantum Theory..130
        6.8.1   Atomic Orbital...130
        6.8.2   Quantum Numbers...130
                 6.8.2.1   The Principal Quantum Number..130
                 6.8.2.2   The Angular Momentum Quantum Number....................................131
                 6.8.2.3   The Magnetic Quantum Number...131
                 6.8.2.4   The Electron Spin Quantum Number...132
        6.8.3   Electronic Configuration..133
                 6.8.3.1   The Aufbau Principle...133
                 6.8.3.2   Hund's Rule..134
                 6.8.3.3   Pauli's Exclusion Principle...135
                 6.8.3.4   Electronic Configuration of Atoms and Ions...................................135
                 6.8.3.5   Electronic Configuration of Transition Metals and Transition Metal Cations..........136
                 6.8.3.6   Exceptions in the Electronic Configuration of some Transition Metals.........137
                 6.8.3.7   Isoelectronic Configuration...137
                 6.8.3.8   Valence Shell Electrons..137
                 6.8.3.9   Short Cut for Writing Electronic Configurations.............................138
Check Your Reading..139
Summary of Chapter 6..139
Practice on Chapter 6...142
Answers to Questions..147
Key Explanations..149

Chapter 7   Periodic Table of Elements and Properties of Atoms..157
7.1   Objectives..157
7.2   The Periodic Table of Elements...157

	7.3	Metals, Non-metals, and Metalloids	159
		7.3.1 Alkali Metals	160
		7.3.2 Alkaline Earth Metals	161
		7.3.3 Transition Metals	161
		7.3.4 Halogen Family	161
		7.3.5 Noble Gases	161
		7.3.6 Lanthanides and actinides	161
	7.4	Properties of Atoms	162
		7.4.1 Effective Nuclear Charge and Shielding Effect	162
		7.4.2 Atomic Size	162
		7.4.3 Ionization Energy	164
		7.4.4 Electron Affinity	165
		7.4.5 Electronegativity	165
		7.4.6 Acidic and Basic Trends of Oxides and Hydrides	166
	Check Your Reading		168
	Summary of Chapter 7		168
	Practice on Chapter 7		170
	Answers to Questions		174
	Key Explanations		175

Chapter 8 Chemical Bonding and Molecular Geometry ... 179

	8.1	Objectives	179
	8.2	Chemical Bonding	179
	8.3	Lewis Dot Representations of Atoms	180
	8.4	Lewis Structures	182
		8.4.1 Octet Rule	182
		8.4.2 Rules for Drawing Lewis Structures	182
		8.4.3 Lewis Structures of Ionic Compounds	184
		8.4.4 Formal Charge	185
		8.4.5 Resonance Formulas	186
	8.5	Molecular Geometry	186
		8.5.1 Electron Groups and Molecule Notation	186
		8.5.2 Basic Geometries and Derivatives	187
		8.5.3 Valence Shell Electron Pair Repulsion Theory	187
	8.6	Molecular Orbital Theory	189
	8.7	Orbital Hybridization Theory	192
		8.7.1 Hybridized sp^3 Orbitals	192
		8.7.2 Hybridized sp^2 Orbitals	192
		8.7.3 Hybridized sp Orbitals	193
		8.7.4 Hybridized sp^3d Orbitals	194
		8.7.5 Hybridized sp^3d^2 Orbitals	195
		8.7.6 Hybridized sp^3d^3 Orbitals	195
		8.7.7 Hybridized dsp^2 Orbitals	195
	Check Your Reading		196
	Summary of Chapter 8		196
	Practice on Chapter 8		200
	Answers to Questions		201
	Key Explanations		205

Chapter 9 Intermolecular Forces and Properties of Matter ... 217

	9.1	Objectives	217
	9.2	Molecular Forces	217
		9.2.1 Dipole Moment and Polarizability	217
		9.2.2 Intermolecular Interactions	219

9.2.2.1   van der Waals Forces 219
9.2.2.2   Hydrogen Bond 221
9.2.2.3   Ion–Dipole Interaction 222
9.2.2.4   Ion-induced Dipole Interaction 223
9.3   Properties of Matter..223
9.3.1   Vapor Pressure...223
9.3.2   Boiling Point 226
9.3.3   Viscosity...227
9.3.4   Surface Tension 228
Check Your Reading...228
Summary of Chapter 9 228
Practice on Chapter 9...230
Answers to Questions 235
Key Explanations 236

Chapter 10 Gases 241
10.1   Objectives 241
10.2   Kinetic Molecular Theory of Gases 241
10.3   Boyle's Law 242
10.4   Charles's Law 243
10.5   Avogadro's Law 243
10.6   Gay-Lussac's Law 244
10.7   Ideal Gas Law 244
10.8   Standard Temperature and Pressure 246
10.9   Density of Gases 246
10.10   Diffusion and Effusion 247
10.11   Dalton's Law of Partial Pressures 249
Check Your Reading...250
Summary of Chapter 10 250
Practice on Chapter 10...251
Answers to Questions 254
Key Explanations 255

Index...261

Preface

This book is a student's book and a homework book. It contains the most complete presentation of the fundamentals of general chemistry, which are made easy to understand through step-by-step worked exercises in every chapter and for every topic. The goal of this book is to show the excitement and relevance of chemistry to contemporary issues in a simple but a pedagogically approachable manner. This book uses powerful methods in easy steps through a wide range of topics, with simple and accurate interpretations, and engaging applications to help chemistry students to understand complex principles of general chemistry. This book is a highly reliable, adaptive learning tool for general chemistry, to improve the results of students all over the world, who can use it to understand even the more complex topics of chemistry. Furthermore, it provides all the material needed for chemistry teachers to teach their courses, and to refine their know-how and understanding.

This book covers the following topics: Matter; Measurements; Atoms, Molecules, Ions and Moles; Calculations in Chemistry; Thermochemistry; Quantum Theory and the Electronic Structure of Atoms; Periodic Properties of the Elements; Chemical Bonding and Molecular Geometry; Intermolecular Forces; and Properties of Liquids, Solids and Gases.

The first chapter is dedicated to matter. By the end of this chapter, the student will be able to define the states of matter and compare their properties, differentiate between the types of matter (elements, compounds and mixtures), and differentiate between chemical change and physical change. The second chapter is dedicated to measurements. After completing this chapter, the student will be able to select the appropriate units in the international system (SI) of measurement units, convert units, differentiate between a fundamental quantity and a derived quantity and define and apply the basic methods and tools used in measurements, such as the significant figures and rounding off.

Chapter 3 concerns the study of atoms, ions, molecules, and moles. By the end of this chapter, the student will be able to describe the components of the atom and differentiate between the properties of each component, differentiate between the atom, the molecule and the ion, recognize the isotopes of an element, define the mole, distinguish between the atom, the molecule and the mole, recognize the different types of chemical formulas and define the ionic compounds. Chapter 4 is dedicated to calculations in chemistry. After completing this chapter, the student will be able to distinguish between the molar mass and the atomic or molecular mass, calculate the number of moles of a substance, calculate the number of atoms or molecules in n moles (or in a mass) of a substance, distinguish between the molarity, the molality, and the solubility of a solution, calculate the mass and the mass percentage of an element in a compound, determine the empirical and molecular formulas from their molar masses, from the masses of the constituent elements of the compound and from the products of a chemical reaction, recognize a balanced chemical equation and show the relationships between the number of moles of the reactants and the number of moles of the products. The fifth chapter is dedicated to thermochemistry. On finishing this chapter, the student will be able to define energy, work and heat, distinguish between the system and its surroundings, define the exothermic and endothermic processes, define the internal energy and the enthalpy of a system, perform calculations involving internal energy change and enthalpy change, define the calorimetry and distinguish between specific heat and heat capacity, define and perform calculations on the standard enthalpy of formation, the enthalpy of reaction and bond enthalpy, recognize and apply Hesse's law, define the Born-Haber cycle and calculate the lattice enthalpy change for ionic compounds.

Chapter 6 is an introduction to quantum theory and properties of atoms. After completing this chapter, the student will be able to recognize the wave particle duality of light and the wave particle duality of matter, explain the atomic line spectrum of a hydrogen atom and the photoelectric effect, define the different rules for the electronic configuration and write the electronic configuration of any atom. Chapter 7 is dedicated to the periodic table of elements and the properties of atoms. When this chapter has been completed, the student will be able to explain how elements are arranged in the periodic table of elements, determine the position of an element in the periodic table, recognize the different categories of elements in the periodic table, describe the properties of atoms, such as electronegativity, affinity, ionization energy and atomic radius, and describe how these properties vary along a period or along a group in the periodic table of elements.

The eighth chapter describes the chemical bonds and the different molecular theories that define the geometry of molecules. By the end of this chapter, the student will be able to recognize the different types of chemical bonding, draw the Lewis structure of different atoms and ions, recognize the different molecular geometries and determine the geometry of molecules. Chapter 9 emphasizes the relationship between the intermolecular forces and the properties of matter. On completion of this chapter, the student will be able to identify the different types of interactions between molecules, differentiate between intramolecular forces and intermolecular forces and interpret the properties of liquids and solids on the basis of intermolecular forces. The last chapter (Chapter 10) is entirely dedicated to gases. By the end of this chapter, the student will be able to interpret the temperature and the pressure of a gas, recognize the different gas laws and derive the general law for an ideal gas, calculate the gas density from the ideal gas law, define the partial pressure of a gas in a mixture of gases and describe the relationship between the partial pressure and the total pressure of gases, apply the ideal gas law to balanced chemical equations between gases, define diffusion and effusion of gases and, finally, differentiate between ideal gases and real gases.

This book provides the main objectives at the beginning of each chapter, "Get Smart" sections throughout every chapter, and a "Check your Reading" section at the end of each chapter. Each chapter is filled with examples and practice examples that illustrate the concepts at hand. In addition, a summary and extensive MCQs, exercises and problems, with the corresponding answers and explanations provided, are readily available at the end of every chapter. The availability of these items enables students to customize their personal tools, improving their skill levels and increasing their chances of success. The author recognizes that students appreciate features like these, and he hopes he has reached an appropriate balance between simplicity, clarity and accuracy in the proposed items.

Atef Korchef, M.S., Ph.D.
Associate Professor
Joint Programs, College of Sciences,
King Khalid University, Abha, Saudi Arabia

About the Author

For the past twenty years, Dr Atef Korchef has been teaching both general chemistry and inorganic chemistry at King Khalid University, Abha, Saudi Arabia and at the University of Sfax, Tunisia. He has participated in several training courses, including application of quality standards QM to teaching courses, building courses and student assessment. He earned his M.S. and Ph.D. in inorganic chemistry at the University of Sfax with highest distinction. Dr Korchef was a research scientist at several institutions in Tunisia, including the Institute of Scientific and Technical Research, the Centre of Research and Technology of Water, and the National Centre of Research in Materials Sciences, and, in France, at the Centre of Metallurgical Chemistry Studies. His main areas of research interest are water treatment by precipitation reactions and characterization of nanomaterials. Dr Korchef has published two books and a number of research papers in highly regarded international journals. Awards and honours include the Wiley award for the top downloaded paper in the *Water and Environment Journal*, the School of Chemistry award for outstanding performance at King Khalid University, the Vice Rector's award in recognition and appreciation of contribution to students at King Khalid University and the General Supervisor of Joint Programs award in recognition of his contribution to engineering students at King Khalid University.

Atef Korchef, M.S., Ph.D.
Associate Professor
Joint Programs, College of Sciences,
King Khalid University, Abha, Saudi Arabia

Matter

1

1.1 OBJECTIVES

At the end of this chapter, the student will be able to:

1. Define the states of matter and compare their properties.
2. Differentiate between the types of matter (elements, compounds, mixtures).
3. Differentiate between chemical properties and physical properties.
4. Differentiate between chemical change and physical change.

1.2 CHEMISTRY AND MATTER

Chemistry is the study of matter and the changes that matter undergoes. **Matter** is defined as anything that has mass, occupies space and is made up of particles.

1.3 STATES OF MATTER

The states of matter differ from each other in terms of the **distance** and the **attractive forces** between their constituent particles.

Solids exhibit high attractive forces and very short distances between particles. As a result, the particles are close together and show little freedom of motion. A solid has a fixed volume and shape, and changes very little as temperature and pressure change slightly.

However, the attractive forces in **gases** are significantly lower than in solids and the distance between the particles is markedly greater. In a gas, the particles are randomly spread apart and have complete freedom of movement. A gas sample conforms to both the shape and volume of the container it fills. The volume of a gas varies considerably in response to either temperature or pressure.

The attractive forces in **liquids** are greater than those in gases but lower than those in solids. In a liquid, the particles are close together but not held rigidly in position; these particles are free to move past one another. A liquid sample has a fixed volume but assumes the shape of the container it fills.

1.4 TYPES OF MATTER

Matter is classified as either a **substance** or a **mixture** of substances.

1.4.1 SUBSTANCES

A substance can be either an **element** or a **compound** and has a definite composition and distinct properties.

DOI: 10.1201/9781003257059-1

An **element** cannot be separated into simpler substances by chemical means. All the elements are cited in the periodic table of elements. **The chemical symbol of an element is represented by one capital (uppercase) letter, or a capital letter followed by a small (lowercase) one.**

> **Examples:** hydrogen (H), carbon (C), nitrogen (N), oxygen (O), sulfur (S), iron (Fe), mercury (Hg), chlorine (Cl) and calcium (Ca).

Note that the building up unit (i.e., the smallest unit) of an element that has all the properties of the element is the atom.

A **compound** consists of two or more different elements that are chemically combined in definite ratios. The properties of the compounds are different from the properties of their elements.

> **Examples:** $NaCl$, H_2O, CO_2, $CaCO_3$, $Fe(NO_3)_2$, $Mg_3(PO_4)_2$, $MgC1_4H_{10}O_4$ and $CaSO_4 \cdot 2H_2O$ are formed of different elements, so all of these are compounds.

GET SMART

A MOLECULE OR A COMPOUND?

When two or more atoms of the same or different elements are chemically combined, they form a molecule. A molecule formed by atoms from different elements, such as $NaCl$ and H_2O, is the building up unit (i.e., the smallest part) of a compound that still has all the properties of that compound. On the other hand, a molecule formed of two or more atoms of the same element, such as H_2, O_2 or O_3, cannot be the smallest part of a compound since a compound consists of two or more different elements that are chemically combined in definite ratios.

1.4.2 MIXTURES

A **mixture** is a physical combination of two or more substances (elements or compounds). Substances retain distinct identities. A mixture can be either homogeneous or heterogeneous.

A **homogeneous mixture** is one in which the components are uniformly distributed. The composition of the mixture is uniform throughout

> **Examples:** air, seawater, alloys, sugar dissolved in water, chemical solutions (e.g., aqueous solutions of $NaCl$, $NaOH$, H_3PO_4 or $KMNO_4$, etc.).

A **solution** is a homogenous mixture of two or more substances which are chemically unreactive. A solution is composed of the solute and the solvent. The **solute** is the substance present in the smaller amount and the **solvent** is the substance present in the larger amount. When a liquid and a solid form a solution, the liquid is the solvent whatever the amounts of the substances in the mixture. Water is considered to be the most common inorganic solvent, while benzene is the most common organic solvent. **All solutions are homogeneous mixtures.**

Solubility is the property of a solute (solid, liquid, or gaseous chemical substance) to dissolve in a solid, liquid, or gaseous solvent. In thermodynamics, solubility is a physical quantity denoted by "s", designating the maximum mass concentration of the solute in the solvent, at a given temperature, at which point the solution obtained is said to be "saturated".

A **heterogeneous mixture** is made up of two or more different substances that remain physically separate. No chemical bonds are formed between the mixed substances. Heterogeneous mixtures have more than one phase and the composition of the mixture is not uniform throughout.

Examples:

- Mixture of two or more elements: Fe and Al; Zn and Fe; Fe, Al, and Cu powders.
- Mixture of two or more compounds: $NaCl$ and sugar; $CaCO_3$ and $CaSO_4$ solids; sand and chalk powders in water; $CaCO_3$ and $NaCl$ mixed with sugar; oil and water.
- Mixture of elements and compounds: iron filings mixed with water.

GET SMART

WHAT IS THE DIFFERENCE BETWEEN A MIXTURE OF ELEMENTS AND A COMPOUND?
In a mixture of elements, there are no chemical bonds between the elements. Elements remain separate and retain their distinct identities and can be separated by physical means. In the compound, the elements are chemically combined. New chemical bonds are formed, and the properties of the compound obtained are different from those of the constituent elements. Compounds can be separated into their constituent elements by chemical means.

1.4.3 METHODS FOR SEPARATING COMPOUNDS AND MIXTURES

1.4.3.1 SEPARATION OF COMPOUNDS

Compounds can be separated into their elements by chemical means. For example, mercuric oxide (HgO) can be separated into its components mercury and oxygen by **heat decomposition:**

$$HgO_{(sd)} \xrightarrow{\text{Heat}} Hg_{(g)} + \frac{1}{2}O_{2(g)}$$

Furthermore, water (H_2O) can be resolved into hydrogen (H_2) and oxygen (O_2) by **electrolysis**. Electrolysis consists of splitting water with electricity to produce oxygen and hydrogen (Figure 1.1), according to the following reaction:

$$H_2O_{(lq)} \rightarrow H_{2(g)} + \frac{1}{2}O_{2(g)}$$

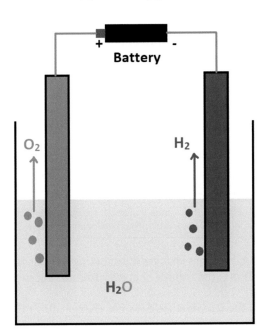

FIGURE 1.1   Electrolysis of water

1.4.3.2 SEPARATION OF MIXTURES

Different methods can be used to separate mixtures into their components by physical means, such as **filtration** or **distillation**. Filtration is used to separate a heterogeneous solid–liquid mixture, e.g., a mixture of water and sand. Distillation is used to resolve homogenous mixtures, e.g., NaCl (table salt) in water, seawater, etc.

1.5 PROPERTIES OF MATTER

1.5.1 PHYSICAL PROPERTIES

Physical properties are the characteristics that can be observed or measured without changing the composition of the substance.

> **Examples:** temperature, color, volume, mass, area, energy, pressure, melting point, boiling point, etc.

Physical properties are either intensive or extensive:

> **Intensive properties** are properties which do not depend on the amount of matter.

> **Example:** temperature, density, boiling point, concentration, solubility.

> **Extensive properties** are properties which depend on the amount of matter.

> **Example:** mass, energy, momentum, volume, number of moles.

GET SMART

HOW TO DETERMINE IF A PHYSICAL PROPERTY IS INTENSIVE OR EXTENSIVE

First, take two beakers containing two identical solutions, then, mix the two solutions in one beaker (Figure 1.2). If, after mixing, the physical property changes and becomes the sum of those in the two solutions, then it is an extensive property. However, if the property remains unchanged, this means that it is an intensive property.

Practice 1.1 Classify each of the following as being intensive or extensive properties: temperature, volume, mass, color, density and solubility.

Answer:

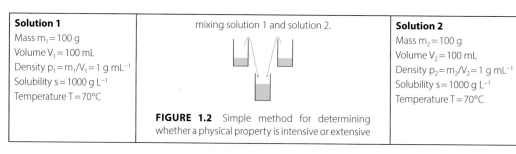

Solution 1	mixing solution 1 and solution 2.	Solution 2
Mass $m_1 = 100$ g		Mass $m_2 = 100$ g
Volume $V_1 = 100$ mL		Volume $V_2 = 100$ mL
Density $\rho_1 = m_1/V_1 = 1$ g mL^{-1}		Density $\rho_2 = m_2/V_2 = 1$ g mL^{-1}
Solubility $s = 1000$ g L^{-1}		Solubility $s = 1000$ g L^{-1}
Temperature $T = 70°C$	**FIGURE 1.2** Simple method for determining whether a physical property is intensive or extensive	Temperature $T = 70°C$

Mass $m = m_1 + m_2 = 200$ g; the mass changes after mixing so mass is an **extensive** property.
Volume $V = V_1 + V_2 = 200$ mL; the volume changes after mixing so volume is an **extensive** property.
Temperature $T = 70°C$; the temperature does not change after mixing, so temperature is an **intensive** property.
Color does not change after mixing the two solutions, so color is an intensive property.
Solubility $s = 1000$ g L^{-1}; the solubility does not change after mixing, so it is an **intensive** property. Note that the dissolved mass changes but also the volume of the solution changes. As a result, the ratio $s = m/V$ remains constant.
Density $\rho = m/V = 200/200 = 1$ g mL^{-1}; the density does not change after mixing so density is an **intensive** property.

GET SMART

THE RATIO OF TWO EXTENSIVE PROPERTIES OF THE SAME OBJECT OR SYSTEM IS AN INTENSIVE PROPERTY

For example, the ratio of an object's mass and volume, which are two extensive properties, is density, which is an intensive property.

Also, concentration, which is the ratio of the number of moles (n) and the volume (V) of a solution (n and V are two extensive properties) is an intensive property.

1.5.2 CHEMICAL PROPERTIES

Chemical properties are the ability of a substance to combine with or change into one or more other substances

Examples: Heat of combustion, enthalpy of formation, toxicity, electronegativity, oxidation.

1.6 CHEMICAL AND PHYSICAL CHANGES

1.6.1 CHEMICAL CHANGE

A chemical change involves making or breaking chemical bonds to create new substances. Chemical changes include the following:

- **Oxidation–reduction (Redox) reactions:** loss or gain of electrons.
  For example, the oxidation of iron (II) to iron (III) by hydrogen peroxide in an acidic medium:

  - $Fe^{2+} \leftrightarrow Fe^{3+} + 1e^-$
  - $H_2O_2 + 2\,e^- \leftrightarrow 2OH^-$
  - Overall equation (Redox reaction): $2Fe^{2+} + H_2O_2 + 2H^+ \rightarrow 2Fe^{3+} + 2H_2O$

- **Reaction of base and acid (neutralization reaction):**

$$HCl_{sol} + NaOH_{sol} \rightarrow H_2O_{lq} + NaCl_{aq}$$

- **Heat decomposition:** $HgO_{(sd)} \rightarrow Hg_{(g)} + \tfrac{1}{2}\,O_{2\,(g)}$
- **Electrolysis:** $H_2O_{(lq)} \rightarrow H_{2(g)} + \tfrac{1}{2}\,O_{2(g)}$
- **Precipitation reaction:**

$$MgCl_2\left(aq\right) + Ca\left(OH\right)_2\left(aq\right) \rightarrow Mg\left(OH\right)_2\left(s\right) + CaCl_2\left(aq\right)$$

- **Cellular respiration** (glucose oxidation to CO_2):

$$C_6H_{12}O_6 + 6O_2 \rightarrow 6CO_2 + 6H_2O$$

- **Other examples:** iron rusting, metabolism of food, burning of wood and all chemical reactions.

1.6.2 PHYSICAL CHANGE

A physical change alters a substance without changing its chemical identity. No new substance is created and no formation of new chemical bonds occurs during a physical change. Physical changes occur when substances are mixed but do not chemically react. Physical changes include the following:

- **Phase changes** such as vaporization, condensation, freezing, sublimation, melting and boiling.

Example: $H_2O_{(lq)} \rightarrow H_2O_{(g)}$

- ○ **Dissolving** sugar or salts in water.
- ○ **Mixing** sand with water.
- ○ **Crushing.**

GET SMART

HOW TO DETERMINE WHETHER A CHANGE IS A CHEMICAL OR A PHYSICAL CHANGE

Ask yourself the following question: What do I have before the change and what do I have after the change?

If you have the same matter before and after the change, then it is a physical change, but, if you have different matters before and after the change, then it is a chemical change.

EXAMPLES

- Combustion of ethane C_2H_6

Before combustion, we have ethane (C_2H_4) but, after combustion, we have carbon dioxide (CO_2) and water (H_2O). We have different matters before and after the change, showing that the combustion of ethane is a chemical change.

- Dissolution of salt

When dissolving salt in water, we have the same matters (salt and water) before and after the change, so the dissolution of salt is a physical change.

CHECK YOUR READING

What is matter and what are the different types of matter?
What is the difference between a mixture and a substance?
How can water and mercuric oxide be separated into their elements?
How can dissolved salts be separated from water?
What is the difference between a chemical property and a physical property?
What is an intensive property and what is an extensive property?
What is the difference between a chemical change and a physical change?

SUMMARY OF CHAPTER 1

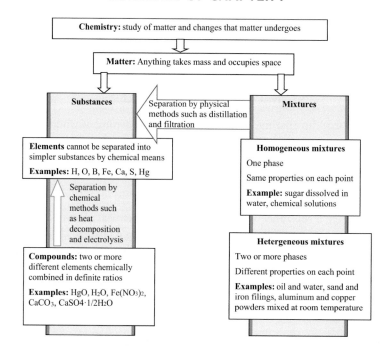

- o **Physical properties:** characteristics that can be observed or measured without changing the composition of the substance.
  - **Intensive properties** are properties which do not depend on the amount of matter
    - **Examples:** Temperature, concentration, density and color.
  - **Extensive properties** are properties which depend on the amount of matter.
    - **Examples:** Mass, volume, energy, momentum and number of moles.
- o **Chemical properties**: the ability of a substance to combine with or change into one or more other substances.
  - **Examples:** Heat of combustion, enthalpy of formation, toxicity and electronegativity.
- o **Chemical change** involves making or breaking chemical bonds to create new substances.
  - **Examples:** Redox reactions, respiration, neutralization and combustion.
- o **Physical change** alters a substance without changing its chemical identity. No new chemical bonds are formed, and no new substances are created. Physical changes occur when substances are mixed but do not chemically react.
  - **Examples:** Phase changes, dissolution processes and crushing.
- o **Solution** is a homogenous mixture of two or more substances which do not chemically react. The solution is composed of the solute and the solvent. The **solubility** is the maximum amount of a solute dissolved in a given volume of a solvent at a given temperature.

PRACTICE ON CHAPTER 1

Q1.1 **Complete the following sentences:**

Matter is defined as anything which has _____ and occupies _____

Matter is either classified as a _____ or a _____ of substances. A substance can be either an _____ or a _____

Compounds can be separated into their elements by _____ methods, such as _____ and _____

Mixture can be separated into their components by _____ methods, such as _____ and _____

Q1.2 **Match the following:**

a. Element   **1.** is a physical combination of two or more substances.

b. Compound   **2.** cannot be separated into simpler substances by chemical means.

c. Mixture   **3.** two or more different elements chemically combined in definite ratios

Q1.3 **Choose the correct answer:**
1. **Zinc (Zn) is**
   a) a heterogeneous mixture
   b) an element
   c) a homogeneous mixture
   d) a compound
2. **Sodium hydroxide (NaOH) is**
   a) a compound
   b) an element
   c) a heterogeneous mixture
   d) an atom
3. **Sodium hydroxide (NaOH) solution is**
   a) a compound
   b) an element
   c) a heterogeneous mixture
   d) a homogeneous mixture
4. **Water (H_2O) is**
   a) a heterogeneous mixture
   b) a compound
   c) a homogeneous mixture
   d) an element

5. **6 g L⁻¹ NaCl solution is**
   a) an element
   b) a compound
   c) a heterogeneous mixture
   d) a homogeneous mixture

6. **Oil, vinegar, and water form**
   a) a heterogeneous mixture
   b) an element
   c) a homogeneous mixture
   d) a compound

7. **A mixture contains**
   a) compounds chemically combined
   b) only one substance
   c) only one element
   d) contains two or more substances

8. **Calcium carbonate ($CaCO_3$) and calcium sulfate ($CaSO_4$) solids form**
   a) a heterogeneous mixture
   b) an element
   c) a homogeneous mixture
   d) a compound

9. **Gypsum ($CaSO_4 \cdot 2H_2O$) is**
   a) a heterogeneous mixture
   b) a compound
   c) a homogeneous mixture
   d) an element

10. **Oxygen (O_2) is**
    a) a compound
    b) a homogeneous mixture
    c) a molecule
    d) a heterogeneous mixture

11. **Blood is**
    a) a compound
    b) an element
    c) a homogeneous mixture
    d) a heterogeneous mixture

12. **A sample composed of isotopes $^{12}_6C$, $^{13}_6C$, and $^{14}_6C$ only is considered to be**
    a) a compound
    b) an element
    c) a homogeneous mixture
    d) a heterogeneous mixture

13. **Water (H_2O) can be resolved to its components oxygen and hydrogen by**
    a) heat decomposition
    b) electrolysis
    c) distillation
    d) filtration

14. **Mercuric oxide (HgO) can be separated into its components mercury and oxygen by**
    a) electrolysis
    b) heat decomposition
    c) distillation
    d) filtration

15. **Seawater can be resolved into its components by**
    a) electrolysis
    b) heat decomposition
    c) filtration
    d) distillation

16. **Sand and water can be separated by**
    a) heat decomposition

    b) electrolysis

    c) distillation

    d) filtration

17. **All the following are chemical properties except**

    a) heat of combustion

    b) electronegativity

    c) pressure

    d) oxidation

18. **Which of the following is a chemical property?**

    a) temperature

    b) mass

    c) pressure

    d) enthalpy of formation

19. **Burning of wood is a**

    a) physical change

    b) chemical change

    c) reduction process

    d) heat decomposition process

20. **Electrolysis of water is a**

    a) physical change

    b) chemical change

    c) distillation process

    d) filtration process

21. **Heat decomposition of mercuric oxide is a**

    a) physical change

    b) chemical change

    c) distillation process

    d) burning process

22. **Dissolving sugar in water is**

    a) a heterogeneous mixture

    b) a compound

    c) a chemical change

    d) a physical change

23. **The transformation of H_2O vapor to liquid (condensation) is**

    a) a chemical change

    b) a physical change

    c) a distillation process

    d) a filtration process

24. **Boiling point is an intensive property because it _____ of the substance**

    a) depends on the amount

    b) depends on the temperature

    c) depends on the solubility

    d) does not depend on the amount

25. **Which of the following is an extensive property?**

    a) temperature

    b) density

    c) melting point

    d) volume

26. **Which of the following is an intensive property?**

    a) temperature

    b) density

    c) concentration

    d) all the above

27. **Solubility is a**

    a) chemical property which depends on temperature

    b) chemical property which does not depend on temperature

    c) physical property which depends on temperature

    d) physical property which does not depend on temperature

28. **If 1 kg of ethanol boils at 78°C, then 500 g of ethanol will boil at**
    a) 156°C
    b) –78°C
    c) 39°C
    d) 78°C
29. **All phase changes are physical changes**
    a) True
    b) False
30. **All chemical solutions are heterogeneous mixtures**
    a) True
    b) False
31. **An atom is a heterogeneous mixture**
    a) True
    b) False
32. **Sand is a homogeneous mixture**
    a) True
    b) False
33. **Heterogeneous mixtures can be separated into their components by physical means**
    a) True
    b) False
34. **Filtration is used to separate the components of a heterogeneous solid–liquid mixture**
    a) True
    b) False
35. **Distillation is used to separate the components of a homogeneous mixture**
    a) True
    b) False
36. **Iron (II) nitrate [Fe(NO₃)₂] is a heterogeneous mixture**
    a) True
    b) False

Q1.4   If different kinds of atoms are represented by different colored dots, match the following pictures on the left, below, with the type of matter on the right, below, which they are most likely to represent

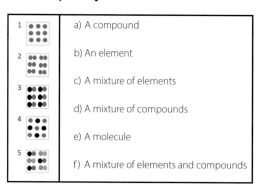

ANSWERS TO QUESTIONS

Q1.1

Matter is defined as anything which has **mass** and occupies **space**
Matter is either classified as a **substance** or a **mixture** of substances. Substance can be either an **element** or a **compound**
Compounds can be separated into their elements by **chemical** methods such as **heat decomposition** and **electrolysis.**
Mixtures can be separated into their components by **physical** methods, such as **distillation** and **filtration.**

Q1.2

(a, 2)

(b, 3)

(c, 1)

Q1.3

1. b
2. a
3. d
4. b
5. d
6. a
7. d
8. a
9. b
10. c
11. d
12. c
13. b
14. b
15. d
16. d
17. c
18. d
19. b
20. b
21. b
22. d
23. b
24. d
25. d
26. d
27. c
28. d
29. a
30. b
31. a
32. b
33. a
34. a
35. a
36. b

Q1.4

1. b
2. e
3. a
4. c
5. d

KEY EXPLANATIONS

Q1.3

1. Zinc (Zn) cannot be separated into simpler substances by chemical means. Zn is cited in the periodic table of elements. The chemical symbol of zinc (Zn) contains a capital letter followed by a small one. Zn is an element.

2. NaOH is formed of the elements Na, O and H chemically combined. The properties of NaOH are different from the properties of Na, O and H. NaOH is a compound.

3. All solutions are homogeneous mixtures.

4. H_2O is formed of the elements H and O chemically combined. The properties of water are different from those of hydrogen and oxygen. H_2O is a compound.

5. All solutions are homogeneous mixtures.

6. Oil, vinegar, and water form distinct phases that remain physically separate.

7. A mixture contains two or more unreacted substances (not chemically combined).

8. Calcium carbonate ($CaCO_3$) and calcium sulfate ($CaSO_4$) solids form a heterogeneous mixture of two compounds.

9. Gypsum ($CaSO_4 \cdot 2H_2O$) is a hydrated compound formed of Ca, S, O and H chemically combined. The properties of gypsum are different from those of the constituent elements. Examples of other calcium compounds: $CaSO_4$, $CaSO_4 \cdot 1/2H_2O$, $Ca(OH)_2$.

10. Oxygen (O_2) is formed of two atoms of the same element. O_2 is not a compound, although it is a molecule. Other examples of molecules which are not compounds: N_2, Cl_2, H_2 and O_3.

11. Blood is a heterogeneous mixture because the blood cells are physically separate from the blood plasma. The blood cells have properties different from those of the plasma. The cells can be separated from the plasma by centrifugation.

12. A mixture of isotopes is very uniform since isotopes of the same element are chemically indistinguishable from one another.

13. Water (H_2O) is a compound which can be resolved to its components oxygen and hydrogen by electrolysis.

14. Mercuric oxide HgO is a compound which can be separated into its components mercury and oxygen by heat decomposition.

15. Seawater is a homogeneous mixture of water and various salts. Seawater can be resolved into its components by distillation.

16. Sand and water form a heterogeneous mixture which can be separated by filtration

17. Heat of combustion, electronegativity and oxidation are chemical properties, but pressure (= force/surface area) is a physical property.

18. Temperature, mass and pressure are physical properties; however, enthalpy of formation is a chemical property.

19. During burning of wood, new substances are created, such as CO_2, so burning of wood is a chemical change.

20. Electrolysis of water is a chemical change since bonds in water are broken down and new substances, namely H_2 and O_2, are formed.

21. Heat decomposition of mercuric oxide (HgO) is a chemical change since bonds are broken down, HgO disappears and new substances, Hg and O_2, form.

22. Sugar dissolved in water forms a homogeneous mixture which can be separated easily by physical means such as distillation. The dissolution of sugar in water is a physical change.

23. The transformation of H_2O vapor to liquid called condensation is a phase change and all phase changes are physical changes.

24. An intensive property is a property that does not depend on the amount of the substance.

25. An extensive property is a property that depends on the amount of the substance. Volume depends on the amount of matter, so it is an extensive property. Temperature, density and melting point are intensive properties because they do not depend on the amount of the substance.

26. Density is the ratio between mass and volume ($d = m/V$); mass and volume are extensive properties. The ratio of two extensive properties of the same object or system is an intensive property. Concentration and temperature are intensive properties because they do not depend on the amount of the substance.

27. Solubility is the analytical composition of a saturated solution expressed as a concentration of a designated solute in a designated solvent at a given temperature. The solute is the substance present in the smaller amount and the solvent is the substance present in the higher amount. The solubility of a substance is a physical property which fundamentally depends on the physical and chemical properties of the solute and solvent, as well as on the temperature, pressure and the pH of the solution.

28. The boiling point is an intensive property (does not depend on the quantity of the substance), so 1 kg and 500 g of ethanol boil at the same temperature of 78°C.

29. All phase changes are physical changes.

30. All chemical solutions are homogeneous mixtures.

31. An atom is a heterogeneous mixture of various parts, such as protons, electrons and neutrons. Atoms can be split apart into their components.

32. Sand is a heterogeneous mixture of rock, shells, metals and other elements, which can be separated from each other by physical methods.

33. Heterogeneous mixtures, such as seawater, can be separated into their components by physical means.

34. Filtration is used to separate a heterogeneous solid–liquid mixture, such as sand and water.

35. Distillation is used to separate a homogeneous mixture such as a potassium chloride (KCl) solution.

36. Iron (II) nitrate $Fe(NO_3)_2$ is not a heterogeneous mixture since it forms only one phase. Fe, N and O are chemically combined in definite ratios to form the $Fe(NO_3)_2$ compound with properties that differ from those of the constituent elements (Fe, N and O).

Q1.4

1. Picture 1 shows that the smallest building unit of matter is an atom of one type (represented by one red dot). The atoms are separated (not bonded together). Picture 1 represents an element.

2. Picture 2 shows that the smallest building unit of matter is two atoms of the same type bonded together (represented by two red dots). Picture 2 represents a homonuclear molecule (containing atoms of the same element).

3. Picture 3 shows that the smallest building unit of matter is two atoms of different types bonded together (represented by one red dot and one black dot). Picture 3 represents a compound.

4. Picture 4 shows that the smallest building unit of matter is two atoms of different types but not bonded together (represented by one red dot and one black dot). Picture 4 represents a mixture of elements.

5. Picture 5 shows that the two atoms of different types bonded together (represented by one red dot and one black dot) represent one compound. Another group of two atoms of different types bonded together (represented by one green dot and one orange dot) represents a second compound. Picture 5 represents a mixture of compounds.

Measurements

2

2.1 OBJECTIVES

At the end of the present chapter, the student should be able to:

1. Select the appropriate units in the International System of Units (known by the international abbreviation SI) of measurement.
2. Differentiate between fundamental quantity and derived quantity.
3. Convert units.
4. Define and use the basic methods and tools of measurements, such as significant figures and rounding off.

2.2 MEASUREMENTS

Measurement is the process of comparing an unknown quantity with another known quantity of the same kind to find out how many times the first includes the second.

Physical quantities are either fundamental (basic) quantities or derived quantities.

Fundamental quantities cannot be defined in terms of other physical quantities.

Examples: length, time, mass, temperature and amount of a substance.

Quantity consists of a number telling us how much, and a unit which shows what the scale of measurement is. Examples of base units in SI are given in Table 2.1. SI base units of mass, temperature and amount of a substance (kg, K and mol, respectively) are widely used by chemists (Table 2.1).

TABLE 2.1   Examples of SI base units

Base quantity		SI base unit	
Name	**Symbol**	**Name**	**Symbol**
Length	l, x, r, etc	meter	m
Mass	m	kilogram	kg
Time	t	second	s
Temperature	T	Kelvin	K
Amount of a substance	n	mole	mol

Note that:

- The SI unit of mass is kg (with small or lowercase k), not Kg (with capital K).
- The SI unit of the amount of a substance (mole) is denoted by mol (without an "e"), not mole. Also, kg is not the SI base unit for the amount of a substance but the SI unit of mass.
- The SI unit of time is second (s) not hour (h).
- The SI unit of temperature Kelvin is denoted by K (no degree sign) not °K.

Derived quantities can be defined in terms of the fundamental physical quantities.

Examples: volume, surface area, density, energy and concentration.

Examples of derived units in SI expressed in terms of base units and examples of derived units in SI with special names and symbols are given in Table 2.2 and Table 2.3, respectively.

DOI: 10.1201/9781003257059-2

Fundamental and derived quantities could be **scalar quantities** or **vector quantities**.

Scalar quantities can be fully defined by only their magnitude (without direction).

Examples: temperature, mass, density, distance, time and energy.

Vector quantities are fully defined by magnitude and direction.

Examples: displacement, dipole moment, velocity, acceleration and force.

TABLE 2.2   Examples of SI derived units

Base quantity		SI base unit	
Name	**Symbol**	**Name**	**Symbol**
Area	A	square meter	m^2
Volume	V	cubic meter	m^3
Density	ρ	kilogram per cubic meter	$kg\ m^{-3}$
Concentration	c	mole per cubic meter	$mol\ m^{-3}$
Mass concentration	γ	kilogram per cubic meter	$kg\ m^{-3}$

Note that:

- The SI derived unit of volume is m^3 since volume equals (length × width × height). The unit liter (L or l) is more commonly used. One liter equals one decimeter cubed (dm^3).
- The SI derived unit of density is kg/m^3 since density equals m/V and the SI unit of mass (m) is kg and the SI derived unit of volume (V) is m^3. There are other common units for density such as $g\ cm^{-3}$, $g\ mL^{-1}$ or $g\ L^{-1}$. For gases, densities are expressed in $g\ L^{-1}$.
- The SI derived unit of concentration is $mol\ m^{-3}$ since concentration equals n/V and the SI unit of the amount of a substance is mol and the SI derived unit of volume is m^3. Concentration could be also expressed by $mol\ L^{-1}$, which is also denoted by M (molarity).

TABLE 2.3   Examples of SI derived units with special names and symbols

Derived quantity	Name	Symbol	Expressed in terms of other SI units	Expressed in terms of SI base units
Frequency	Hertz	Hz	–	s^{-1}
Force	Newton	N	–	$kg\ m\ s^{-2}$
Pressure	Pascal	Pa	$N\ m^2$	$m^{-1}\ kg\ s^{-2}$
Energy	Joule	J	Nm	$m^2\ kg\ s^{-2}$

Note that:

- The SI derived unit of pressure is the Pascal (Pa). Pressure equals force/area and, since the SI derived unit of force is Newton (N) and the SI derived unit of area is m^2, then pressure is expressed, in terms of SI units, as $N\ m^{-2}$ and in terms of SI base units as $m^{-1}\ kg\ s^{-2}$.
- SI units can be used with prefixes (Table 2.4).

TABLE 2.4   Prefixes used with SI units

Derived quantity	Prefix	Symbol	Meaning	Example
multiplication	Tera-	T	$1 \times 10^{12} = 1,000,000,000,000$	1 TL = 1,000,000,000,000 L
	Giga-	G	$1 \times 10^{9} = 1,000,000,000$	1 Gg = 1,000,000,000 g
	Mega-	M	$1 \times 10^{6} = 1,000,000$	1 MHz = 1,000,000 Hz
	Kilo-	k	$1 \times 10^{3} = 1,000$	1 km = 1,000 m
division	deci-	d	$1 \times 10^{-1} = 0.1$	1 dL = 0.1 L
	centi-	c	$1 \times 10^{-2} = 0.01$	1 cL = 0.01 L
	milli-	m	$1 \times 10^{-3} = 0.001$	1 mg = 0.001 g
	micro-	μ	$1 \times 10^{-6} = 0.000001$	1 μL = 0.000001 L
	nano-	n	$1 \times 10^{-9} = 0.000000001$	1 nm = 0.000000001 m
	pico-	p	$1 \times 10^{-12} = 0.000000000001$	1 pL = 0.000000000001 L

Note that the prefixes used with the SI unit related to multiplication are denoted by capital (upper-case) letters, except for kilo- which is denoted by lowercase "k", whereas the prefixes related to division are denoted by small (lowercase) letters. For example, 1 Mg (megagram) equals 1,000,000 g and 1 mg (milligram) equals 0.001 g.

You can use Table 2.5 to convert a unit.

TABLE 2.5   Table that can be used to convert mass units

Tg			Gg			Mg			kg			g	dg	cg	mg			µg			ng

Note that the unit (g) can be replaced by any other unit in the table, i.e., m, s or L.

To use Table 2.5, follow the following steps:

Step 1: write the quantity to be converted in the table, one digit in each column. The last non- decimal digit should be written in the column of the unit of the quantity to be converted. If the quantity contains decimals, then the (dot) should be not written in the table and the last non- decimal digit should be written in the column representing the unit of the original value.

Example: how should one write 23 Gg or 13.45 g in the table?

Tg			Gg			Mg			kg			g	dg	cg	mg			µg			ng
	2	3																			
										1	3	4	5								

Step 2: if necessary, place zeros in all empty columns up to the column representing the desired unit.

Example: convert 23 Gg to dg, 13.45 g to Mg and 4,579 µg to mg

Tg			Gg			Mg			kg			g	dg	cg	mg			µg			ng
	2	3	0	0	0	0	0	0	0	0	0	0	0								
						0	0	0	0	0	1	3	4	5							
															4	5	7	9			

Step 3: if necessary, add a decimal point after the digit written in the column of the desired unit.

Tg			Gg			Mg			kg			g	dg	cg	mg			µg			ng
						0.	0	0	0	0	1	3	4	5							
															4.	5	7	9			

To summarize, 23 Gg = 230000000000 dg = 23×10^{10} dg,
13.45 g = 0.00001345 Mg = 1.345×10^{-5} Mg,
and 4,579 µg = 4.579 mg

GET SMART

HOW DO YOU CONVERT A UNIT?

Before converting a unit, you need to know how much larger one unit is than the other. First, convert the value to the basic unit (i.e., the unit without the prefix), then convert that to the desired unit.

Example: To convert 15.6 km to µm, you first convert **km** to **m**:

$$15.6\,km = 15.6 \times 1\,km \text{ or } 1\,km = 10^3\,m, \text{ so } 15.6\,km = 15.6 \times 10^3\,m.$$

Then, convert **m** to **µm**: (1 m = 10^6 µm)

$$15.6 \times 10^3 \ \mathbf{m} = 15.6 \times 10^3 \times \mathbf{10^6} \ \mathbf{µm} = 15.6 \times 10^9 \ \text{µm} = 156 \times 10^8 \ \text{µm}$$

Practice 2.1 Convert the followings to the desired unit:

a) 23 Gg = _____ dg
b) 126 Mg = _____ mg
c) 15.6 kg = _____ µg
d) 2.126 mg = _____ ng
e) 2 ng = _____ mg
f) 1,827 ng = _____ µg
g) 45 mg = _____ g
h) 456.89 g = _____ kg

Answer:

a) 23 Gg = 23×10^9 g = $23 \times 10^9 \times 10$ dg = $\mathbf{23 \times 10^{10}}$ **dg**
b) 126 Mg = 126×10^6 g = $126 \times 10^6 \times 10^3$ mg = $\mathbf{126 \times 10^9}$ **mg**
c) 15.6 kg = 15.6×10^3 g = $15.6 \times 10^3 \times 10^6$ µg = $\mathbf{156 \times 10^8}$ **µg**
d) 2.126 mg = 2.126×10^{-3} g = $2.126 \times 10^{-3} \times 10^9$ ng = $\mathbf{2126 \times 10^3}$ **ng**
e) 2 ng = 2×10^{-9} g = $2 \times 10^{-9} \times 10^3$ mg = $\mathbf{2 \times 10^{-6}}$ **mg**
f) 1,827 ng = $1,827 \times 10^{-9}$ g = $1,827 \times 10^{-9} \times 10^6$ µg = **1.827 µg**
g) 45 mg = $\mathbf{45 \times 10^{-3}}$ **g**
h) 456.89 g = 456.89×10^{-3} kg = **0.45689 kg**

Or use the following table of conversion:

Tg			Gg			Mg			kg			g	dg	cg	mg			µg			ng
		2	3	0	0	0	0	0	0	0	0	0	0								
				1	2	6	0	0	0	0	0	0	0	0	0						
								1	5	6	0	0	0	0	0	0	0	0			
															2	1	2	6	0	0	0
															0.	0	0	0	0	0	2
																		1.	8	2	7
												0.	0	4	5						
									0.	4	5	6	8	9							

2.3 EXAMPLES OF PHYSICAL PROPERTIES AND THEIR UNITS

2.3.1 VOLUME

Volume is an extensive physical property and is a derived property. The SI unit for volume is m^3 (Figure 2.1) but the unit liter (L) is commonly used.

1 m^3 is the volume of a cube with a 1 m length on each side. **1 m^3 = 1,000 L**
1 dm^3 is the volume of a cube with a 0.1 m = 10 cm length on each side. **1 dm^3 = 1 L**
1 cm^3 is the volume of a cube with a 0.01 m = 1 cm length on each side. **1 cm^3 = 1 mL**
1 mm^3 is the volume of a cube with a 0.001 m = 1 mm length on each side. **1 mm^3 = 1 µL**

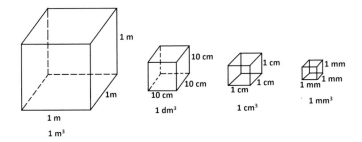

FIGURE 2.1   Different units used for volume

You can use Table 2.6 to convert a volume unit.

TABLE 2.6   Table that can be used to convert volume units

m³		dm³			cm³			mm³		

Practice 2.2 Convert the following examples to the desired unit:

Answer:

a) 25 m³ = 25,000 dm³
b) 3.2 m³ = 3,200,000 cm³ = 3.2 × 10⁶ cm³
c) 15 cm³ = 0.015 dm³
d) 2.5 cm³ = 2,500 mm³
e) 1,543 mm³ = 1.543 cm³
f) 73 dm³ = 73 L = 73,000 mL (= 73000 cm³)
g) 45 m³ = 45,000 dm³ = 45,000 L
h) 45.8 cm³ = 45.8 mL = 0.458 L

Or use Table 2.6 for conversion.

m³		dm³			cm³			mm³		
2	5	0	0	0						
	3	2	0	0	0	0	0			
				0.	0	1	5			
							2	5	0	0
							1.	5	4	3
			7	3	0	0	0			
4	5	0	0	0						
				0.	0	4	5	8		

2.3.2 TEMPERATURE

The SI unit of temperature is K, but temperature can also be expressed in degrees Celsius (°C) based on the freezing point of water as 0°C and boiling point of water as 100°C or in degrees Fahrenheit (°F), as in the US system.

○ The equation for temperature conversion from °C to K is:

$$T(K) = T(°C) + 273$$

○ The equation for temperature conversion from °F to °C is:

$$T(°F) = \left(\frac{9}{5} \times T(°C)\right) + 32 = \left(1.8 \times T(°C)\right) + 32$$

GET SMART

HOW TO DETERMINE AN APPROXIMATE VALUE OF THE TEMPERATURE IN °F?

When temperature values in (°C) are not very high (<100°C), an approximate value of the temperature in °F can be rapidly determined by multiplying the value of temperature in °C by 2 and adding 30

Example: For a temperature of 30°C, T (°F) = (1.8 × T (°C)) + 32
$$= (1.8 \times 30 °C) + 32 = 86°F$$

T (approximate) = 30 × 2 + 30 = 90°F which differs slightly from the correct value (86°F).

Practice 2.3 A clock in your town reported a temperature reading of 35 °C. What is this temperature in K and in °F?

Answer:

$$T(°C) = 35°C + 273 = 308 \text{ K}$$

$$T(°F) = \left(1.8 \times 35°C\right) + 32 = 95°F$$

When converting temperature from K to °F, or conversely from °F to K, convert first to °C.

$$T(°C) = T(K) - 273$$

$$T(°C) = \frac{\left(T(°F) - 32\right)}{1.8}$$

Practice 2.4 A temperature equals 300 K. What is this temperature in °F?

Answer:
First, convert 300 K to °C: $300 \text{ K} = \left(300 - 273\right)°C = 27 \text{ °C}$

Then, convert the obtained temperature in °C to °F: 27°C = (1.8 × 27) + 32°F = 80.6°F
Therefore, 300 K = 80.6°F

Practice 2.5 A temperature equals 150°F. What is this temperature in K?

Answer:
First, convert 150°F to °C: 150°F = (150 − 32)/1.8 = 65.6°C

Then, convert the obtained temperature in °C to K: 65.6°C = 65.6 + 273 = 338.6 K
Therefore, 150°F = 338.6 K

2.3.3 DENSITY AND SPECIFIC GRAVITY

The density of a substance (ρ) is defined as its mass per unit volume:

$$\rho = \frac{m}{V}$$

where ρ is the density, m is the mass and V is the volume. For example, the density of water is 1 g/cm^3. The density of ethanol is 0.79 g/cm^3 which is lower than the density of water. For this reason, ethanol floats on water. The density of iron is 7.9 g/cm^3 which higher than the density of water. For this reason, iron sinks in water.

Density is a derived quantity and is an intensive property (it does not depend on the amount of matter).

The specific gravity (or relative density) is the ratio of the density of a substance to the density of a given reference substance. Generally, water is considered to be the reference substance.

$$specific\ gravity = \frac{\rho_{(substance)}}{\rho_{(reference\ substance)}}$$

Specific gravity is a dimensionless (without a unit) quantity because it represents the ratio of two quantities having the same unit. For example, the density of copper is 8.96 g/cm^3 and the specific gravity of copper is 8.96 (without a unit).

Practice 2.6 A sample of copper has a volume of 0.478 mL and a mass of 4.28 g. Calculate the density of copper in g cm^{-3} and in the SI derived unit kg m^{-3}.

Answer:
To calculate the density of copper in g cm^{-3}, the mass should be expressed in g and the volume in cm^3:

$$1\ mL = 1\ cm^3,\ the\ volume\ V = 0.478\ mL = 0.478\ cm^3$$

$$g = \frac{m}{V} = \frac{4.28\ g}{0.478\ cm^3} = 8.95\ g\ cm^{-3}.$$

To calculate the density of copper in kg m^{-3}, the mass should be expressed in kg and the volume in m^3:

$$the\ mass\ of\ copper\ m = 4.28 \times 10^{-3}\ kg = 0.00428\ kg$$

The volume of copper V = 0.478 × 10^{-3} L = 0.478 × 10^{-3} × 10^{-3} m^3 = 0.478 × 10^{-6} m^3

$$g = \frac{m}{V} = \frac{4.28 \times 10^{-3}\ kg}{0.478 \times 10^{-6}\ m^3} = 8,950\ kg\ m^{-3}$$

Note that: 1 mL = 10^{-3} L and 1 L = 10^{-3} m^3, so 1 mL = 10^{-3} × 10^{-3} m^3 = 10^{-6} m^3

Practice 2.7 A cube of metal has a mass of 4 g and a length of 2 cm on each side. Calculate its density.

Answer:
The metal has a cubic shape with a length of 2 cm on each side (Figure 2.2).

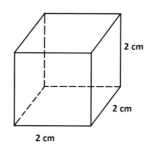

FIGURE 2.2 A cube with a length of 2 cm on each side

The volume of the cube is $V = 2 \times 2 \times 2 = 8\,\text{cm}^3$

The density $g = \dfrac{m}{V} = \dfrac{4\,\text{g}}{8\,\text{cm}^3} = 0.5\,\text{g cm}^{-3}$

2.4 UNCERTAINTY IN MEASUREMENT

2.4.1 EXACT AND INEXACT NUMBERS

Numbers can be **exact** or **inexact:**

- ○ Exact numbers are numbers with defined values.
  - • **Examples:** counted numbers, conversion factors.
- ○ Inexact numbers are numbers obtained by any method other than counting.
  - • **Example:** measured values.

GET SMART

HOW TO DETERMINE IF A NUMBER IS AN EXACT NUMBER OR A MEASURED NUMBER?

A measured number is followed by a unit which shows the scale of measurement; an exact number does not show a unit. For example, 3 is an exact number and 3 mL is a measured number.

2.4.2 SIGNIFICANT FIGURES

The last digit in a measured value is an uncertain digit. The certain digits and the last uncertain digit in a measurement are called **significant figures (SF)**. Significant figures are used to express the uncertainty of inexact numbers obtained by a measurement. For example, when a pen is measured using three different graduated rules, three different measurements are obtained (Figure 2.3):

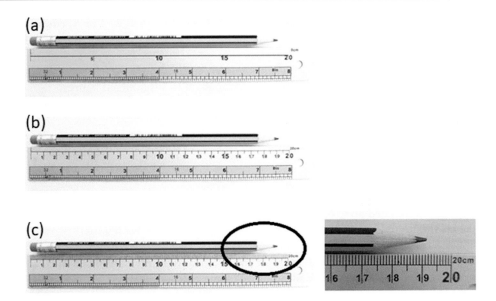

FIGURE 2.3   The last digit in a measured value is an uncertain digit: (a) *measurement 1. The length of the pen is 19 or 20 cm. The last digit is uncertain (9 or 10)*; (b) *measurement 2. The length of the pen is 19.1 or 19.2 cm. Now 9 becomes a certain digit and the last digit is uncertain (1 or 2)*; (c) *measurement 3. The length of the pen is 19.18 or 19.19 cm. Now 9 and 1 are certain digits and the last digit is uncertain (8 or 9)*

2.4.3 RULES DETERMINING THE SIGNIFICANT FIGURES

Rule 1. Any non-zero digit is significant.
   3,455 cm → 4 significant figures (4 SF)
   28.7 cm → 3 significant figures (3 SF)
Rule 2. Zeros between non-zero digits are significant.
   12,051 m → 5 significant figures (5 SF)
   108.036 cm → 6 significant figures (6 SF)
   3.000001 cm → 7 significant figures (7 SF)
Rule 3. Zeros to the left of the first non-zero digit are not significant.
   0.8 mL → 1 significant figure (1 SF)
   0.58 g → 2 significant figures (2 SF)
   0.00246 L → 3 significant figures (3 SF)
Rule 4. Zeros to the right of the last non-zero digit are significant if a decimal point is present.
   251.00 mL → 5 significant figures (5 SF)
   30.0 g → 3 significant figures (3 SF)
   0.2300 dL → 4 significant figures (4 SF)
Rule 5. Zeros to the right of the last non-zero digit are not significant if no decimal point is present.
   30 mL → 1 significant figure (1 SF)
   2,500 mL → 2 significant figures (2 SF)
   12,3000 kg → 3 significant figures (3 SF)

2.4.4 RULES FOR ROUNDING-OFF NUMBERS

Rounding off means making a **number** simpler but keeping its value close to what it was. The result is less accurate, but easier to use. For example, 32 **rounded** off to the nearest ten is 30, because 32 is closer to 30 than to 40.

Rule 1. If the number you are rounding off is followed by 0, 1, 2, 3 or 4, round the number down.

Examples:

8.242310 is rounded off to 8.24231 when rounded off to 6 significant figures

8.242310 is rounded down to 8.2423 when rounded off to 5 significant figures

8.242310 is rounded down to 8.242 when rounded off to 4 significant figures

8.242310 is rounded down to 8.24 when rounded off to 3 significant figures

8.242310 is rounded down to 8 when rounded off to 1 significant figure

Rule 2. If the number you are rounding off is followed by 6, 7, 8 or 9, round the number up.

Examples:

6.298687 is rounded up to 6.29979 when rounded off to 6 significant figures

6.298687 is rounded up to 6.2998 when rounded off to 5 significant figures

6.298687 is rounded up to 6.299 when rounded off to 4 significant figures

6.298687 is rounded up to 6.30 when rounded off to 3 significant figures

Rule 3. Concerning digit 5, if the number you are rounding off is followed by 5, and 5 is not the last digit, round the number up (same rule as rule 2).

Examples:

3.298357 is rounded up to 3.2984 when rounded off to 5 significant figures

6.798587 is rounded up to 6.799 when rounded off to 4 significant figures

1.395687 is rounded up to 1.40 when rounded off to 3 significant figures

If the number you are rounding off is followed by 5, and 5 is the last digit or followed by zeros, make the preceding digit even (exception to rule 2).

Examples:

Where the preceding digit is odd (1, 3, 5, 7 and 9), make it even:

3.29835 is rounded up to 3.2984 when rounded off to 5 significant figures

3.2975 is rounded up to 3.298 when rounded off to 4 significant figures

3.295 is rounded up to 3.30 when rounded off to 3 significant figures

3.29500 is rounded up to 3.30 when rounded off to 3 significant figures

Where the preceding digit is even (0, 2, 4, 6 and 8), do not change it:

3.2985 is rounded down to 3.298 when rounded off to 4 significant figures

3.29845 is rounded down to 3.2984 when rounded off to 5 significant figures

3.125 is rounded down to 3.12 when rounded off to 3 significant figures

3.05 is rounded down to 3.0 when rounded off to 2 significant figures.

(Note that 0 is an even digit)

2.5 CALCULATIONS WITH MEASURED NUMBERS

2.5.1 ADDITION AND SUBTRACTION

The answer cannot have more digits to the right of the decimal point than any of original numbers.

Examples:

○ 102.50 + 0.231 = ?

The number 102.50 contains two digits after the decimal point and the number 0.231 contains three digits after the decimal point. Therefore, the result should contain two digits after the decimal point.

Calculator answer: 102.731 rounded off to 102.73

$$102.50 + 0.231 = 102.73$$

○ 21 − 13.8 = ?

The number 21 (here we considered that 21 is not an exact number, e.g., is 21 cm) contains no digits after the decimal point and the number 13.8 contains one digit after the decimal point. Therefore, the result should contain no digits after the decimal point.

Calculator answer: 7.2 rounded off to 7

$$21 - 13.8 = 7$$

2.5.2 MULTIPLICATION AND DIVISION

The final answer contains the smallest number of significant figures

Examples:

o $1.4 \times 8.011 = ?$

The number 1.4 contains two significant figures (2 SF) and the number 8.011 contains four significant figures (4 SF). Therefore, the answer should contain two significant figures (limited by 1.4 to two significant figures in the answer).

Calculator answer: 11.2154 which is rounded off to two significant figures, resulting in 11
   Therefore, $1.4 \times 8.011 = 11$

o $\dfrac{13.9}{4.0215} = ?$

The number 13.9 contains three significant figures (3 SF) and the number 4.0215 contains five significant figures (5 SF). Therefore, the result should contain three significant figures (limited by 13.9 to three significant figures in the answer).

Calculator answer: 3.4564217332 rounded off to three significant figures results in 3.46
   Therefore, $\dfrac{13.9}{4.0215} = 3.46$

2.5.3 EXACT NUMBERS IN CALCULATION

Exact numbers do not limit the answer because they have an infinite number of significant figures.

Examples:

o A coin has a mass of 3.5 g. If we have six such coins, the total mass is

$$6 \times 3.5 \text{ g} = 21 \text{ g}$$

In this case, 6 is an exact number and it does not limit the number of significant figures in the result.

o A student repeated a neutralization experiment three times in the laboratory. He found the following values of titration volume (10.5 mL, 11.35 mL, and 10.8 mL). The average titration volume is:

$$\frac{10.5 + 11.35 + 10.8}{3} = 10.9$$

In this case, 3 is an exact number and it does not limit the number of significant figures in the result.

2.5.4 SCIENTIFIC NOTATION IN CALCULATION

Scientific notation is a special way of writing numbers that are too big or too small to be conveniently written in decimal form. In scientific notation, numbers are written in the form ($n \times 10^m$) where n is a number written with only one digit before the decimal point.

Examples: 2.0×10^5, 1.003×10^2, 3.12×10^{-7}

Scientific notation is very useful for writing results of calculations with respect to the rules of significant figures.

Practice 2.8 Calculate the following (8.9 × 56.1) to the correct number of significant figures.

Answer:

$$(8.9 \times 56.1) = ?$$

The number 8.9 contains two significant figures (2 SF) and the number 56.1 contains three significant figures (3 SF). Therefore, the result should contain two significant figures (limited by 8.9 to <u>two</u> significant figures in the answer).

Calculator answer: 499.29 (5 SF)

    Note that if we write 8.9 × 56.1 = 499, the result is correct but does not respect the rule of significant figures in multiplication and division since the result should contain two significant figures and the number 499 contains three significant figures. If we write 8.9 × 56.1 = 49, for example, the result respects the rule of significant figures in multiplication and division, but it is not correct.

    Scientific notation is mandatory for writing the correct result of calculation:

        8.9 × 56.1 = 499.29

    With respect to the rules of significant figures:

        $499.29 = 4.\underline{9}929 \times 10^2$ rounded off to two significant figures results in 5.0×10^2

    Therefore, using the scientific notation, we have: $8.9 \times 56.1 = 5.0 \times 10^2$.

GET SMART

HOW TO CALCULATE AN ARITHMETIC OPERATION WHEN IT IS A COMBINATION OF MULTIPLICATION (OR DIVISION) AND ADDITION (OR SUBTRACTION)

Calculate the addition and subtraction operations separately, and the multiplication and division operations separately, then calculate the overall operation using the results obtained separately.

Example: Calculate the following operation to the correct number of significant figures:

$$\frac{1.200 - 0.06711 + 0.00040}{0.01 + 3.2326} = ?$$

Calculate separately the arithmetic operations (1.200 – 0.06711 + 0.00040) and (1.0 + 3.2346) and then calculate the overall operation:

- (1.200 – 0.06711 + 0.00040) = **1.133** (calculator answer is 1.13329)
- (0.<u>0</u>1 + 3.2326) = **3.24** (calculator answer is 3.<u>2</u>426)

- $\dfrac{1.200 - 0.06711 + 0.00040}{0.01 + 3.2326} = \dfrac{1.133}{3.24} = 0.350$

- (Calculator answer is 0.<u>3</u>49691358)

Do not calculate the overall arithmetic operation using a calculator, since, in that case, what rule should we apply? Should we apply the rule concerning addition and subtraction or the rule concerning multiplication and division?

If we calculate the overall arithmetic operation using a calculator, we found the answer to be 0.3492849658.

- If we apply the rule concerning addition and subtraction, the answer (which should contain two digits after the decimal point) is 0.35. The uncertainty of the measurement lies from the second decimal. However, in the correct answer (0.350) containing three significant figures, the uncertainty of the measurement lies from the third decimal.
- If we apply the rule concerning multiplication and division, the answer (which should contain one significant figure) is 0.3. This result is different from both the correct answer (0.350) and the answer obtained when applying the rule concerning addition and subtraction (0.35).

2.6 ACCURACY AND PRECISION

The quality of a set of measured numbers is determined on the bases of accuracy and precision.

Accuracy is how close a measurement is to the accepted value (or true value).
Precision is how closely measurements of the same thing are to one another.

Practice 2.9 The true value of a neutralization volume is 12.370 mL. Describe the accuracy and precision for each set of measurements:

Set A (12.345 mL, 12.346 mL, 12.343 mL)
Set B (12.357 mL, 12.337 mL, 12.393 mL)
Set C (12.369 mL, 12.370 mL, 12.371 mL)

Answer:
The values in set A are close to one another but are not close to the true value so set A is precise but not accurate.

The values in set B are not close to one another and are not close to the true value so set B is neither precise nor accurate.

The values in set C are close to one another and close to the true value so set C is both precise and accurate.

CHECK YOUR READING

What is a measurement?
What is the difference between a fundamental quantity and a derived quantity?
What is the difference between precision and accuracy?
What are the SI basic units used for mass, temperature and amount of a substance?
What are the SI derived units used for volume, density and pressure?
How should temperature be converted from °C to K and °F?
How should volume be converted from m³ or cm³ to L?
How should significant figures be determined and what are the rules of significant figures in calculations?
What are the rules for rounding off?

SUMMARY OF CHAPTER 2

Measurement is the process of comparing an unknown quantity with another known quantity of its kind.

Fundamental physical quantities cannot be defined in terms of other physical quantities.

Examples: length, time, mass, temperature and amount of a substance.

Derived physical quantities can be defined in terms of the fundamental physical quantities.

Examples: volume, area (surface), density, energy and concentration.

Scalar quantities can be fully defined by only their magnitude (without direction).

Examples: temperature, mass, density, distance, time and energy.

Vector quantities are fully defined in terms of magnitude and direction.

Examples: displacement, velocity, acceleration and force.

The SI basic unit of mass is kg, the SI basic unit of the amount of substance is mol and the SI basic unit of temperature is denoted by K.

The SI derived unit of volume is m^3. The unit liter (L) is more commonly used. One liter equals one decimeter cubed (dm^3) and 1 m^3 equals 1000 L.

The SI derived unit of density is kg m^{-3} since density equals m/V.

The prefixes used with the SI unit related to multiplication are denoted by a capital letter, except for kilo- which is denoted by a lowercase "k": 1 Mg (megagram) equals 1,000,000 g.

The prefixes used with the SI unit related to division are denoted by small letters: 1 mg (milligram) equals 0.001 g.

Temperature conversion:

The equation for temperature conversion from °C to K is: **T(K) = T(°C) + 273**.

The equation for temperature conversion from °F to °C is: **T(°F) = [1.8 × T(°C)] + 32**.

To convert temperature from °F to K, firstly convert temperature from °F to °C, then convert from °C to K.

Accuracy is how close a measurement is to the accepted value (or true value).

Precision is how close measurements of the same thing are to one another.

Rules for determining the significant figures:

o Any non-zero digit is significant: <u>342156798</u> contains 9 SF.
o Zeros between non-zero digits are significant: <u>2008.0305</u> contains 8 SF.
o Zeros to the left of the first non-zero digit are not significant: 0.000<u>211</u> contains 3 SF.
o Zeros to the right of the last non-zero digit are significant if a <u>decimal point is present</u>: <u>20.0</u> contains 3 SF.
o Zeros to the right of the last non-zero digit are not significant if a decimal point is not present: <u>45</u>000 contains 2 SF.

Rules for rounding off numbers:

o If the number is less than 5, round down (example: 3.1<u>62</u> is rounded down to 3.16 when rounded off to 3 SF).
o If the number is greater than 5, round up (example: 2.23<u>26</u> is rounded off to 2.233 when rounded off to 4 SF).
o When the first digit deleted is 5, and 5 is the last digit or followed by zeros, make the preceding digit even:
   3.2<u>95</u> is rounded up to 3.30 when rounded to 3 significant figures.
   3.1<u>25</u> is rounded down to 3.12 when rounded to 3 significant figures.
   3.<u>05</u> is rounded down to 3.0 when rounded to 2 significant figures (0 is an even digit).
   3.1<u>25</u>000 is rounded down to 3.12 when rounded to 3 significant figures (5 is followed by zeros).
   3.2<u>85</u>687 is rounded up to 3.29 when rounded to 3 SF (5 is not the last digit).

Significant figures in calculation:

Addition and subtraction:
The answer cannot have more digits to the right of the decimal point than is exhibited by any of the original numbers.

Examples: 102.50 + 0.231 = 102.73
21 − 13.8 = 7

Multiplication and division:
The final answer contains the smallest number of significant figures.

Examples: $1.4 \times 8.011 = 11$
13.9 / 4.0215 = 3.46

Exact numbers (counting numbers, conversion factors, etc.) do not limit the answer because they have an infinite number of significant figures.

Example: If the mass of one coin is 2.55 g, then three coins have the mass 3×2.55 g = 7.65 g (3 SF)

Scientific notation is very useful for writing results of calculations with respect to the rules of significant figures. In scientific notation, numbers are written in the form $(n \times 10^m)$ where n is a number written with only one digit before the decimal point.

Example: $8.9 \times 56 = \underline{5.0} \times 10^2$ (2 SF)

PRACTICE ON CHAPTER 2

Q2.1 **Complete the following sentences:**
Measurement is the process of _____ an _____ with another known _____ of its kind to find out how many times the first includes the second.
Physical quantities are either _____ or _____ quantities.
_____ cannot be defined in terms of other physical quantities and _____ can be defined in terms of the fundamental physical quantities.
Fundamental and derived quantities could be _____ or _____ quantities _____ can be fully defined by their magnitude only and _____ can be fully defined by both magnitude and direction.
_____ is how close a measurement is to the accepted value (or true value).
_____ is how close measurements of the same thing are to one another.

Q2.2 **Match the following:**

a. kg m⁻³	**1.** is a fundamental quantity
b. Amount of substance	**2.** is a derived quantity
c. K	**3.** is the SI derived unit of density
d. Force	**4.** is the SI basic unit of temperature
e. Accuracy	**5.** is how close measurements are to one another.
f. Precision	**6.** is how close a measurement is to the accepted value

Q2.3 **Choose the correct answer:**
  1. **Temperature is**
     a) a fundamental quantity
     b) a vector quantity
     c) a derived quantity
     d) an SI unit

2. **Density is**
    a) a fundamental quantity
    b) a vector quantity
    c) a derived quantity
    d) a SI unit

3. **Volume is**
    a) expressed in m³ in SI units
    b) a scalar quantity
    c) a derived quantity
    d) all of the above

4. **Mass is**
    a) a fundamental quantity
    b) a vector quantity
    c) a derived quantity
    d) a SI derived unit

5. **Amount of a substance is**
    a) a fundamental quantity
    b) a vector quantity
    c) a derived quantity
    d) an SI basic unit

6. **Pressure is**
    a) a fundamental quantity
    b) expressed in atmospheres in SI units
    c) a derived quantity
    d) an SI unit

7. **Energy is**
    a) a fundamental quantity
    b) a vector quantity
    c) a derived quantity
    d) none of the above

8. **Length is**
    a) expressed in cm in SI units
    b) a fundamental quantity
    c) a derived quantity
    d) none of the above

9. **Time is**
    a) expressed in h in SI units
    b) a derived quantity
    c) a fundamental quantity
    d) none of the above

10. **Concentration is**
    a) expressed in mol m⁻³ in SI units
    b) a derived quantity
    c) a scalar quantity
    d) all of the above

11. **Which of the following is not a fundamental quantity?**
    a) temperature
    b) mass
    c) pressure
    d) length

12. **Which of the following is a fundamental quantity?**
    a) temperature
    b) volume
    c) concentration
    d) density

13. **Which of the following is a derived quantity?**
    a) force
    b) temperature
    c) time
    d) amount of a substance
14. **Temperature, mass, length, and amount of a substance are fundamental quantities**
    a) True
    b) False
15. **Density, volume, and pressure are derived quantities**
    a) True
    b) False
16. **Force is a scalar quantity**
    a) True
    b) False

Q2.4   Choose the correct answer:
1. **The International System (SI) unit of temperature is**
   a) °C
   b) °F
   c) K
   d) atm
2. **The International System (SI) unit of volume is**
   a) L
   b) m^3
   c) mL
   d) cm^3
3. **The International System (SI) unit of the amount of substance is**
   a) kg
   b) g
   c) dg
   d) mol
4. **The International System (SI) unit of pressure is**
   a) atm
   b) mmHg
   c) Pa
   d) K
5. **The International System (SI) unit of mass is**
   a) kg
   b) g
   c) dg
   d) mol
6. **The International System (SI) unit of density is**
   a) $kg\ m^{-3}$
   b) $g\ cm^{-3}$
   c) $kg\ L^{-1}$
   d) $g\ L^{-1}$
7. **$1\ m^3$ equals**
   a) 10 L
   b) 1 L
   c) 1,000 L
   d) 1,000 mL
8. **10 μm equals**
   a) 10^{-2} mm
   b) 10^{-2} m
   c) 10^{-3} m
   d) 10^{-3} mm

9. **1000 nm equals**
   a) 10^{-2} m
   b) 1 cm
   c) 10 dm
   d) 1 μm

10. **1 L equals**
    a) 100 mL
    b) 1,000 mL
    c) 10 dm^3
    d) 1,000 m^3

11. **1 cm^3 equals**
    a) 1 mL
    b) 10 mL
    c) 100 mL
    d) 1 L

12. **Which of the following is correct?**
    a) 1 mg = 10^{-2} g
    b) 10 ng = 10^{-5} g
    c) 10 kg = 10^3 g
    d) 100 μg = 10^{-7} kg

13. **Which of the following is incorrect?**
    a) 1 mL = 10^{-3} L
    b) 10 mL = 10 cm^3
    c) 10 m^3 = 10^4 dm^3
    d) 1 dL = 10 cm^3

14. **Which of the following is correct?**
    a) 1 Mg = 10^{-3} g
    b) 1 Tg = 10^9 kg
    c) 1 pg = 10^3 g
    d) 10 Gg = 10^6 kg

15. **Which of the following is correct?**
    a) 100 ng = 10^{-4} mg
    b) 1 mg = 10^{-3} g
    c) 1 μg = 10^{-9} kg
    d) All of the above are correct

16. **A temperature of 350°C, when expressed in °F, equals**
    a) 630°F
    b) 662°F
    c) 32°F
    d) 700°F

17. **A temperature of 20°C, when expressed in °F, equals**
    a) 68°F
    b) 90°F
    c) 150°F
    d) 10°F

18. **A temperature of 150°C, when expressed in K, equals**
    a) 302 K
    b) 150 K
    c) 423 K
    d) 0 K

19. **A temperature of 180 K, when expressed in °C, equals**
    a) 82°F
    b) 356°F
    c) 453°C
    d) −93°C

20. **A temperature of 313°F, when expressed in °C, equals**
    a) 156°C
    b) 149°C
    c) 50°C
    d) 0°C

21. **A temperature of 300 K, when expressed in °F, equals**
    a) 573°F
    b) 150°F
    c) 80.6°F
    d) 27°F

22. **A temperature of 315°F, when expressed in K, equals**
    a) 430 K
    b) 150 K
    c) 327 K
    d) 573 K

Q2.5   **Choose the correct answer:**

1. **Precision refers to**
   a) the meaningful digits obtained in a measurement
   b) the proximity of a measurement to the true value of a quantity
   c) the proximity of several measurements to each other
   d) the difference between a measurement and the true value

2. **Accuracy refers to**
   a) the meaningful digits obtained in a measurement
   b) the proximity of a measurement to the true value of a quantity
   c) the proximity of several measurements to each other
   d) the difference between a measurement and the true value

3. **The true value of a neutralization volume is 9.870 mL. The set of measurements (9.345 mL, 9.346 mL, 9.343 mL) is:**
   a) precise and accurate
   b) precise but not accurate
   c) not precise but accurate
   d) not precise and not accurate

4. **The true value of a neutralization volume is 9.870 mL. The set of measurements (9.872 mL, 9.870 mL, 9.871 mL) is**
   a) precise and accurate
   b) precise but not accurate
   c) not precise but accurate
   d) not precise and not accurate

5. **The true value of a neutralization volume is 9.870 mL. The set of measurements (9.345 mL, 9.349 mL, 9.340 mL) is**
   a) precise and accurate
   b) precise but not accurate
   c) not precise but accurate
   d) not precise and not accurate

6. **The true value of a neutralization volume is 11.870 g. Which of the following sets of measurements is accurate but not precise?**
   a) 10.873 g, 10.870 g, 10.877 g
   b) 11.873 g, 11.870 g, 11.877 g
   c) 11.852 g, 11.850 g, 11.851 g
   d) 11.870 g, 11.871 g, 11.872 g

7. **Which of the following measurements is the least precise?**
   a) 10.8 g
   b) 10.87 g
   c) 10.877 g
   d) 10.8776 g

8. **Which of the following measurements is the most precise?**
    a) 10.8 g
    b) 10.87 g
    c) 10.877 g
    d) 10.8776 g

9. **The number of significant figures (SF) in 129.345 is**
    a) 2 SF
    b) 3 SF
    c) 5 SF
    d) 6 SF

10. **The number of significant figures in 109.04 is**
    a) 2 SF
    b) 3 SF
    c) 5 SF
    d) 6 SF

11. **The number of significant figures in 1,008 is**
    a) 2 SF
    b) 3 SF
    c) 5 SF
    d) 4 SF

12. **The number of significant figures in 0.003 is**
    a) 1 SF
    b) 3SF
    c) 4 SF
    d) 2 SF

13. **The number of significant figures in 1,500 is**
    a) 1 SF
    b) 2 SF
    c) 3 SF
    d) 4 SF

14. **The number of significant figures in 1.5×10^3 is**
    a) 1 SF
    b) 2 SF
    c) 3 SF
    d) 10^3 SF

15. **The number of significant figures in the result of (1.5 − 1.0089) is**
    a) 1 SF
    b) 2 SF
    c) 5 SF
    d) 4 SF

16. **The number of significant figures in the result of (3.5 − 1.0077) is**
    a) 1 SF
    b) 2 SF
    c) 5 SF
    d) 4 SF

17. **The number of significant figures in the result of (2.51 + 1.0077) is**
    a) 1 SF
    b) 2 SF
    c) 3 SF
    d) 4 SF

18. **The number of significant figures in the result of (11.5 × 3.0089) is**
    a) 1 SF
    b) 2 SF
    c) 3 SF
    d) 5 SF

19. The number of significant figures in the result of (1.5/2.1067) is
    a) 1 SF
    b) 2 SF
    c) 3 SF
    d) 5 SF
20. The number of significant figures in the result of (12.5 × 2.0089)/2.089 is
    a) 1 SF
    b) 2 SF
    c) 3 SF
    d) 5 SF
21. Rounding off the following number (2.143) to 3 SF results in
    a) 2.14
    b) 2.15
    c) 2.13
    d) 2.00
22. Rounding off the following number (1.3479) to 4 SF results in
    a) 1.347
    b) 1.348
    c) 1.349
    d) 1.350
23. Rounding off the following number (1.3497) to 4 SF results in
    a) 1.349
    b) 1.350
    c) 1.3497
    d) 1.450
24. Rounding off the following number (3.0015) to 4 SF results in
    a) 3.000
    b) 3.001
    c) 3.002
    d) 3.006
25. Rounding off the following number (1.025) to 3 SF results in
    a) 1.03
    b) 1.04
    c) 1.025
    d) 1.02
26. Rounding off the following number (11.205) to 4 SF results in
    a) 11.20
    b) 11.21
    c) 11.22
    d) 11.2
27. The correct answer to express the arithmetic operation
    (9.41 − 8.0)/(2.00 × 2.0090) is
    a) 0.34
    b) 0.35
    c) 0.348
    d) 0.4
28. The correct answer to express the arithmetic operation (9.03 cm × 88 cm) is
    a) 795
    b) 794.64
    c) 7.9×10^3
    d) 8.0×10^3
29. The mass of one chocolate bar is 22 g. The mass of four chocolate bars (4 × 22 g) is
    a) 90 g
    b) 0.88 g
    c) 9.0×10 g
    d) 88 g

30. **A student had the following marks in general chemistry: 19.5, 19, 18.5. The correct answer to express the average mark obtained by this student is (19.5 + 19 + 18.5)/3 = ?**
    a) 20
    b) 1.9
    c) 2×10
    d) 19

Calculations

Q2.6

An alloy cube with edge length of 20 cm has a mass of 80 g.
a) Calculate its density in $g\ cm^{-3}$ and in $kg\ m^{-3}$.
b) The cube is cut into two pieces. Does its density change? Why?

Q2.7

Calculate the following to the correct number of significant figures
a) $(3.234 - 0.06) \times 2.05/(1.0 + 1.2346) = ?$
b) $(86 \times 23) = ?$
c) $(20.0 \times 100.0) = ?$
d) $(50.00 \times 140.0) = ?$
e) $(4.05 - 0.2 + 2.123) / (0.0501 \times 0.40) = ?$

ANSWERS TO QUESTIONS

Q2.1

Measurement is the process of **comparing** an **unknown quantity** with another, known **quantity** of its kind to find out how many times the first includes the second.
Physical quantities are either **fundamental** or **derived** quantities.
Fundamental quantities cannot be defined in terms of other physical quantities and **derived quantities** can be defined in terms of the fundamental physical quantities.
Fundamental and derived quantities could be **scalar** or vector quantities. **Scalar quantities** can be fully defined by only their magnitude, whereas **derived quantities** are fully defined by their magnitude and direction.
Accuracy is how close a measurement is to the accepted value (or true value).
Precision is how close measurements of the same thing are to one another.

Q2.2 (a, 3); (b, 1); (c, 4); (d, 2); (e, 6); (f, 5).

Q2.3

1. a
2. c
3. d
4. a
5. a
6. c
7. c
8. b
9. c
10. d
11. c
12. a
13. a
14. a
15. a
16. b

Q2.4

1. c
2. b

3. d
4. c
5. a
6. a
7. c
8. a
9. d
10. b
11. a
12. d
13. d
14. b
15. d
16. b
17. a
18. c
19. d
20. a
21. c
22. a

Q2.5

1. c
2. b
3. b
4. a
5. d
6. b
7. a
8. d
9. d
10. c
11. d
12. a
13. b
14. b
15. a
16. b
17. c
18. c
19. b
20. c
21. a
22. b
23. b
24. c
25. d
26. a
27. b
28. c
29. d
30. d

Q2.6

1. The density $\rho = m/V = 80/8000 = 0.01$ g cm^{-3}
    The density in kg/m^3 is $\rho = m/V = 0.08/0.0080 = 10$ kg m^{-3}
2. When the cube is cut into two pieces, the density does not change because density is
    an intensive property (it does not depend on the amount of matter)

Q2.7

    a. $(3.234 - 0.067) \times 2.05 / (1.0 + 1.2346) = 3.0$

    b. $86 \times 23 = 2.0 \times 10^3$

    c. $(20.0 \times 100.0) = 2.00 \times 10^3$

    d. $50.00 \times 140.0 = 7.000 \times 10^3$

    e. $(4.05 - 0.2 + 2.123) / (0.0501 \times 0.40) = 3.0 \times 10^2$

KEY ANSWER EXPLANATIONS

Q2.3

1. Temperature cannot be defined in terms of other fundamental physical quantities. Temperature is a fundamental quantity and its SI unit is K.
2. Density (ρ) is the ratio of mass (m) and volume (V) of a substance ($\rho = m/V$), so it can be defined in terms of the fundamental physical quantities mass and length (volume = length $\times$ length $\times$ length).
3. Volume (V) is a derived quantity (volume = length $\times$ length $\times$ length). It can be fully defined by magnitude only (without direction), so volume is a scalar quantity. The SI unit of volume is m^3.
4. Mass is a fundamental quantity, and its SI basic unit is kg.
5. Amount of substance is fundamental quantity, and its SI basic unit is mol.
6. Pressure is a derived quantity, and its SI unit is Pa.
7. Energy is a derived quantity, and its SI unit is J.
8. Length is a fundamental quantity, and its SI unit is m.
9. Time is a fundamental quantity, and its SI unit is s.
10. Concentration (C = n/V, where n is the number of moles of the substance and V is the volume) is a derived scalar quantity. Its SI unit is mol m^{-3}.
11. Temperature, mass and length are fundamental quantities. Pressure is a derived quantity.
12. Volume, concentration and density are derived quantities. Temperature is a fundamental quantity.
13. Force is a derived quantity and its SI unit is N (= kg m s^{-2}). Temperature, time and amount of a substance are fundamental quantities.
14. Temperature, mass, length and amount of a substance are fundamental quantities.
15. Density, volume and pressure are derived quantities.
16. Force is a vector quantity. It needs to be fully defined by magnitude and direction.

Q2.4

1. The International System (SI) unit of temperature is K. °C and °F are not the SI units of temperature.
2. The International System (SI) unit of volume is m^3 (V = length $\times$ length $\times$ length).
3. The International System (SI) unit of amount of substance is mol.
4. The International System (SI) unit of pressure is Pa. The units (atm and mmHg) are also used but they are not SI units.
5. The International System (SI) unit of mass is kg (with lowercase or small k).
6. The density $\rho = m/V$; the mass m is in kg and the volume V is in m^3, so the SI unit of density is kg m^{-3}. The units g cm^{-3}, kg L^{-1} and g L^{-1} are commonly used for density, but they are not the SI unit of density.
7. $1\ m^3$ is the volume of a cube with a 1-m edge and equals 1,000 L; 1 L = 1,000 mL.
8. $1\ \mu m = 10^{-6}\ m = 10^{-3}\ mm$ since $1\ m = 10^3\ mm$ and $10\ \mu m = 10 \times 10^{-3}\ mm = 10^{-2}\ mm$.
9. $1\ nm = 10^{-3}\ \mu m$ so $1,000\ nm = 1000 \times 10^{-3}\ \mu m = 1\ \mu m$.
10. 1 L equals 1,000 mL.
11. $1\ cm^3$ is the volume of a cube with a 1-cm edge. $1\ cm^3 = 1\ mL$.
12. For each equation, convert to the same unit to find the correct answer:
    a) $1\ mg = 10^{-2}\ g$ is incorrect because $1\ mg = 10^{-3}\ g$.
    b) $10\ ng = 10^{-5}\ g$ is incorrect because $1\ ng = 10^{-9}\ g$ so $10\ ng = 10 \times 10^{-9}\ g = 10^{-8}\ g$.
    c) $10\ kg = 10^3\ g$ is incorrect because $10^3\ g = 1\ kg$.
    d) $100\ \mu g = 10^{-7}\ kg$ is correct because $1\ \mu g = 10^{-6}\ g$,

so 100 µg = 100 × 10^{-6} g = 10^{-4} g; 1 kg = 10^3 g, and
10^{-7} kg = 10^{-7} × 10^3 g = 10^{-4} g.

13. For each equation, convert to the same unit to find the correct answer:
    a) 1 mL = 10^{-3} L is correct.
    b) 10 mL = 10 cm^3 is correct because 1 cm^3 = 1 mL so 10 cm^3 = 10 mL.
    c) 10 m^3 = 10^4 dm^3 is correct because 1 m^3 = 1,000 L = 1,000 dm^3 (1 L = 1 dm^3) so
       10 m^3 = 10 × 1,000 L = 10^4 L = 10^4 dm^3.
    d) 1 dL = 10 cm^3 is incorrect because 1 dL = 0.1 L
       and 10 cm^3 = 10 mL = 10 × 10^{-3} mL = 0.01 L.

14. For each equation, convert to the same unit to find the correct answer:
    a) 1 Mg = 10^{-3} g is incorrect because 1 Mg = 10^6 g.
    b) 1 Tg = 10^9 kg is correct because 1 Tg = 10^{12} g and 10^9 kg = 10^9 × 10^3 g = 10^{12} g.
    c) 1 pg = 10^3 g is incorrect because 1 pg = 10^{-12} g.
    d) 10 Gg = 10^6 kg is incorrect because 10 Gg = 10 × 10^9 g = 10^8 g
       and 10^6 kg = 10^6 × 10^3 g = 10^9 g.

15. For each equation, convert to the same unit to find the correct answer:
    a) 100 ng = 10^{-4} mg is correct because 100 ng = 100 × 10^{-9} g = 10^{-7} g and
       10^{-4} mg = 10^{-4} × 10^{-3} g = 10^{-7} g.
    b) 1 mg = 10^{-3} g is correct.
    c) 1 µg = 10^{-9} kg is correct because 1 µg = 10^{-6} g and 10^{-9} kg = 10^{-9} × 10^3 g = 10^{-6} g.
       Therefore, all of the above are correct.

16. T (°F) = (1.8 × T(°C)) + 32 = (1.8 × 350°C) + 32 = 662°F
17. T (°F) = (1.8 × T(°C)) + 32 = (1.8 × 20°C) + 32 = 68°F
    When temperature values in °C are not very high (<100°C), an approximate value
    of the temperature in °F can be rapidly determined by multiplying the value of tem-
    perature in °C by 2 and adding 30:
    Example: T (approximate) = 20 × 2 + 30 = 70°F which differs slightly from the correct
    value (68°F).
18. T (K) = T(°C) + 273
    T = 150 °C + 273 = 423.15 K
19. T (°C) = T(K) − 273
    T (°C) = 180 K − 273 = − 93 °C
20. T (°F) = (1.8 × T(°C)) + 32
    T (°C) = (T (°F) − 32)/1.8
    T (°C) = (313 − 32)/1.8 = 156°C
21. First, convert temperature from K to °C, then from °C to °F:
    T (°C) = T(K) − 273 = 300 − 273 = 27°C
    T (°F) = (1.8 × T(°C)) + 32 = (1.8 × 27°C) + 32 = 80.6°F
22. First, convert temperature from °F to °C, then from °C to K:
    T (°C) = (T (°F) − 32)/1.8 = (315 − 32)/1.8 = 157°C
    T (K) = T(°C) + 273 = 157.2°C + 273 = 430 K

Q2.5
1. Precision refers to the proximity of several measurements to each other.
2. Accuracy refers to the proximity of a measurement to the true value (accepted) of a
   quantity.
3. The true value of a neutralization volume is 9.870 mL. The values of the set of measurements (9.345 mL, 9.346 mL, 9.343 mL) are very different from the true value, so
   this set is not accurate; however, they are close to one another, so the set is precise.
4. The true value of a neutralization volume is 9.870 mL. The values of the set of measurements (9.872 mL, 9.870 mL, 9.871 mL) are very close to the true value so this set
   is accurate, and they are close to one another, so the set is precise.
5. The true value of a neutralization volume is 9.870 mL. The values of the set of measurements (9.345 mL, 9.349 mL, 9.340 mL) are very different from the true value so
   this set is not accurate, and they are not close to one another, so the set is not precise.
6. The true value of a neutralization volume is 11.870 g. The set of measurements
   (11.873 g, 11.870 g, 11.877 g) is accurate but not precise since values are close to the
   true value but different from one another.

The set of measurements (11.852 g, 11.850 g, 11.851 g) is precise but not accurate.
The set of measurements (10.873 g, 10.870 g, 10.877 g) is not precise and not accurate.
The set of measurements (11.870 g, 11.871 g, 11.872 g) is precise and accurate.

7. The measurement 10.8 g shows the lowest number of significant figures (3 SF), so it is the least precise.

8. The measurement 10.8776 g shows the highest number of significant figures (6 SF), so it is the most precise.

9. Any non-zero digit is significant; the number of significant figures in 129.345 is then 6.

10. Zeros between non-zero digits are significant. The number of significant figures in 109.04 is 5 SF.

11. Zeros between non-zero digits are significant; the number of significant figures in 1,008 is 4 SF.

12. Zeros to the left of the first non-zero digit are not significant; the number of significant figures in 0.003 is 1 SF.

13. Zeros to the right of the last non-zero digit are significant if a decimal point is present. 1,500 is not a decimal number so zeros in 1,500 are not significant. The number of significant figures in 1,500 is 2.

14. The number 1.5×10^3 is a scientific notation. The number of significant figures in 1.5×10^3 is 2 (1.5 shows 2 SF).

15. In both addition and subtraction calculations, the answer cannot have more digits to the right of the decimal point than any of original numbers. 1.5 contains one decimal place and 1.0089 contains 4 decimal places so the result of (1.5 − 1.0089) should contain one decimal place.
    The calculator answer of (1.5 − 1.0089) is 0.4911
    The correct answer to express the arithmetic operation (1.5 − 1.0089) is 0.5 which contains 1 SF.

16. The correct answer to express the arithmetic operation (3.5 − 1.0077) should contain one decimal place.
    The calculator answer of (3.5 − 1.0077) is 2.4923.
    The correct answer to express the arithmetic operation (3.5 − 1.0077) is 2.5 which contains 2 SF.

17. The correct answer to express the arithmetic operation (2.51 + 1.0077) should contain two decimal places.
    The calculator answer of (2.51 + 1.0077) is 3.5177.
    The correct answer to express the arithmetic operation (3.5 − 1.0077) is 2.52, which contains 3 SF.

18. In multiplication (and division), the correct answer contains the lowest number of significant figures; 11.5 contains 3 SF and 3.0089 contains 5 SF so the answer of the calculation (11.5 × 3.0089) should contain 3 SF.

19. In multiplication and division, the correct answer contains the lowest number of significant figures; 1.5 contains 2 SF and 2.1067 contains 5 SF so the answer of the calculation (1.5/2.1067) should contain 2 SF.

20. In multiplication (and division), the correct answer contains the lowest number of significant figures; 12.5 contains the lowest number of SF which is 3, so the answer of the calculation (12.5× 2.0089)/2.089 should contain 3 SF.

21. If the number is less than 5, round down. (3 < 5), so rounding down (2.14<u>3</u>) to 3 SF results in 2.14.

22. If the number is greater than 5, round up. (9 > 5), so rounding up (1.34<u>79</u>) to 4 SF results in 1.348.

23. If the number is greater than 5, round up. (7 > 5), so rounding up (1.34<u>97</u>) to 4 SF results in 1.350. Note that the number after 9 is 10 so that 9 is replaced by 0 and 1 is added to 4 which becomes 5.

24. When the first digit deleted is 5, and 5 is the last digit or is followed by zeros, the preceding digit should be made to be even. In the value 3.00<u>15</u>, the preceding digit is 1 (odd digit) so rounding off 3.0015 to 4 SF results in 3.002 (2 is an even digit).

25. When the first digit deleted is 5, and 5 is the last digit or followed by zeros, the preceding digit should be made to be even. In the value 1.0<u>2</u>5, the preceding digit is 2 (an even digit) so rounding off 1.025 to 3 SF results in 1.02.

26. When the first digit deleted is 5, and 5 is the last digit or is followed by zeros, the preceding digit should be made even. In the value 11.2<u>0</u>5, the preceding digit is 0 (an even digit) so rounding off 11.205 to 4 SF results in 11.20.

27. In an arithmetic operation which contains both addition (or subtraction) and division (or multiplication) first, carry out the addition (or subtraction) operation and apply the rounding-off rules, then carry out the multiplication (or division) operation and apply the rounding-off rules. Finally, carry out the operation.

    9.41- 8.0 = ?

    Calculator answer: 1.41, but the result should contain one digit to the right of the decimal point

    The correct answer is 9.41 − 8.0 = 1.4

    2.00 × 2.0090 = ?

    Calculator answer: 4.018

    Because the result should contain 3 SF, the correct answer is 4.02.

    Finally, (9.41 − 8.0)/(2.00 × 2.0090) = 1.4/4.02

    1.4/4.02 = ?

    Calculator answer: 0.3482587065, but the result should contain 2 SF, so the correct answer is 0.35.

28. (9.03 cm × 88 cm) = ?

    Calculator answer: 794.64

    In the arithmetic operation (9.03 cm × 88 cm), 9.03 contains 3 SF and 88 contains 2 SF; therefore, the result should contain 2 SF. In this case, use the scientific notation.

    $794.64 = \underline{7.9}464 \times 10^3$

    4<5, so round down to obtain the correct answer of 7.9×10^3.

29. The mass of one chocolate pack is 22 g. The mass of 4 chocolate packs is (4 × 22 g). In the arithmetic operation (4 × 22 g), the number 4 is an exact number (counting number), so do not consider this number in giving the correct result.

    (22) contains 2 SF so the result should contain 2 SF. The correct answer is:

    (4 × 22 g) = 88 g.

30. In the arithmetic operation (19.5 + 19 + 18.5)/3, the number 3 is an exact number (counting number), so do not consider this number in giving the correct result.

    (19.5) contains 3 SF, (19) contains 2 SF and (18.5) contains 3 SF so the result should contain 2 SF. The correct answer is 19.

Q2.6

An alloy cube with an edge length of 20 cm has a mass of 80 g.

1. The mass of the alloy cube is m = 80 g.

   The volume of the alloy cube is V = 20 cm × 20 cm × 20 cm = 8,000 cm^3.

   The density ρ = m/V = 80/8,000 = 0.01 g/cm^3.

   The mass of the alloy cube in kg is m = 80 g = 80 × 10^{-3} kg = 0.08 kg.

   1 cm = 0.01 m, so 20 cm = 0.2 m

   The volume of the alloy cube in m^3 is V = 0.20 m × 0.20 m × 0.20 m = 0.0080 m^3.

   The density in kg/m^3 is ρ = m/V = 0.08 / 0.0080 = 10 kg/m^3.

2. Density is an intensive property which means that it does not depend on the amount of a substance. Therefore, when the alloy cube is cut into two pieces, the density does not change.

Q2.7

(3.234 − 0.067) × 2.05/ (1.0 + 1.2346) = ?

3.234 − 0.06 = ?

Calculator answer: 3.174

The result should contain two digits to the right of the decimal point.

The correct answer is 3.234 − 0.06 = 3.17.

(3.234 − 0.067) × 2.05 = 3.17 × 2.05 = ?

Calculator answer: 6.4885

The result should contain 3 SF, so the correct answer is 6.49.

(1.0 + 1.2346) = ?

Calculator answer: 2.2346

The result should contain one digit to the right of the decimal point.

The correct answer is 1.0 + 1.2346 = 2.2.

Finally, (3.234 − 0.067) × 2.05/ (1.0 + 1.2346) = 6.49/2.2

6.49/2.2 = ?

Calculator answer: 2.9̲5

The result should contain 2 SF, so the correct answer is:

(3.234 − 0.067) × 2.05/ (1.0 + 1.2346) = 3.0

b. (86 × 23) = ?

   (2 SF)(2 SF)(2 SF)

   Calculator answer: 1978 (4 SF), so the result should contain 2 SF.

   1,978 = 1.9̲78 × 10^3

   1.9̲78 is rounded off to 2.0 (2 SF) .

$$86 \times 23 = 2.0 \times 10^3 \left(\text{scientific notation} \right)$$

c. (20.0 × 100.0) = ?

   (2 SF)(2 SF)(2 SF)

   Calculator answer: 2,000

$$\left(20.0 \times 100.0 \right) = 2.00 \times 10^3 \left(\text{scientific notation} \right)$$

d. (50.00 × 140.0) = ?

   (4 SF)(4 SF)(4 SF)

   Calculator answer: 7000

$$50.00 \times 140.0 = 7.000 \times 10^3 \left(\text{scientific notation} \right)$$

e. (4.05 − 0.2 + 2.123) / (0.0501 × 0.40) = ?

   4.05 − 0.2̲ + 2.123 = ?

   Calculator answer: 5.9̲73

   4.0̲5 contains 2 digits after the decimal point, 0.2̲ contains one digit after the decimal point and 2.1̲23 contains 3 digits after the decimal point. Therefore, the answer should contain one digit to the right of the decimal point.

   The correct answer is 34.05 − 0.2 + 2.123 = 6.0.

   (0.0501 × 0.40) = ?

   Calculator answer: 0.02̲004

   The result should contain 2 SF.

   The correct answer is 0.020.

   Finally, (4.05 − 0.2 + 2.123) / (0.0501 × 0.40) = 6.0/0.020

   6.0/0.020 = ?

   Calculator answer: 300

   The result should contain 2 SF so, the correct answer is:

$$\left(4.05 - 0.2 + 2.123 \right) / \left(0.0501 \times 0.40 \right) = 3.0 \times 10^2 \left(\text{scientific notation} \right)$$

Atoms, Molecules, Ions and Moles

3

3.1 OBJECTIVES

At the end of the present chapter, the student will be able to

1. Describe the components of the atom and differentiate between the properties of each component.
2. Differentiate between the atom, the molecule and the ion.
3. Recognize the different types of chemical formulas.
4. Recognize the isotopes.
5. Define the ionic compounds.
6. Define the mole.

3.2 A BRIEF HISTORY OF THE ATOM

○ **Democritus** (was born around 460 BC, although there are disagreements about the exact year) named the smallest piece of matter 'atomos', which means uncuttable or indivisible structure.

○ **Dalton** (1803) suggested that an element is composed of tiny particles called atoms. The atom can be defined as the smallest particle of an element that can enter into a chemical reaction. According to Dalton's theory, all atoms of a given element show the same chemical properties, while atoms of different elements show different properties. Compounds are formed when atoms of two or more elements combine one with another. The relative numbers of atoms of each kind in a compound are definite and constant.

○ **Thomson** (1897) discovered the **electrons**, during a cathode ray experiment. He suggested that the atom was like a plum pudding which is is an English dessert like a blueberry muffin. According to Thomson's plum pudding model of the atom, the negatively charged electrons were surrounded by a volume of positive charge. Indeed, electrons were embedded in a positively charged sphere like blueberries stuck into a muffin. Thomson determined the electron's charge-to-mass ratio (q/m) to be 1.7588×10^8 Coulomb g^{-1}. The value of the ratio q/m is high which indicates that the electron charge is very low.

○ **Robert A. Millikan and Harvey Fletcher** (1909) determined the electron charge (q = 1.6022×10^{-19} Coulomb), using the oil drop experiment. The mass of the electron was then determined since the ratio q/m was known. The mass of the electron is m = 9.1096×10^{-31} kg.

○ Using the golden foil experiment, **Rutherford** (1911) showed that the atom is mostly empty space and contains a nucleus. The nucleus contains positively charged particles called **protons**. Rutherford noted that atomic mass is equal to almost twice the total mass of protons and electrons in the atom. Atom contains particles other than protons and electrons, with these particles being neutral.

○ **Chadwick** (1932) noted that bombarding atoms of some elements, such as beryllium, with alpha particles caused these atoms to release small particles, different from protons and electrons, and which are electrically neutral. Chadwick called these particles **neutrons**. Chadwick determined the mass of the neutron to equal 1.6749×10^{-24} kg.

DOI: 10.1201/9781003257059-3

3.3 ATOMIC STRUCTURE

An atom consists of three subatomic particles (protons, neutrons and electrons). Protons and neutrons form the nucleus of the atom:

- A proton (p^+) has a positive charge (+) and has a mass nearly equal to that of a hydrogen atom. The mass of a proton equals 1.672×10^{-27} kg and its charge is 1.6×10^{-19} Coulomb.
- A neutron (n^0) is uncharged and has a mass slightly greater than that of the proton (1.675×10^{-27} kg).
- An electron (e^-) has a negative charge (–) and a tiny mass. The mass of the electron equals 9.11×10^{-31} kg and its charge is equal to -1.6×10^{-19} Coulomb.

Note that

- The proton and neutron have comparable masses.
- The mass of the proton (or the mass of the neutron) is significantly greater than that of the electron.

GET SMART

IS AN ION SMALLER IN MASS THAN THE NEUTRAL ATOM FROM WHICH IS FORMED?

The mass of a proton (or the mass of a neutron) is significantly greater than the mass of an electron. Consequently, the mass of electrons can be ignored in calculating the mass of an atom, especially when the atom contains few electrons; this means that the mass of the atom is approximately the sum of the masses of the protons and neutrons. Since the proton and neutron have comparable mass, **the mass of an atom is approximately twice the mass of protons in the atom.** This also means that, **when an atom loses or gains electrons, its atomic mass does not change markedly.**

Example 1: A lithium (Li) atom contains three protons, three neutrons and three electrons. The approximate mass of a Li atom is
$m = 2 \times (3 \times 1.672 \times 10^{-27}$ kg$) = 1.003 \times 10^{-26}$ kg.
The real mass of Li atom is 1.152×10^{-26} kg $\pm 0.0003 \times 10^{-26}$ kg.

Example 2: The atomic mass of iron is 56 amu (atomic mass unit). When an iron atom loses three electrons, it becomes Fe^{3+}. The atomic mass of Fe^{3+} is 56 amu which is the same as that of Fe.

3.4 SYMBOL OF AN ELEMENT

The full chemical *symbol* for an *element* shows its atomic number at the bottom and its *mass number* at the top:

$$^A_Z X$$

where X represents the element chemical symbol, A is the mass number and Z is the atomic number.

The atomic number (Z) is the number of protons in the atom.

The mass number (A) is the sum of the number of protons and the number of neutrons (N)

$$A = N + Z$$

The number of neutrons (N) is

$$N = A - Z$$

For a neutral atom, the number of protons equals the number of electrons:

$$\text{Number of protons (Z)} = \text{number of electrons (Ne}^-)$$

Practice 3.1 How many protons, neutrons and electrons are there are there in $^{31}_{15}P$?

Answer:

The atomic number Z = 15, therefore the number of protons in a phosphorus (P) atom is 15.

The mass number A = 31, therefore the total number of protons and neutrons is 31.

The number of neutrons N = 31–15 = 16 neutrons.

P is a neutral atom (there is no charge sign in the atom symbol), so the number of protons is equal to the number of electrons (Ne$^-$): Ne$^-$ = Z = 15 electrons.

3.5 ISOTOPES

Atoms of a given element that have the same number of protons and **differ in the number of neutrons** are called isotopes.

Examples:

- ○ **Isotopes of hydrogen:** ^1_1H **(hydrogen)**, ^2_1H **(deuterium)**, ^3_1H **(tritium)**
  - • ^1_1H contains one proton, one electron and zero neutrons.
  - • ^2_1H contains one proton, one electron and one neutron.
  - • ^3_1H contains one proton, one electron and two neutrons.
- ○ **Isotopes of carbon:** $^{12}_6C$, $^{13}_6C$, $^{14}_6C$
  - • $^{12}_6C$ contains 6 protons, 6 electrons and 6 neutrons.
  - • $^{13}_6C$ contains 6 protons, 6 electrons and 7 neutrons.
  - • $^{14}_6C$ contains 6 protons, 6 electrons and 8 neutrons.
- ○ **Isotopes of oxygen:** $^{16}_8O$, $^{17}_8O$, $^{18}_8O$
  - • $^{16}_8O$ contains 8 protons, 8 electrons and 8 neutrons.
  - • $^{17}_8O$ contains 8 protons, 8 electrons and 9 neutrons.
  - • $^{18}_8O$ contains 8 protons, 8 electrons and 10 neutrons.
- ○ **Examples of isotopes of cobalt:** $^{56}_{27}Co$, $^{58}_{27}Co$, $^{60}_{27}Co$
  - • $^{56}_{27}Co$ contains 27 protons, 27 electrons and 29 neutrons.
  - • $^{58}_{27}Co$ contains 27 protons, 27 electrons and 31 neutrons.
  - • $^{60}_{27}Co$ contains 27 protons, 27 electrons and 33 neutrons.

3.6 MOLECULES

A molecule is an electrically neutral group of two or more atoms bonded together that may or may not be the same element. The molecule is the smallest unit ("building unit") of a compound and has all the properties of the substance.

Homonuclear molecules are molecules formed from one type of atom

Examples: O_2 (oxygen), O_3 (ozone), H_2 (hydrogen) and N_2 (nitrogen).

Heteronuclear molecules are molecules formed from more than one type of atom.

Examples: H_2O (water), NH_3 (ammonia), NaCl (sodium chloride), Na_2CO_3 (sodium carbonate), $CaSO_4 \cdot 2H_2O$ (gypsum) and $Fe(NO_3)_3$ (iron(III) nitrate).

GET SMART

WHAT ARE THE BUILDING UNITS OF AN ELEMENT AND A COMPOUND?

The atom is the smallest (building) unit of an element and has all the properties of the substance.

The heteronuclear molecule is the smallest (building) unit of a compound and has all the properties of the substance.

Example: The building unit of the aluminum substance is the Al atom and the building unit of water is the H_2O molecule.

3.7 CHEMICAL FORMULAS

A molecular formula is the formula that represents the actual number and types of atoms in a molecule.

Examples: H_2O_2 (hydrogen peroxide), C_2H_6 (ethane), C_6H_6 (benzene) and $C_6H_{12}O_6$ (glucose).

An empirical ("simplest") formula is the formula that gives only the relative number of atoms of each type in a molecule. It gives rise to the smallest set of whole numbers of atoms.

Examples:

The empirical formula of hydrogen peroxide (H_2O_2) is HO
The empirical formula of ethene (C_2H_4) is CH_2
The empirical formula of glucose ($C_6H_{12}O_6$) is CH_2O

Note that, for several compounds, the empirical formula can be the same as the molecular formula, e.g., water (H_2O), sodium chloride (NaCl), calcium carbonate ($CaCO_3$).

A structural formula is the formula that indicates the attachment of atoms (Figure 3.1).

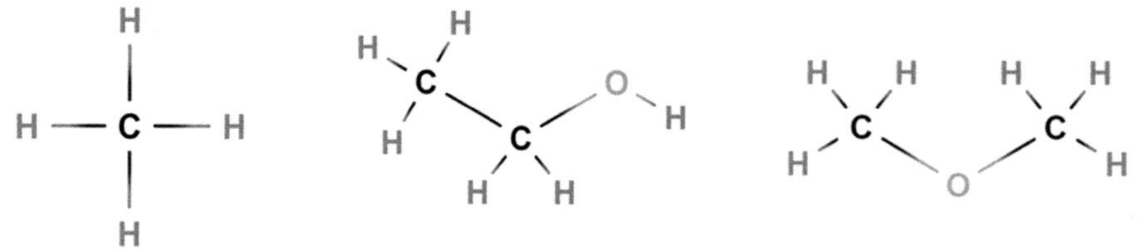

FIGURE 3.1 Structural formulas of methane (CH_4), ethanol (C_2H_5OH) and ether (C_2H_6O)

GET SMART

HOW IS THE EMPIRICAL FORMULA DETERMINED FROM THE MOLECULAR FORMULA?

Find the **greatest common divisor (GCD)** of the subscripts of the molecular formula of the compound. Then, divide all subscripts by the GCD to find the empirical formula.

Example: The molecular formula of the lactic acid is $C_3H_6O_3$. The greatest common divisor of 3, 6 and 3 is 3. Divide all numbers by 3 and you find the empirical formula of lactic acid: CH_2O.

HOW CAN A STRUCTURAL FORMULA BE RECOGNIZED?

The attachment of atoms is indicated by continuous lines between the atoms that constitute the molecule. Therefore, any chemical formula that contains continuous lines is a structural formula.

Example: The structural formula of carbon dioxide CO_2 is O=C=O.

3.8 IONS

An ion is an atom or a group of atoms (molecule) in which the number of electrons differs from the number of protons. Therefore, an ion has a net positive or a net negative electric charge.

A neutral atom or a molecule becomes an ion either by losing electron(s) (to form a cation) or by gaining electron(s) (to form an anion) (Figure 3.2).

Examples: Cations: H^+, Ca^{2+}, Fe^{3+}, H_3O^+.

Anions: Cl^-, O^{2-}, HCO_3^-, CO_3^{2-}

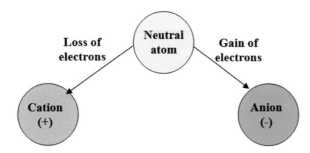

FIGURE 3.2 A neutral atom becomes a cation by losing electron(s) and an anion by gaining electron(s).

Practice 3.2 Calculate the number of protons, electrons and neutrons in $^{9}_{4}Be^{2+}$ and $^{17}_{8}O^{2-}$.

Answer:

The numbers of protons, electrons, and neutrons in $^{9}_{4}Be^{2+}$ and $^{17}_{8}O^{2-}$ are summarized in Table 3.1.

TABLE 3.1 Numbers of protons, electrons, and neutrons in $^{9}_{4}Be^{2+}$ and $^{17}_{8}O^{2-}$

Atom	Number of protons	Number of electrons	Number of neutrons	Key explanation
$^{9}_{4}Be^{2+}$	4	2	5	The neutral atom (Be) has four electrons and loses two electrons.
$^{17}_{8}O^{2-}$	8	10	9	The neutral atom (O) has eight electrons and gains two electrons.

3.9 IONIC COMPOUNDS

Since metals tend to lose electrons and non-metals tend to gain electrons, metals and non-metals form compounds with each other. These compounds are called ionic compounds.

Examples:

- Sodium (Na) is a metal which loses one electron (Na^+), and chlorine (Cl) is a non-metal which gains an electron (Cl^-). The ionic compound formed between Na and Cl is NaCl.

- Magnesium (Mg) is a metal which loses two electrons (Mg^{2+}), and sulfur (S) is a non-metal which gains two electrons (S^{2-}). The ionic compound formed between Mg and S is MgS.
- Aluminum (Al) is a metal which loses three electrons (Al^{3+}), and chlorine (Cl) is a non-metal which gains an electron (Cl^-). The ionic compound formed between Al and Cl is $AlCl_3$.
- The ionic compound formed between Mg^{2+} and nitrate (NO_3^-) is $Mg(NO_3)_2$.
- The ionic compound formed between Co^{2+} and chloride (Cl^-) is $CoCl_2$.

GET SMART

HOW CAN YOU DETERMINE THE CHEMICAL FORMULA OF AN IONIC COMPOUND?

The **cross-over rule** can be used to determine the chemical formula of ionic compounds. This rule suggests that you write down the symbols of the elements (metal and non-metal), then you take the charge of the metal (without the sign) and put it down in the nonmetal symbol and you take the charge of the nonmetal (without the sign) and put it down in the metal symbol.

Example: aluminum oxide

$$Al^{3+} \diagdown\diagup O^{2-} \quad \rightarrow \quad Al_2O_3$$

3.10 MOLES

A mole is the amount of a substance that contains the same number of entities as there are atoms in exactly 12 g of carbon-12. This number is equal to $\mathbf{6.022 \times 10^{23}}$. This number is called Avogadro's number and is abbreviated to N_A.

One mole (1 mol) contains $N_A = \mathbf{6.022 \times 10^{23}}$ **entities.** Entities are atoms (e.g., H, C, O, Cl, Ca, Fe, ...), molecules (e.g., H_2, O_2, $CaCl_2$, NaOH, ...) or ions (e.g., H^+, Cl^-, OH^-, SO_4^{2-}, NH_4^+, ...).

Examples:
One mole of chlorine (Cl) contains $\mathbf{6.022 \times 10^{23}}$ **atoms** of Cl
One mole of calcium carbonate ($CaCO_3$) contains $\mathbf{6.022 \times 10^{23}}$ **molecules** of $CaCO_3$
One mole of carbonate (CO_3^-) contains $\mathbf{6.022 \times 10^{23}}$ **ions** of CO_3^-

---

GET SMART

WHAT IS THE DIFFERENCE BETWEEN ONE MOLE OF A GIVEN ELEMENT (OR A MOLECULE) AND ONE MOLE OF ANOTHER ELEMENT (OR A MOLECULE)?

They have the same number of atoms or molecules but they have different masses.

Examples: One mole of calcium (Ca) (or sodium fluoride NaF) and one mole of chlorine (Cl) (or calcium carbonate $CaCO_3$) have the same number of atoms (or molecules) but they have different masses.

CHECK YOUR READING

What are the constituents of the atom?
What do the atomic number and mass number mean?
What are isotopes?
What is a molecule? What is an ion? What is a mole?
What is the difference between a molecule and an atom?

What are the different types of a chemical formula?
How can the molecular formula of an ionic compound be determined?

SUMMARY OF CHAPTER 3

o **Thomson** (1897) discovered **electrons** when carrying out a cathode ray experiment.
o **Robert A. Millikan and Harvey Fletcher** (1909) determined the electron charge (**q = 1.6022 × 10^{-19} coulomb**), using the oil drop experiment. The mass of the electron was then determined (**m$_e$ = 9.1096 × 10^{-28} g**).
o Using the golden foil experiment, **Rutherford** (1911) showed that the atom is mostly empty space and contains positively charged particles called **protons**. The mass of the proton equals **m$_p$ = 1.672 × 10^{-27} kg**.
o **Chadwick** (1932) discovered electrically neutral particles called neutrons by bombarding atoms of some elements with alpha particles. The mass of the neutron equals **m$_n$ = 1.674 × 10^{-24} kg**).
o The full chemical *symbol* for an *element* (X) shows its **atomic number (Z)** at the bottom and its *mass number* **(A)** at the top: $^A_Z X$
The atomic number (Z) is the number of protons in the atom
The mass number (A) is the sum of the number of protons and the number of neutrons (N)

$$A = N + Z$$

For a neutral atom: number of protons (Z) = number of electrons (Ne$^-$)
Example: carbon $^{14}_6C$ contains six protons, eight (=14−6) neutrons and six electrons; carbon $^{14}_6C$ is a neutral atom since there is no positive or negative charge assigned to the symbol.
o **Isotopes** are atoms of a given element that have the same number of protons but **differ with respect to the number of neutrons**
Example: isotopes of carbon $^{12}_6C$, $^{13}_6C$ and $^{14}_6C$.
Molecule is an electrically neutral group of two or more atoms bonded together that may or may not be the same element.
Homonuclear molecules: molecules formed with one type of atom
Examples: oxygen (O_2), ozone (O_3), hydrogen (H_2), nitrogen (N_2)
Heteronuclear molecules: molecules formed with more than one type of atom. The heteronuclear molecule is the smallest (building up) unit of a compound that has all the properties of the substance.
Examples: water (H_2O), ammonia (NH_3), sodium chloride (NaCl), sodium carbonate (Na_2CO_3), gypsum ($CaSO_4·2H_2O$), iron (III) nitrate [$Fe(NO_3)_3$]
The molecular formula is the formula that represents the actual number and types of atoms in a molecule
Examples: hydrogen peroxide (H_2O_2), ethane (C_2H_6), benzene (C_6H_6), glucose ($C_6H_{12}O_6$)
The empirical (or **simplest) formula** is the formula that gives only the relative number of atoms of each type in a molecule (giving rise to the smallest set of whole numbers of atoms).
Examples: HO is the empirical formula of hydrogen peroxide H_2O_2, (CH_3) is the empirical formula of ethane (C_2H_6), CH is the empirical formula of benzene (C_6H_6); CH_2O is the empirical formula of glucose ($C_6H_{12}O_6$), (H_2O) is the empirical formula of water H_2O.
The structural formula is the formula that indicates the attachment of atoms.
Examples: The structural formula of carbon dioxide CO_2 is O = C = O and the structural formula of methane CH_4 is:

A neutral atom or a molecule becomes **an ion** either by losing electron(s) (to form a **cation**) or by gaining electron(s) (**anion**)

Cations: H^+, Ca^{2+}, Fe^{3+}, H_3O^+, ...
Anions: Cl^-, O^{2-}, HCO_3^-, CO_3^{2-}, ...
Metals and non-metals form compounds with each other. These compounds are called **ionic compounds.**
Examples: Aluminum (Al) is a metal which loses three electrons (Al^{3+}) and chlorine (Cl) is a non-metal which gains one electron (Cl^-). The ionic compound formed between Al and Cl is $AlCl_3$, to balance the + and −.
A mole is the amount of a substance that contains the same number of entities as there are atoms in exactly 12 g of carbon-12. This number is $N_A = 6.022 \times 10^{23}$. The number N_A is called Avogadro's number.
One mole (1 mol) contains 6.022×10^{23} entities (atoms, molecules, or ions).

PRACTICE ON CHAPTER 3

Q3.1 Complete the following sentences
An _____ is formed by _____ with a positive charge, _____ which have no charge and _____ with a negative charge. The atom is the _____ of the element and has all the properties of the substance. When a neutral atom loses electrons, it becomes a _____. When a neutral atom gains electrons, it becomes an _____.
When two or more atoms, that may or may not be the same element, bond together they form a _____. The molecule is the smallest (building) unit of _____ and has all the properties of the substance. _____ molecules are molecules formed with one type of atoms and _____ molecules are molecules formed with more than one type of atoms.
There are three types of chemical formulas: the _____ formula which gives the actual number and types of atoms in the molecule, the _____ that gives the relative number of atoms in the molecule and the structural formula which indicates the _____ of atoms in the molecule.

Q3.2 Match the followings

a. An ion	**1.** is formed between metal and nonmetal.
b. Isotopes	**2.** contains N_A entities (atoms, molecules or ions).
c. An ionic compound	**3.** is the number of protons.
d. A mole	**4.** are different atoms of a given element that differ in the number of neutrons.
e. The mass number	**5.** is the number of protons and neutrons.
f. The atomic number	**6.** is obtained when a neutral atom loses or gains electrons.

Q3.3 Choose the correct answer
1. **The cathode ray experiment allowed Thomson to discover the**
    a) electron
    b) proton
    c) neutron
    d) atom
2. **Using the golden foil experiment, Rutherford showed that the atom was mostly empty space and contains**
    a) electrons
    b) protons
    c) neutrons
    d) atoms
3. **Chadwick discovered the _____ by bombarding atoms of some elements with alpha particles.**
    a) electrons
    b) protons
    c) neutrons
    d) atoms

4. **The mass of the proton is approximately the same as the mass of the**
   a) electron
   b) alpha particle
   c) neutron
   d) atom
5. **The mass of the electron is significantly lower than the mass of the proton or the neutron**
   a) True
   b) False
6. **The atomic mass of iron ($_{26}$Fe) is 56. The atomic mass of the ferric ion (Fe^{3+}) is**
   a) 26
   b) 56
   c) 29
   d) 59

Q3.4 Choose the correct answer
1. **Which of the following is a homonuclear molecule?**
   a) H_2O
   b) CaF_2
   c) KNO_3
   d) Cl_2
2. **Which of the following is a heteronuclear molecule?**
   a) H_2
   b) F_2
   c) $CaCO_3$
   d) O_3
3. **An ion is obtained when a neutral atom gains or loses electrons.**
   a) True
   b) False
4. **Cations form when**
   a) neutrons are added to or removed from an atom
   b) electrons are removed from an atom
   c) electrons are added to an atom
   d) an atom gains or loses electrons
5. **When a neutral atom gains electrons, it becomes**
   a) a proton
   b) a neutron
   c) a cation
   d) an anion
6. **When the atom Ti loses four electrons, it becomes**
   a) Ti^+
   b) Ti^{4+}
   c) Ti^{2+}
   d) Ti^{4-}
7. **When the cation Ti^{3+} gains an electron, it becomes**
   a) Ti^+
   b) Ti^{4+}
   c) Ti^{2+}
   d) Ti^-
8. **When the oxygen atom (O) gains two electrons, it becomes**
   a) O^+
   b) O^{2+}
   c) O_2
   d) O^{2-}
9. **The atomic number of an element is the number of**
   a) protons and electrons
   b) neutrons and electrons

c) protons

d) protons and neutrons

10. **In the symbol of an atom $_Z^A X$, A is the**

a) atomic number and represents the number of protons in the atom

b) mass number and represents the number of protons and neutrons in the atom

c) mass number and represents the number of protons and electrons in the atom

d) chemical symbol of the atom

11. **The numbers of protons (p^+), neutrons (n^0) and electrons (e^-) in $_{55}^{133}$Cs are**

a) 55 p^+, 133 n^0 and 55 e^-

b) 55 p^+, 78 n^0 and 78 e^-

c) 55 p^+, 78 n^0 and 55 e^-

d) 133 p^+, 78 n^0 and 55 e^-

12. **The numbers of protons (p^+), neutrons (n^0) and electrons (e^-) in radium ion $_{88}^{226}$Ra^{2+} are**

a) 88 p^+, 226 n^0 and 88 e^-

b) 86 p^+, 138 n^0 and 88 e^-

c) 88 p^+, 138 n^0 and 86 e^-

d) 88 p^+, 138 n^0 and 90 e^-

13. **Isotopes are different atoms of a given element that differ in the number of**

a) protons

b) electrons

c) neutrons

d) protons and neutrons

14. **^{12}C, ^{13}C and ^{14}C are called**

a) molecules

b) compounds

c) ions

d) isotopes

15. **The formula that indicates the attachment of atoms in a molecule is the**

a) structural formula

b) ionic formula

c) molecular formula

d) simplest formula

16. **The formula that gives the relative number and types of atoms in a molecule is the**

a) structural formula

b) ionic formula

c) molecular formula

d) simplest formula

17. **The molecular formula**

a) gives the relative number and types of atoms in a molecule

b) gives the actual number and types of atoms in a molecule

c) indicates the attachment of atoms in a molecule

d) gives the number of protons in an atom

18. **The empirical (simplest) formula of glucose ($C_6H_{12}O_6$) is**

a) CH_2O

b) $C_3H_4O_3$

c) CHO

d) $C_6H_{12}O_6$

19. **Which of the following is the structural formula of ether (C_2H_6O)?**

a) C_2H_6O

b) CHO

c)

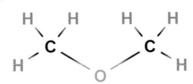

d)

H   H   H   H
 \ /   \ /
  C   C
 / \   / \
H   \   /   H
     O_____

20. **The ionic compound is formed between**
    a) two metals
    b) two non-metals
    c) a metal and a non-metal
    d) two cations
21. **The formula of the ionic compound formed between Fe^{2+} and SO_4^{2-} ions is**
    a) $FeSO_4$
    b) $Fe_3(SO_4)_2$
    c) $Fe_2(SO_4)_3$
    d) $Fe_3(SO_4)$
22. **The formula of the ionic compound formed between Fe^{3+} and O^{2-} ions is**
    a) Fe_2O_3
    b) Fe_3O_2
    c) Fe_2O_5
    d) FeO
23. **A mole**
    a) is the amount of a substance that contains the same number of entities as there are atoms in exactly 12 g of carbon-12
    b) contains N_A entities (atoms, ions or molecules)
    c) is a fundamental quantity
    d) All the above are correct
24. **One mole of calcium ($CaCl_2$) and one mole of calcium carbonate ($CaCO_3$) do not have the same number of molecules.**
    a) True
    b) False
25. **One mole of calcium ($CaCl_2$) and one mole of calcium carbonate ($CaCO_3$) have the same mass.**
    a) True
    b) False

ANSWERS TO QUESTIONS

Q3.1

An **atom** is formed by **protons** with a positive charge, **neutrons** which have no charge and **electrons** with a negative charge. The atom is the **smallest (building) unit** of the element and has all the properties of the substance. When a neutral atom loses electrons, it becomes a **cation**. When it gains electrons, it becomes an **anion.**

When two or more atoms, that may or may not be the same element, bond together they form a **molecule.** The molecule is the smallest (building) unit of **the compound** and has all the properties of the substance. **Homonuclear** molecules are molecules formed from one type of atoms and **heteronuclear** molecules are molecules formed from more than one type of atoms.

There are three types of chemical formulas: the **molecular** formula which gives the actual number and types of atoms in the molecule, the **empirical formula** that gives the relative number of atoms in the molecule and the structural formula which indicates the **attachment** of atoms in the molecule.

Q3.2   (a, 6); (b, 4); (c, 1); (d, 2); (e, 5); (f, 3)

Q3.3
1. a
2. b
3. c
4. c
5. a
6. b

Q3.4
1. d
2. c
3. a
4. b
5. d
6. b
7. c
8. d
9. c
10. b
11. c
12. b
13. c
14. d
15. a
16. d
17. b
18. a
19. d
20. c
21. a
22. a
23. d
24. b
25. b

KEY EXPLANATIONS

Q3.3
3. The mass of the proton (= 1.672×10^{-27} kg) is approximately the same as the mass of the neutron (1.675×10^{-27} kg).
4. The mass of the electron (= 9.11×10^{-31} kg) is significantly lower than the mass of the proton (or the neutron). This means that when an atom loses a few electrons, its atomic mass does not change markedly.

5. When an atom loses few electrons, its mass does not change since the mass of the electron is significantly lower than the mass of the proton or the neutron. Therefore, the atomic mass of iron Fe is 56 amu and when Fe loses three electrons, it becomes Fe^{3+} and its mass does not change but remains equal to 56 amu.

Q3.4

1. Homonuclear molecules are molecules formed from one type of atom. Cl_2 is formed from one type of atoms (Cl). Therefore, Cl_2 is a homonuclear molecule.
2. Heteronuclear molecules are molecules formed from more than one type of atom. $CaCO_3$ is formed from three types of atoms (Ca, C and O). Therefore, $CaCO_3$ is a heteronuclear molecule.
4. Cations have a positive charge. They form when electrons are removed from an atom
5. Anions have a negative charge. They form when a neutral atom gains electrons.
7. The cation Ti^{3+} has a charge (3+). When it gains an electron (charge –1), its charge becomes (+3–1 = +2). Therefore, when the cation Ti^{3+} gains an electron, it becomes Ti^{2+}.
11. In $^{133}_{55}Cs$, the numbers of protons (p^+) is 55, the number of electrons is 55 since Cs is a neutral atom (no sign is assigned to the symbol of the element) and the number of neutrons is (133–55 = 78).
12. In $^{226}_{88}Ra^{2+}$, the numbers of protons (p^+) is 88, the number of electrons (e^-) is 88–2 = 86 since Ra^{2+} means that the neutral atom $_{88}Ra$ loses two electrons, and the number of neutrons (n^0) is 226–88 = 138.
18. The empirical (simplest) formula of glucose ($C_6H_{12}O_6$) is obtained by dividing the subscripts in the molecular formula of glucose (6, 12 and 6) by the greatest common divisor, which is 6. Therefore, the simplest formula of glucose ($C_6H_{12}O_6$) is CH_2O.
19. The structural formula indicates the attachment of atoms in a molecule. When two compounds have the same molecular formula, i. e., ethanol and ether, with the same molecular formula C_2H_6O, they could be distinguished by the structural formula (Figures 3.4 and 3.5):

Ethanol   Ether

22. An ionic compound is formed between a metal and a non-metal. To determine the chemical formula of the ionic compound formed between Fe^{3+} and O^{2-}, use the **cross- over rule.** To apply this rule, follow the following steps:
    1. Write down the symbols of the elements (metal and non-metal)
    2. Take the charge of the metal (without the sign) and put it down in the non-metal symbol
    3. Take the charge of the non-metal (without the sign) and put it down in the metal symbol.

$$Fe^{3+} \diagdown O^{2-} \rightarrow Fe_2O_3$$

24 and 25 One mole of a compound and one mole of another compound necessarily have the same number of molecules ($N_A = 6.022 \times 10^{23}$) but, in most cases, different masses. One mole of $CaCl_2$ and one mole of $CaCO_3$ have the same number of molecules but they have different masses. Indeed, the molar mass of $CaCl_2$ is 111 g mol^{-1} and the molar mass of $CaCO_3$ is 100 g mol^{-1}.

Calculations in Chemistry

4

4.1 OBJECTIVES

At the end of this chapter, the student will be able to:

1. Distinguish between the molar mass and the atomic mass or molecular mass.
2. Calculate the number of moles of a substance.
3. Calculate the number of atoms or molecules in n moles (or in a mass) of a substance.
4. Distinguish between the molarity, the molality and the solubility of a solution.
5. Calculate the mass and the mass percent of an element in a compound.
6. Determine the empirical and molecular formulas from their molar masses, from the masses of the constituent elements of the compound and from the products of a chemical reaction.
7. Recognize a balanced chemical equation and show the relationships between the number of moles of the reactants and the number of moles of the products.
8. Calculate the final concentration after dilution.

4.2 ATOMIC MASS (ATOMIC WEIGHT)

An atom contains protons, electrons, and neutrons. Isotopes are atoms of a given element that have the same number of protons and differ in number of neutrons. The atomic mass (defined as the mass of one atom) is the average mass of all the isotopes of an atom with respect to their abundances. It is commonly expressed in unified atomic mass units (amu) where, by international agreement, 1 atomic mass unit is defined as 1/12 of the mass of a single carbon-12 atom:

$$1\,amu = \frac{1}{12} \times \frac{12g}{N_A} = \frac{1}{N_A}\,g = 1.660 \times 10^{-24}\,g$$

So, by definition, a carbon (^{12}C) atom has a mass of 12 amu. The atomic masses of all elements are given in the periodic table of elements. For example, the atomic mass of hydrogen (H) is 1 amu and the atomic mass of calcium (Ca) is 40 amu.

Practice 4.1 Magnesium has three isotopes: $^{24}_{12}Mg$, $^{25}_{12}Mg$ and $^{26}_{12}Mg$. The abundances of these isotopes are 78.99%, 10.00% and 11.01%, respectively. The atomic masses (in amu) of these isotopes are 23.985, 24.986 and 25.983, respectively. Calculate the average atomic mass of magnesium Mg.

Answer:
First, we calculate the contribution of each isotope to the atomic mass of the element defined by:

$$Contribution = Atomic\,mass\,of\,the\,isotope \times \frac{Abundance}{100}$$

DOI: 10.1201/9781003257059-4

$$^{24}_{12}Mg \text{ contribution} = 23.985 \times (78.99/100) = 18.9458 \text{ amu}$$

$$^{25}_{12}Mg \text{ contribution} = 24.986 \times (10.00/100) = 2.4986 \text{ amu}$$

$$^{26}_{12}Mg \text{ contribution} = 25.983 \times (11.01/100) = 2.8607 \text{ amu}$$

The average atomic mass of magnesium is the sum of the contributions of all isotopes: $18.9458 + 2.4986 + 2.8607 = 24.305$ amu.

GET SMART

The average atomic mass of an element takes into account the abundance of each isotope. For this reason, it is different from the average of the atomic masses of all isotopes of this element. For example, to calculate the average atomic mass of Mg, **do not calculate the average of the atomic masses of all isotopes of Mg** $(23.985 + 24.986 + 25.983)/3 = 24.985$; this value is different from the average atomic mass of Mg (calculated in practice 4.1) and should not be considered as the average atomic mass of Mg.

4.3 MOLECULAR MASS (FORMULA MASS)

A molecule is an electrically neutral group of two or more atoms bonded together that may or may not be the same element.

The molecular mass (defined as the mass of one molecule) is the sum of atomic masses of all the constituent atoms of the molecule. It is commonly expressed in atomic mass units (amu).

Practice 4.2 Calculate the molecular masses of sulfuric acid (H_2SO_4), glucose ($C_6H_{12}O_6$), calcium sulfate hemihydrate ($CaSO_4 \cdot 1/2H_2O$) and iron (II) nitrate ($Fe(NO_3)_2$).

Atomic masses: H = 1, C = 12, N = 14, O = 16, S = 32, Ca = 40 and Fe = 56.

Answer:

- One molecule of sulfuric acid (H_2SO_4) contains two atoms of hydrogen (H), one atom of sulfur (S) and four atoms of oxygen (O). The molecular mass of sulfuric acid (H_2SO_4):
  $= 2 \times \text{(atomic mass of H)} + 1 \times \text{(atomic mass of S)} + 4 \times \text{(atomic mass of O)}$
  $= (2 \times 1) + (1 \times 32) + (4 \times 16) = 98$ amu.
- One molecule of glucose ($C_6H_{12}O_6$) contains six atoms of carbon (C), twelve atoms of hydrogen (H) and six atoms of oxygen (O). The molecular mass of glucose ($C_6H_{12}O_6$):
  $= 6 \times \text{(atomic mass of C)} + 12 \text{ (atomic mass of H)} + 6 \text{ (atomic mass of O)}$
  $= 6 \times 12 + 12 \times 1 + 6 \times 16$
  $= 180$ amu.
- Calcium sulfate hemihydrate ($CaSO_4 \cdot 1/2H_2O$) is a compound. One molecule of this compound contains one atom of Ca, one atom of S, four atoms of O and 1/2 molecule of H_2O. The molecular mass of calcium sulfate hemihydrate ($CaSO_4 \cdot 1/2H_2O$):
  $= 1 \times \text{(atomic mass of Ca)} + 1 \times \text{(atomic mass of S)} + 4 \times \text{(atomic mass of O)} + \frac{1}{2} [2 \times \text{(atomic mass of H)} + 1 \times \text{(atomic mass of O)}]$
  $= 1 \times 40 + 1 \times 32 + 4 \times 16 + \frac{1}{2} [2 \times 1 + 1 \times 16]$
  $= 145$ amu.
- One molecule of iron (II) nitrate ($Fe(NO_3)_2$) contains one atom of Fe, two atoms of N and six ($= 2 \times 3$) atoms of O. The molecular mass of iron (II) nitrate ($Fe(NO_3)_2$):
  $= 1 \times \text{(atomic mass of Fe)} + 2 \times \text{(atomic mass of N)} + 6 \times \text{(atomic mass of O)}$
  $= 1 \times 56 + 2 \times 14 + 6 \times 16$
  $= 180$ amu.

GET SMART

HOW CAN THE MASS OF ONE ATOM OR ONE MOLECULE BE DETERMINED IN GRAMS?

The mass of an atom $\left(\text{or a molecule}\right)$ in grams

$$= \frac{\text{mass of one atom}\left(\text{or molecule}\right)\text{in amu}}{\text{Avogadro's number } N_A}.$$

- A hydrogen (^{1}H) atom has a mass of 1 amu. The mass of hydrogen atom in grams is $(1/N_A)\,g = 1.660 \times 10^{-24}\,g$
- A calcium (Ca) atom has a mass of 40 amu. The mass of calcium atom in grams is $(40/N_A)\,g = 6.642 \times 10^{-24}\,g$.
- The mass of one sulfuric acid (H_2SO_4) molecule is 98 amu. The mass of one molecule of H_2SO_4 in grams is $(98/N_A)\,g = 1.627 \times 10^{-24}\,g$.
- The molecular mass of glucose ($C_6H_{12}O_6$) is 180 amu. The mass of one molecule of glucose in grams is $(180/N_A)\,g = 2.989 \times 10^{-22}\,g$.

4.4 MOLAR MASS

A mole is the amount of substance that contains $N_A = 6.022 \times 10^{23}$ entities (atoms, molecules, ions). The molar mass (defined as the mass of one mole) is abbreviated as M (or M_{wt}).

For monatomic elements, the molar mass is the numerical value on the periodic table expressed in g mol^{-1}. For example, the atomic mass of oxygen (O) is equal to 16 amu and the molar mass of oxygen (O) is equal to 16 g mol^{-1}.

For molecules, the molar mass is the sum of the molar masses of each of the atoms in the molecular formula.

Examples:

- A molecule of H_2O contains two H atoms and one O atom. The atomic mass of O is 16 amu and the atomic mass of H is 1 amu. The molecular mass of H_2O is equal to $16 + 2 \times 1 = 18$ amu and the molar mass of H_2O is equal to 18 g mol^{-1}.
- The molecular mass of sulfuric acid (H_2SO_4) is 98 amu and the molar mass of H_2SO_4 is 98 g mol^{-1}.
- The molecular mass of glucose ($C_6H_{12}O_6$) is 180 amu and the molar mass of $C_6H_{12}O_6$ is 180 g mol^{-1}.
- The molecular mass of calcium sulfate hemihydrate ($CaSO_4 \cdot 1/2H_2O$) is 145 amu and the molar mass of calcium sulfate hemihydrate ($CaSO_4 \cdot 1/2H_2O$) is 145 g mol^{-1}.

GET SMART

WHAT IS THE DIFFERENCE BETWEEN THE ATOMIC MASS, THE MOLECULAR MASS, AND THE MOLAR MASS?

The atomic mass and the molecular mass are expressed in amu. The molar mass is the numerical value obtained for the atomic mass or the molecular mass expressed in g mol^{-1}.

Examples:

- The atomic mass of sulfur (S) is 32 amu. The molar mass of sulfur is 32 g mol^{-1}.
- The molecular mass of water (H_2O) is 18 amu. The molar mass of water is 18 g mol^{-1}.

4.5 NUMBER OF MOLES

The number of moles (n) in a mass of a substance is determined by the following equation:

$$n = \frac{m}{M}$$

where m is the mass of the substance in g and M is the molar mass of the substance expressed in g mol^{-1}.

Practice 4.3 Calculate the number of moles of calcium carbonate CaCO$_3$ in a stick of chalk containing 14 g of calcium carbonate. Atomic masses: C = 12, O = 16, Ca = 40.

Answer:
The molar mass of CaCO$_3$ is M = (1 × 40) + (1 × 12) + (3 × 16) = 100 g mol^{-1}
and the mass of CaCO$_3$ is m = 14 g
The number of moles of CaCO$_3$ is n = m/M = 14/100 = 0.148 mol

Practice 4.4 What is the mass of 0.28 mol of acetylsalicylic acid C$_9$H$_8$O$_4$? (Atomic masses: H = 1, C = 12, O = 16)

Answer:
The molar mass of C$_9$H$_8$O$_4$ is M = (9 × 12) + (8 × 1) + (4 × 16) = 180 g mol^{-1}
The number of moles (n) of C$_9$H$_8$O$_4$ is given by n = m/M where m and M are the mass and the molar mass of C$_9$H$_8$O$_4$, respectively.
Rearranging gives m = n × M
Resolving gives m = 0.28 × 180 = 50.4 g

GET SMART

HOW IS THE EQUATION n = m/M USED CORRECTLY?
The number of moles (n), the mass (m) and the molar mass (M) can be determined for the same element or the same compound. For example, to determine the number of moles of CaCO$_3$, m should be the mass of CaCO$_3$ and M should be the molar mass of CaCO$_3$. To calculate the number of moles of oxygen (O) in 5 g of CaCO$_3$, for example, do not write n(O) = 5/16 since 5 g is the mass of CaCO$_3$ and 16 g mol^{-1} is the molar mass of oxygen. The calculation of the number of moles and the mass of an element in a compound is depicted in Section 4.7.

4.6 NUMBER OF ATOMS, MOLECULES, OR IONS

One mole contains N$_A$ atoms, molecules or ions. Therefore, the number of atoms, molecules or ions (N) in n moles of a substance is given by the following equation:

$$N = n \times N_A$$

where N is the number of atoms, molecules or ions, n is the number of moles (n = m/M) and N$_A$ is Avogadro's number (N$_A$ = 6.022 × 10^{23}).

Practice 4.5 Calculate the number of zinc (Zn) atoms in 1 g of Zn.

Atomic mass: Zn = 65.38

Answer:
The molar mass of Zn = 65.38 g mol^{-1}.
Number of Zn moles: n = m/M = 1/65.38 = 1.529 × 10^{-2} mol

Number of Zn atoms: $N = n \times N_A$
$$= 1.529 \times 10^{-2} \times 6.022 \times 10^{23}$$
$$= 9.207 \times 10^{21} \text{ atoms}$$

Practice 4.6 Calculate the number of molecules of calcium carbonate ($CaCO_3$) in 14 g of calcium carbonate.

Atomic masses: C = 12, O = 16, Ca = 40

Answer:
Molar mass of $CaCO_3$: $M = (1 \times 40) + (1 \times 12) + (3 \times 16) = 100 \text{ g mol}^{-1}$
Number of moles of $CaCO_3$: $n = m/M = 14/100 = 0.14 \text{ mol}$
Number of molecules of $CaCO_3$: $N = n \times N_A = 0.14 \times 6.022 \times 10^{23} = 8.43 \times 10^{22} \text{ molecules}$

4.7 NUMBER OF MOLES, NUMBER OF ATOMS AND MASS OF AN ELEMENT IN A COMPOUND

Follow the following steps to determine the number of moles, the number of atoms and the mass of an element in a compound:

1. Determine the molar mass of the compound.
2. Determine the number of moles of the compound.

$$\left(n = m/M \right)_{compound}$$

3. Determine the number of moles of the element in one mole of the compound from the molecular formula.
4. Determine the number of moles of the element n(element) in n moles of the compound. The number of moles of the compound is determined in step 2.
5. Determine the number of atoms of the element using the equation:

$$N = n\left(element \right) \times N_A$$

or determine the mass of the element using the equation:

$$m\left(element \right) = n\left(element \right) \times M\left(element \right)$$

Practice 4.7 Calculate the number of atoms of oxygen (O) in 0.2 kg of sulfuric acid (H_2SO_4).

Atomic masses: H = 1, O = 16, S = 32

Answer:

1. The molar mass of sulfuric acid (H_2SO_4):

$$M = 2 \times M\left(H \right) + M\left(S \right) + 4 \times M\left(O \right)$$
$$= \left(2 \times 1 \right) + \left(1 \times 32 \right) + \left(4 \times 16 \right) = 98 \text{ g mol}^{-1}$$

2. The mass of H_2SO_4 is 0.2 kg = 200 g. The number of moles of sulfuric acid H_2SO_4:
   $n = m/M = 200/98 = 2.04 \text{ mol}$
3. One mole of sulfuric acid H_2SO_4 contains **4** moles of oxygen (O)
4. The number of moles of oxygen in 2.04 mol of sulfuric acid:

$$n\left(O \right) = 4 \times 2.04 = 8.16 \text{ mol}$$

5. The number of atoms of oxygen in 2.04 mol of sulfuric acid:

$$N(O) = n(O) \times N_A = 8.16 \times 6.022 \times 10^{23} = 4.91 \times 10^{24} \text{ atoms}$$

Practice 4.8 Calculate the mass of iron in 1 kg of limonite ($Fe_2O_3 \cdot 3/2 H_2O$).

Atomic masses: H = 1, O = 16, Fe = 56

Answer:

1. Limonite ($Fe_2O_3 \cdot 3/2 H_2O$) is a compound. The molar mass of limonite:

$$M = (2 \times M(Fe)) + (3 \times M(O)) + 3/2 [2 \times M(H) + 1 \times M(O)]$$
$$= (2 \times 56) + (3 \times 16) + 3/2 (2 \times 1 + 1 \times 16) = 187 \text{ g mol}^{-1}$$

2. The mass of limonite is 1 kg = 1000 g. The number of moles of limonite:

$$n = m/M = 1000/187 = 5.35 \text{ mol}$$

3. One mole of limonite ($Fe_2O_3 \cdot 3/2 H_2O$) contains **2** moles of Fe
4. The number of moles of iron: n(Fe) = **2** $\times$ 5.35 = 10.7 mol
5. The mass of iron in 1 kg of limonite ($Fe_2O_3 \cdot 3/2 H_2O$) is

$$m(Fe) = n(Fe) \times M(Fe) = 10.7 \times 56 = 599.2 \text{ g}$$

4.8 LAW OF DEFINITE PROPORTIONS

The law of definite proportions (also called Proust's law or the law of constant composition) states that a given chemical compound always has the same proportion of its constituent elements by mass and does not depend on its source or method of preparation. This means that **if a compound is broken into its elements, the ratio of their masses will always be the same.** For example, it can be measured that, regardless of the mass of water (H_2O) considered, the ratio between the mass of hydrogen (H) and the mass of oxygen (O) in water is always 1 to 8. This means that 9 g of pure water, for example, decomposes into 1 g of hydrogen and 8 g of oxygen and 100 g of pure water decomposes into 11.1 g of hydrogen and 88.9 g of oxygen.

4.9 MASS PERCENT OF AN ELEMENT IN A COMPOUND

The mass (weight) percent of an element in a molecule shows the amount contribution of the element to the total molecular mass. The mass percent refers to the ratio of the mass of an element to the total mass of the compound.

The mass percent is determined by the following equation:

$$\%(\text{element}) = \frac{n \times \text{atomic mass of the element}}{\text{molecular mass of the compound}} \times 100$$

Where %(element) is the mass percent of the element in the compound and n is the number of atoms of the element in the molecule, n is determined from the molecular formula of the compound.

To determine the mass percent of the constituent elements of a compound, carry out the following steps:

1. Determine the number of atoms of each element in the molecule.
2. Calculate the total mass of each element.
3. Calculate the molecular mass of the molecule.
4. Determine the mass percent of each element.
5. Check the calculation: If you add up all the mass percent compositions, they should reach 100%.

Practice 4.9 Calculate the mass percent of each constituent element in iron (III) nitrate ($Fe(NO_3)_3$).

(Atomic masses: N = 14.0, O = 16.0, Fe = 56.0)

Answer:

1. The number of atoms of each constituent element in the molecule:
   one molecule of iron (III) nitrate ($Fe(NO_3)_3$) contains one atom of iron (Fe), three atoms of nitrogen (N) and (3×3 =) 9 atoms of oxygen (O).
2. The total mass of each constituent element:
   The total mass of iron: $m(Fe) = 1 \times 56.0 = 56.0$ amu
   The total mass of nitrogen mass: $m(N) = 3 \times 14.0 = 42.0$ amu
   The total mass of oxygen: $m(O) = 9 \times 16.0 = 144$ amu
3. The molecular mass M of $Fe(NO_3)_3$ is the sum of the atomic masses of all the constituent atoms of $Fe(NO_3)_3$:

$$M = 1 \times 56.0 + 3 \times 14.0 + 9 \times 16.0 = 242 \text{ amu}$$

4. The mass percent of each constituent element:

$$\text{The mass percent of Fe is } \%(Fe)$$

$$= \left(\text{total mass}(Fe) / \text{molecular mass}\right) \times 100$$

$$= (56.0 / 242) \times 100 = 23.1\%$$

$$\text{The mass percent of N is } \%(N)$$

$$= \left(\text{total mass}(N) / \text{molecular mass}\right) \times 100$$

$$= (42.0 / 242) \times 100 = 17.4\%$$

$$\text{The mass percent of O is } \%(O)$$

$$= \left(\text{total mass}(O) / \text{molecular mass}\right) \times 100$$

$$= (144 / 242) \times 100 = 59.5\%$$

5. Check calculation: $\%(Fe) + \%(N) + \%(O) = 23.1 + 17.4 + 59.5 = 100\%$

Practice 4.10 Calculate the mass percent of iron (Fe) in 1 kg of iron (III) nitrate ($Fe(NO_3)_3$).

Atomic masses: N = 14.0, O = 16.0, Fe = 56.0

Answer:
The molar mass of $Fe(NO_3)_3$ is $M = 1 \times 56.0 + 3 \times 14.0 + 9 \times 16.0 = 242$ g mol^{-1}.
The mass of $Fe(NO_3)_3$ is 1 kg = 1000 g. The number of moles of $Fe(NO_3)_3$ is

$$n = m / M = 1000 / 242 = 4.13 \, \text{mol}$$

From the molecular formula of $Fe(NO_3)_3$, it is easy to determine that one mole of $Fe(NO_3)_3$ contains one mole of Fe. Therefore, 4.13 mol of $Fe(NO_3)_3$ contain 4.13 mol of Fe.

The mass of Fe: $m(Fe) = n(Fe) \times M(Fe) = 4.13 \times 56 = 231.28 \, \text{g}$

The mass percent of Fe is $\% \, (Fe) = (m(Fe)/m(Fe(NO_3)_3) \times 100 = (231.28/1000) \times 100 = 23.1\%$

GET SMART

IS THERE ANY DIFFERENCE BETWEEN CALCULATING THE MASS PERCENT OF AN ELEMENT IN A COMPOUND, CONSIDERING ONE MOLECULE OF THE COMPOUND OR A GIVEN MASS OF THE COMPOUND?

No, there is no difference. When you calculate the mass percent of an element with respect to the molecular mass or a given mass of the compound, you will obtain the same mass percent value. For example, the mass percent of iron calculated using one molecule of $Fe(NO_3)_3$ is 23.1% (Practice 4.12), which is exactly the same as the value calculated for 1,000 g of $Fe(NO_3)_3$ (Practice 4.13).

The mass percent of an element could be used to calculate the mass of the element in any given amount of the compound:

$$\mathbf{m(element) = \%(element) \times m(compound) / 100}$$

For example, the mass percent of Fe in $Fe(NO_3)_3$ is 23.1%. The mass of Fe in 5 kg of $Fe(NO_3)_3$ is $(23.1 \times 5)/100 = 1.155 \, \text{kg}$.

4.10 LAW OF MULTIPLE PROPORTIONS

The law of multiple proportions states that when two elements combine with each other to form more than one compound, the mass of one element that combines with a fixed mass of the second one is in a ratio of small, whole numbers. For example, carbon monoxide (CO) and carbon dioxide (CO_2) contain the same elements carbon (C) and oxygen (O). In CO, 12 g of C combined with 16 g of O. The ratio between the mass of oxygen (O) and the mass of carbon (C) in CO is 1.33 (=16/12). In CO_2, 12 g of C combined with $16 \times 2 = 32$ g of O. The ratio between the mass of oxygen (O) and the mass of carbon (C) in CO_2 is 2.67 (=32/12). The mass of carbon in both compounds is the same and the ratio of the masses of oxygen in CO_2 and CO can be reduced to the ratio of small, whole numbers 2.67/1.33 = 32/16 = 2/1. This means that the mass of oxygen in CO_2 will always be twice as much as the mass of oxygen in CO.

Practice 4.11 The masses of oxygen that combine with 10 g of nitrogen to form 3 compounds (A, B and C) are respectively, 11.41 g, 22.84 g and 28.55 g. Can the oxygen masses be reduced to the ratio of small, whole numbers?

Answer:
The ratio between the mass of oxygen (O) and the mass of nitrogen (N) in:

Compound A is 11.41/10 = 1.144
Compound B is 22.84/10 = 2.284
Compound C is 11.41/10 = 2.855

The mass of nitrogen in all compounds is the same and the ratio of the masses of oxygen in the compounds B and A can be reduced to the ratio of small, whole numbers 2/1 (2.284/1.144 = 2/1).

The ratio of the masses of oxygen in the compounds C and B can be reduced to the ratio of small, whole numbers 5/4 (2.855/2.284 = 1.25 = 5/4).

The ratio of the masses of oxygen in the compounds C and A can be reduced to the ratio of small, whole numbers 5/2 (2.855/1.144 = 2.5 = 5/2).

4.11 DETERMINING THE EMPIRICAL AND THE MOLECULAR FORMULAS

4.11.1 FROM MOLAR MASSES

The empirical (simplest) formula for a compound is the formula that agrees with the elemental analysis and gives rise to the smallest set of whole numbers of atoms.

The molecular formula is the formula of the compound as it exists, and it may be a multiple of the empirical formula:

The molecular formula = D × the empirical formula

where D is the greatest common divisor of the subscripts determining the number of atoms in the molecular formula. D means how many times the empirical formula is repeated to give the molecular formula (D = 1, 2, 3,…).

D could be determined from the following equation:

$$D = \frac{The\ molar\ mass\ of\ the\ molecular\ formula}{The\ molar\ mass\ of\ the\ empirical\ formula}$$

Practice 4.12 The molar mass of a compound is found to be 180 g mol⁻¹. Its empirical (simplest) formula is CH₂O. What is the molecular formula of this compound?

Atomic masses: H = 1, C = 12, O = 16

Answer:
The empirical (simplest) formula is CH_2O. The molar mass of the empirical formula CH_2O is:

$$M = 1\times12 + 2\times1 + 1\times16 = 30\ g\ mol^{-1}$$

The molar mass of the molecular formula is 180 g mol⁻¹.

The number of times D the empirical formula is repeated to give the molecular formula is

$$D = \frac{The\ molar\ mass\ of\ the\ molecular\ formula}{The\ molar\ mass\ of\ the\ empirical\ formula} = \frac{180}{30} = 6$$

The molecular formula of this compound is obtained by multiplying the relative number of atoms in the empirical formula (CH_2O) by 6. The molecular formula of this compound is $C_6H_{12}O_6$ (glucose).

4.11.2 FROM THE MASSES OF THE CONSTITUENT ELEMENTS OF THE COMPOUNDS

To determine the empirical and molecular formulas of a compound from the different masses of the constituent elements of the compound, carry out the following steps:

1. Determine the number of moles of each element from its mass (n = m/M).
2. Divide by the lowest mol amount to find the relative mol ratios.
3. Use the relative mol ratios as subscripts to obtain a preliminary formula.
4. Change to integer subscripts to obtain the empirical formula.

Practice 4.13 Elemental analysis of an orange compound shows that it contains 6.64 g of potassium (K), 8.84 g of chromium (Cr) and 9.60 g of oxygen (O). Find the empirical and molecular formulas of this compound.

(Atomic masses: K = 39, Cr = 52, O = 16)

Answer:

1. The number of moles of each element determined from its mass:
   Number of moles of potassium $n(K) = 6.64/39 = 0.17$ mol
   Number of moles of chromium $n(Cr) = 8.84/52 = 0.17$ mol
   Number of moles of oxygen $n(O) = 9.60/16 = 0.60$ mol
2. Divide each mole number by the smallest number of moles (0.17):
   Relative mole ratio of $(K) = 0.17/0.17 = 1.0$
   Relative mole ratio of $(Cr) = 0.17/0.17 = 1.0$
   Relative mole ratio of $(O) = 0.60/0.17 = 3.5$
3. A preliminary formula of this compound is $KCrO_{3.5}$.
4. Multiply the relative mol ratios by 2 to obtain integer subscripts. The empirical formula of this compound is $K_2Cr_2O_7$.

Note that the empirical formula is $K_2Cr_2O_7$ which is the same as the molecular formula.

4.11.3 FROM THE MASSES OF PRODUCTS

To determine the empirical and molecular formulas of a compound from the masses of products (substances produced in a chemical reaction), carry out the following steps:

1. Determine the number of moles of each product ($n = m/M$)
2. Determine the number of moles and mass of the element from the number of moles of the product that contains this element, using the chemical formula of the product.
3. When an element appears in more than one product, its number of moles cannot be calculated explicitly from the number of moles of the products. In that case, use the total mass of the compound and the masses of the other constituent elements calculated in step 2 to determine the mass and number of moles of this element:

$$m(element) = m(compound) - \left(\sum masses\ of\ the\ other\ elements\right)$$

4. Divide by the lowest mol amount to find the relative mol ratios
5. Use the relative mol ratios as subscripts to obtain a preliminary formula.
6. Change to integer subscripts to obtain the empirical formula.

Practice 4.14 Ethyl alcohol contains the elements C, H and O. When 5 g of ethyl alcohol are burned in air, 9.68 g CO_2 and 5.94 g H_2O are obtained as products.

Find the simplest formula of ethyl alcohol.

(Atomic masses: H = 1, C = 12, O = 16)

Answer:

Steps 1 and 2: Determine the number of moles of each constituent element using the masses of products:
Number of moles of CO_2: $n(CO_2) = m(CO_2)/M(CO_2) = 9.68/(12 + 2 \times 16) = 0.22$ mol
One mole of CO_2 contains one mole of C, therefore

$$n(C) = n(CO_2) = 0.22\ mol$$

The mass of C is $m(C) = n(C) \times M(C) = 0.22 \times 12 = 2.64$ g
Number of moles of H_2O: $n(H_2O) = m(H_2O)/M(H_2O) = 5.94/(2 \times 1 + 16) = 0.33$ mol

One mole of H_2O contains 2 moles of H, therefore

$$n(H) = 2 \times n(H_2O) = 2 \times 0.33 = 0.66 \, mol$$

The mass of H is $m(H) = n(H) \times M(H) = 0.66 \times 1 = 0.66$ g.

Step 3: A sample of 5.00 g ethyl alcohol contains 2.64 g of C and 0.66 g of H

The mass of O in ethyl alcohol is $m(O) = 5.00 - (2.64 + 0.66) = 1.7$ g

The number of moles of O is:

$$n(O) = m(O) / M(O) = 1.7 / 16 = 0.11 \, mol$$

A primary formula of this compound is $C_{0.22}H_{0.66}O_{0.11}$

Step 4: Divide by the lowest mol amount (0.11) to find the relative mol ratios:

Relative mole ratio of C = 0.22/0.11 = 2

Relative mole ratio of H = 0.66/0.11 = 6

Relative mole ratio of O = 0.11/0.11 = 1

Step 5: use the relative mol ratios as subscripts to obtain a preliminary formula.

A preliminary formula of the compound is C_2H_6O

Step 6: Change to integer subscripts to obtain the empirical formula.

Since the mole ratios are integers, so the empirical formula is C_2H_6O.

4.12 BALANCING CHEMICAL EQUATIONS AND STOICHIOMETRY

4.12.1 LAW OF CONSERVATION OF MASS

The law of conservation of mass states that, in an isolated system (closed to all transfers of matter and energy), the mass of the system must remain constant over time, as the system's mass is neither created nor destroyed by chemical reactions or physical transformations.

According to the law of conservation of mass, the mass of products (substances produced) in a chemical reaction must equal the mass of the reactants (substances consumed).

4.12.2 BALANCING CHEMICAL EQUATIONS

A balanced chemical equation is an equation that represents the correct amount of reactants and products in a chemical reaction. Balancing a chemical equation is determining the **coefficients** that provide equal numbers of each type of atom on each side of the equation.

To balance a chemical equation, follow the following steps:

Step 1: Write the correct formulas for the reactants (on the left side of the equation) and the products (on the right side of the equation). Do not change the **subscripts** that indicate the actual numbers of atoms in one molecule).

Example: Butane reacts with oxygen to form carbon dioxide and water.

The correct formula of butane is C_4H_{10}. Do not write for example $2(C_2H_5)$ or $(C_2H_5)_2$

The correct formula of oxygen is O_2 and the correct formula of water is H_2O

$$C_4H_{10} + O_2 \rightarrow CO_2 + H_2O$$

Step 2: Change the **coefficients** (numbers in front of the molecular formulas) to make the number of atoms of each element the same on both sides of the equation. Start by balancing those elements that appear in only one reactant and one product.

In the equation: $C_4H_{10} + O_2 \rightarrow CO_2 + H_2O$

carbon (C) and hydrogen (H) appear in only one reactant. However, oxygen (O) appears in two products, which are carbon dioxide (CO_2) and water (H_2O). Therefore, start by balancing C or H and not O. For example, start by balancing carbon (C) and then by balancing hydrogen (H):

In the equation: $C_4H_{10} + O_2 \rightarrow CO_2 + H_2O$
there are four atoms of carbon on the left side and one atom of carbon on the right side, so multiply CO_2 by 4: $C_4H_{10} + O_2 \rightarrow 4CO_2 + H_2O$
Then, start balancing H:
There are ten atoms of hydrogen on the left side and two atoms of hydrogen on the right side, so multiply H_2O by 5:

$$C_4H_{10} + O_2 \rightarrow 4CO_2 + 5H_2O$$

Finally, balance O.
There are two atoms of oxygen on the left side and 13 atoms of oxygen on the right side so, multiply O_2 by 13/2:

$$C_4H_{10} + 13/2\,O_2 \rightarrow 4CO_2 + 5H_2O$$

Multiply the overall equation by 2 to obtain integers:

$$2C_4H_{10} + 13O_2 \rightarrow 8CO_2 + 10H_2O$$

Step 3: Check to make sure that you have the same number of each type of atom on both sides of the equation:

Reactants (left side)	Products (right side)
8 C	8 C
20 H	20 H
26 O	26 O

GET SMART

HOW TO BALANCE A CHEMICAL EQUATION THAT CONTAINS IONS

Determine the coefficients that provide equal numbers of each type of atom on each side of the equation. Then, check to make sure that the sum of charges on each side of the equation are the same.

Example: $2MnO_4^- + 5H_2O_2 + 6H^+ \rightarrow 2Mn^{2+} + 5O_2 + 8H_2O$

The sum of charges on the left side is $2 \times (-1) + 6 \times (+1) = +4$
The sum of charges on the right side is $2 \times (+2) = +4$

4.12.3 STOICHIOMETRY

Stoichiometry is the study of mass relationships that exist between reactants (substances consumed) and products (substances produced) in a chemical reaction:

$$\text{Reactants} \rightarrow \text{Products}$$

To determine the mass relationships between reactants and products, carry out the following steps:

Step 1: Write the balanced chemical equation.
Step 2: Use coefficients in the balanced chemical equation to determine the relationships between the number of moles of the reactants and the number of moles of the products.
Step 3: Convert quantities of known substances into moles.
Step 4: Use relationships between the number of moles of reactants and products (determined in Step 2) to calculate the number of moles of the desired quantity.
Step 5: Determine the mass of the desired quantity.

Practice 4.15 Butane reacts with oxygen to form carbon dioxide and water. Determine the mass of CO$_2$ obtained when 5.8 g of butane reacts.

Atomic masses: H = 1, C = 12, O = 16.

Answer:

Step 1: Butane reacts with oxygen to form carbon dioxide and water. The corresponding balanced chemical equation is:

$$2C_4H_{10} + 13O_2 \rightarrow 8CO_2 + 10H_2O$$

Step 2: relationships between the number of moles (n) of reactants and products:

$$\frac{n(C_4H_{10})}{2} = \frac{n(O_2)}{13} = \frac{n(CO_2)}{8} = \frac{n(H_2O)}{10}$$

Note that these relationships are obtained simply by dividing the number of moles of the substance by its coefficient in the balanced chemical equation.
Step 3: The number of moles of butane C$_4$H$_{10}$

$$n(C_4H_{10}) = m/M = 5.8 / (4 \times 12 + 10 \times 1) = 0.1 \, mol$$

Step 4: n(C$_4$H$_{10}$)/2 = n(CO$_2$)/8 so n(CO$_2$) = 4 × n(C$_4$H$_{10}$) = 4 × 0.1 = 0.4 mol
Step 5: The mass of (CO$_2$) obtained is

$$m(CO_2) = n(CO_2) \times M(CO_2) = 0.4 \times (1 \times 12 + 2 \times 16) = 17.6 \, g$$

GET SMART

HOW TO CALCULATE THE MASS OF A SUBSTANCE IN A BALANCED CHEMICAL EQUATION, KNOWING ANOTHER QUANTITY

First, read the question carefully and find how the two substances are related in this question. Then, from the balanced chemical equation, find the relationship between the number of moles of the two substances and convert it to a relationship between masses (m = n × M). Finally, use the cross- multiplication method to find the unknown quantity.

Example: From the following balanced chemical equation:

$$2C_4H_{10} + 13H_2O \rightarrow 8CO_2 + 10H_2O$$

determine the mass of H$_2$O obtained when 5.8 g of butane (C$_4$H$_{10}$) reacts.
Atomic masses: H = 1, C = 12, O = 16.

- The question related between butane (**C$_4$H$_{10}$**) and water (**H$_2$O**)
- From the above balanced chemical equation:

$$\begin{array}{cc} 2 \text{ moles } (C_4H_{10}) & 10 \text{ moles } (H_2O) \\ \text{Or} \quad 2 \times M(C_4H_{10}) = 116 \, g & 10 \times M(H_2O) = 180 \, g \end{array}$$

- Use the cross-multiplication method:

$$116 \, g \text{ of } (C_4H_{10}) \rightarrow 180 \, g \text{ of } (H_2O)$$

$$5.8 \, g \text{ of } (C_4H_{10}) \rightarrow m(H_2O) = ?$$

The mass of water obtained is:
$$m(H_2O) = (180 \times 5.8) / 116 = 8.8 \, g$$

4.13 SOLUTIONS

4.13.1 TYPES OF SOLUTIONS

A solution is a homogeneous mixture which consists of a smaller amount of a substance, the solute, dissolved in a larger amount of another substance, the solvent. When a solid is dissolved in a liquid, the liquid is the solvent whatever the amounts of the mixed substances. A solution can exist as a solid solution, such as the Ni_3Al alloy, where a smaller amount of a solid is dissolved in another solid present in a larger amount. A gas can also be dissolved in a liquid or in another gas. For example, CO_2 gas can be dissolved in water or can be mixed with different gases in atmospheric air. Finally, a solution can be obtained by dissolving a liquid in another liquid, e.g., ethylene glycol $(CH_2OH)_2$ in water.

A solution is said to be saturated if it reaches its limits in dissolving the solute at a given temperature. But, if the solution can still dissolve more solute, it is said to be undersaturated. By raising the temperature, a saturated solution can dissolve more solute which becomes present in an excess amount. In that case, the solution is said to be supersaturated. The excess amount of the solute can precipitate when the temperature decreases.

4.13.2 MOLE RATIO AND MOLE PERCENT

The mole ratio of a solute X_{solute} in a solution is defined by the following equation:

$$X_{solute} = \frac{n_{solute}}{n_{total}} = \frac{n_{solute}}{n_{solute} + n_{solvent}}$$

Where n is the number of moles. The mole percent of the solute is defined by:

$$mol\left(\%\right)_{solute} = X_{solute} \times 100\%$$

Similarly, we can define the mole ratio and the mol percent of the solvent as follows:

$$X_{solvent} = \frac{n_{solvent}}{n_{total}} = \frac{n_{solvent}}{n_{solute} + n_{solvent}}$$

And $mol\left(\%\right)_{solvent} = X_{solvent} \times 100\%$

Note that $X_{solute} + X_{solute} = 1$

And $mol\left(\%\right)_{solute} + mol\left(\%\right)_{solvent} = 100\%$

Practice 4.16 Calculate the mole ratio and mole percent of ethylene glycol $(CH_2OH)_2$ in a solution containing 44.6 g of $(CH_2OH)_2$ and 81 g of water.

Atomic masses: H = 1, C = 12 and O = 16.

Answer:
The molar mass of $(CH_2OH)_2$ is $M = 2 \times (12 + 3 \times 1 + 16) = 62$ g mol^{-1}
The number of moles of $(CH_2OH)_2$ is $n = m/M = 45/62 = 0.72$ mol
The molar mass of H_2O is $M = 2 \times 1 + 16 = 18$ g/mol
The number of moles of H_2O is $n = m/M = 81/18 = 4.5$ mol
Ethylene glycol $(CH_2OH)_2$ is present in the solution in a smaller amount than water. Therefore, $(CH_2OH)_2$ is the solute and water is the solvent. The mole ratio X_{solute} of $(CH_2OH)_2$ is given by:

$$X_{solute} = \frac{n_{solute}}{n_{total}} = \frac{n_{solute}}{n_{solute} + n_{solvent}}$$

Resolving gives $X_{solute} = \dfrac{0.72}{0.72 + 4.5} = 0.13$

And the mole percent of the solute is given by:

$$mol\left(\%\right)_{solute} = X_{solute} \times 100\%$$

Resolving gives $mol\left(\%\right)_{solute} = 0.13 \times 100\% = 13\%$

4.13.3 MOLARITY

The concentration of a solution is expressed as the amount of solute dissolved in an amount of solution. The term most used is **molarity (M)**, defined as the number of moles of solute per liter of solution:

$$M = \frac{n}{V}$$

where n is the number of moles of the solute (n = m/M) and V is the volume of the solution. A common unit of molarity is mol L^{-1}, designated also by M.

Practice 4.17 Calculate the molarity of 460 cm³ hydrochloric acid (HCl) solution containing 5.0 g of HCl.

Atomic masses: H = 1, Cl = 35.5.

Answer:
The molar mass of HCl is M = 35.5 + 1 = 36.5 g mol^{-1}
The number of moles of HCl is n = m/M$_{wt}$ = 5.0/36.5 = 0.13 mol
The volume of the solution is 460 cm³ = 460 mL = 0.460 L
The molarity M = n/V = 0.13/0.460 = 0.28 mol L^{-1} = 0.28 M

Practice 4.18 Calculate the mass percent of ethylene glycol (CH₂OH)₂ in a 1.25 M solution. The density of the solution is 1.03 g cm⁻³.

Atomic masses: H = 1, C = 12 and O = 16.

Answer:
The density of the solution is $\rho = \frac{m}{V} = 1.03\ \frac{g}{cm^3} = 1030\ g\ L^{-1}$

The mass of 1 L of the solution is $m = \rho\ \frac{g}{L} \times V = 1030\ g$

The molarity of the solution is $M = \frac{n}{V} = 1.25\ M$

This means that 1 L of the solution contains 1.25 mol of (CH₂OH)₂.

The molar mass of (CH₂OH)₂ is M = 2 × (12 + 3 × 1 + 16) = 62 g mol^{-1}
The mass of (CH₂OH)₂ in 1L is m = n × M = 1.25 × 62 = 77.5 g

The mass percent of (CH₂OH)₂ is $\%\left(CH_2OH\right)_2 = \frac{m_{(CH_2OH)_2}}{m_{solution}} \times 100 = \frac{77.5\ g}{1030\ g} \times 100 = 7.5\%$

4.13.4 MOLALITY

The molality is defined as the number of moles of the solute per kilogram of the solvent. In the International System of Units (SI units), molality is expressed in mol kg^{-1}.

Practice 4.19 Calculate the molality of NaCl solution prepared by dissolving 32.5 g of NaCl in 300 g of water.

Atomic masses: Na = 23, Cl = 35.5.

Answer:
The molar mass of solute NaCl is M = 23 + 35.5 = 58.5 g mol^{-1}
The number of moles of solute is n(NaCl) = m/M = 32.5/58.5 = 0.55 mol
The mass of the solvent m(H_2O) = 300 g = 0.3 kg
The molality is m = n(solute)/m(solvent) = n(NaCl)/m(H_2O) = 0.55/(0.3) = 1.8 mol kg^{-1}.

4.13.5 SOLUBILITY

Solubility is the property of a solute (a solid, liquid or gaseous chemical substance) to dissolve in a solid, liquid or gaseous solvent. The solubility designates the maximum mass concentration of the solute in the solvent, at a given temperature. This means that solubility depends strongly on temperature. The solution thus obtained is saturated. In thermodynamics, the solubility is a physical quantity, denoted by s, that may be expressed in various units of concentration such as g L^{-1}, mol L^{-1} or g 100 g^{-1} of solvent.

Practice 4.20 The solubility of potassium nitrate (KNO_3) in water is 2460 g L^{-1} at 100°C and 270 g L^{-1} at 20°C. Calculate:

a) **The mass of water required to dissolve 100 g KNO_3 at 100°C (Solution A).**
b) **The mass of KNO_3 that remains when Solution A is cooled to 20°C.**

Answer:

a) At 100°C, the solubility of KNO_3 in water is 2460 g/L water. The mass of 1L of water is 1000 g. The mass of water required to dissolve 100 g KNO_3 at 100°C is

$$m\left(water\right) = \left(100 \times 1000\right)/2460 = 40.7 \text{ g}$$

b) Before calculating the mass of KNO_3 that remains when the solution in (a) is cooled to 20 ° C, we should calculate the mass of KNO_3 dissolved in 40.7 g of water at 20°C. At 20°C, the solubility of KNO_3 is 270 g L^{-1}. The mass dissolved in 40.7 g of water at 20°C is

$$m\left(dissolved\ KNO_3\right) = \left(40.7 \times 1000\right)/270 = 15 \text{ g}$$

The mass of dissolved KNO_3 at 100°C is 100 g. When the solution is cooled to 20°C, the mass of dissolved KNO_3 is 13 g. Therefore, the remaining amount (m) of KNO_3 after cooling is m = 100 − 15 = 85 g

4.13.6 DILUTION

Dilution is the addition of a solvent to the solution that decreases the concentration of the solute in the solution. The number of moles of the solute before and after the dilution remains constant. Consequently, if we consider M_1 and V_1 are, respectively, the molarity and the volume of the solution before dilution, and M_2 and V_2 are, respectively, the molarity and the volume of the solution after dilution, we have:

$$M_1 = \frac{n}{V_1} \text{ and } M_2 = \frac{n}{V_2}$$

Rearranging gives $n = M_1 \times V_1 = M_2 \times V_2$

Practice 4.21 If 30 mL of 2.3 M hydrochloric acid solution are diluted to 82.5 mL, calculate the final concentration.

Answer:
We consider $M_1 = 2.3$ M and $V_1 = 30$ mL, respectively, to be the molarity and the volume of the solution before dilution, respectively, and M_2 and $V_2 = 82.5$ mL, respectively, to be the molarity and the volume of the solution after dilution. As a result, we have:

$$M_1 \times V_1 = M_2 \times V_2$$

Rearranging gives $M_2 = \dfrac{M_1 \times V_1}{V_2}$

Resolving gives $M_2 = \dfrac{2.3 \times 30}{82.5} = 0.83$ M

CHECK YOUR READING

How is the number of moles of an element or a compound determined?
How is the number of atoms (or molecules) of an element (or a compound) determined?
How is the number of moles and the mass of an element in a compound determined?
How is the mass (weight) percent of an element in a compound determined?
How are the empirical and molecular formulas determined?
What is a balanced chemical equation and how can a chemical equation be balanced?
How is the relationship between the number of moles of reactants and products determined?
In a balanced equation, how is the mass of a substance calculated, knowing another quantity?
What are the differences between molarity, molality, and solubility?
How to calculate the final concentration after dilution?

SUMMARY OF CHAPTER 4

A mole is the amount of a substance that contains the same number of entities as there are atoms in exactly 12 g of carbon-12. This amount is $N_A = 6.022 \times 10^{23}$. The number N_A is called Avogadro's number. **One mole (1 mol) contains 6.022×10^{23} entities (atoms, molecules, ions).**

The atomic mass (the mass of one atom) is the average mass of all the isotopes of an atom with respect to their abundances. The atomic masses of all elements are given in the periodic table of elements. The atomic mass is commonly expressed in atomic mass units **(amu).**

$$\textbf{1 amu} = \left(\textbf{1/N}_{\textbf{A}}\right)\textbf{g} = 1.660 \times 10^{-24}\,\textbf{g}$$

Example: A calcium (Ca) atom has a mass of 40 amu. The mass of calcium atom in g is $(40/N_A)$ g $= 6.642 \times 10^{-23}$ g

The molecular mass is the sum of the atomic masses of all the atoms in the molecule. It is commonly expressed in atomic mass units **(amu).**

Example: One molecule of sulfuric acid (H_2SO_4) contains two atoms of H, one atom of S and four atoms of O. The molecular mass M of sulfuric acid is

$$M = 2 \times \left(\text{atomic mass of H}\right) + 1 \times \left(\text{atomic mass of S}\right) + 4 \times \left(\text{atomic mass of O}\right)$$

$$= \left(2 \times 1\right) + \left(1 \times 32\right) + \left(4 \times 16\right) = 98\text{ amu}$$

The mass of one sulfuric acid molecule in g is $(98/N_A)g = 1.627 \times 10^{-22}$ g

For monatomic elements, the molar mass (mass of one mole, abbreviated as M or M_{wt}) is the numerical value on the periodic table expressed in g mol^{-1}.

> **Example:** The atomic mass of O is equal to 16 amu and the molar mass of O is equal to 16 gmol^{-1}.

For molecules, the molar mass is the sum of the molar masses of each of the atoms in the molecular formula.

> **Example:** A molecule of H_2O contains two H atoms and one O atom. The atomic mass of O is 16 amu and the atomic mass of H is 1 amu. The molecular mass of H_2O is equal to $16 + 2 \times 1 = 18$ amu and the molar mass of H_2O is equal to 18 gmol^{-1}.

The number of moles (n) is determined by the following equation: **n = m/M** where m is the mass of the substance in g and M is the molar mass of the substance in g mol^{-1}.

The number of entities (atoms, molecules or ions) (N) is determined by the following equation: $\mathbf{N = n \times N_A}$, where n is the number of moles of the substance and N_A is Avogadro's number.

> **Example:** 5 g of $CaCO_3$ contains n = m/M = 5/100 = 0.05 mol of $CaCO_3$ and $N = 0.05 \times N_A = 3.011 \times 10^{21}$ molecules of $CaCO_3$.

To determine the **number of atoms, moles, and the mass of an element in a compound**, first determine the number of moles of the compound and then found the relationship between the number of moles of the compound and the number of moles of the element, using the molecular formula of the compound.

> **Example:** 5 g of $CaCO_3$ contains 0.05 mol of $CaCO_3$. One mole of $CaCO_3$ contains **3** moles of oxygen (O). Therefore, 0.05 mol of $CaCO_3$ contains $3 \times 0.05 = 0.15$ mol of O and $N = 0.15 \times N_A = 9 \times 10^{22}$ atoms of O.

The mass of oxygen in 5 g of $CaCO_3$ is $m(O) = n(O) \times M(O) = 0.15 \times 16 = 2.4$ g

$$\text{The molecular formula} = D \times \text{the empirical formula}$$

where D is the greatest common divisor of the subscripts determining the number of atoms in the molecular formula. D means how many times the empirical formula is repeated to give the molecular formula. D could be determined from the following equation:

$$D = \frac{\textit{The molar mass of the molecular formula}}{\textit{The molar mass of the empirical formula}}$$

The molecular and empirical formulas could be also determined from the masses of the elements in the compound or from the masses of the products of a chemical reaction.

The mass (or **weight**) **percent** refers to the ratio of the mass of one element to the total mass of a compound. The mass percent of an element, % (element), is determined by the following equation:

$$\%(\text{element}) = \frac{n \times \text{atomic mass of the element}}{\text{molecular mass of the compound}} \times 100$$

where n is the number of atoms of the element in the compound.

A balanced chemical equation is an equation that represents the correct amounts of **reactants** and **products** in a chemical reaction. The number of atoms of each element is the same on both sides of the equation. To balance achemical equation, **start by balancing those elements that appear in only one reactant and one product.**

> **Example of a balanced chemical equation:** $2HgO \rightarrow 2Hg + 1/2O_2$

Stoichiometry is the study of mass relationships that exist between reactants (substances consumed) and products (substances produced) in a chemical reaction. Use coefficients in the balanced equation to determine the relationships between the number of moles of reactants and products. Then, calculate the number of moles and the mass of the desired quantity.

A **solution** consists of a smaller amount of a substance, the **solute**, dissolved in a larger amount of another substance, the **solvent.**

The **concentration** of the solution is expressed as the amount of solute dissolved in a given amount of solution. The term most used is **Molarity (M)**, defined as number of moles of solute per liter of solution ($M = n/V$). A common unit of molarity is $M (= mol\ L^{-1})$.

The **molality** is defined as the number of moles of the solute per kilogram of the solvent. The SI unit of molality is $mol\ kg^{-1}$.

The **solubility** designates the maximum mass concentration of the solute in the solvent, at a given temperature. The solution thus obtained is saturated. The solubility depends on temperature and may be expressed in $g\ L^{-1}$, $mol\ L^{-1}$ and $g\ 100\ g^{-1}$ of solvent.

Dilution is the addition of a solvent to the solution that decreases the concentration of the solute in the solution. The number of moles of the solute before and after the dilution remains constant.

PRACTICE ON CHAPTER 4

Q4.1   **Choose the correct answer**
Atomic masses: H = 1, C = 12, N = 14, O = 16, Na = 23, Mg = 24, S = 32, Cl = 35.5, Ca = 40 and Fe = 56.
Avogadro's number $N_A = 6.022 \times 10^{23}$

1. **1 amu equals**
   a) 1 g
   b) N_A g
   c) $1/N_A$ g
   d) $2 \times N_A$ g

2. **The atomic mass of calcium (Ca) is**
   a) 40 g mol^{-1}
   b) 40 amu
   c) $40 \times N_A$ g
   d) 6.022×10^{-23} g

3. **The mass in g of one atom of chlorine (Cl) is**
   a) 35.5 g
   b) 5.895×10^{-23} g
   c) $35.5 \times N_A$ g
   d) 6.022×10^{-23} g

4. **The molecular mass of calcium (CaCl$_2$) is**
   a) 111 g mol^{-1}
   b) 111 amu
   c) $111 \times N_A$ g
   d) 6.022×10^{-23} g

5. **The mass in g of one molecule of sodium carbonate (Na$_2$CO$_3$) is**
   a) 106 g mol^{-1}
   b) 106 amu
   c) $106/N_A$ g
   d) 6.022×10^{-23} g

6. **The mass of one molecule of calcium carbonate (CaCO$_3$) is**
   a) 100 g mol^{-1}
   b) 100 amu
   c) $100/N_A$ g
   d) both b) and c) are correct

7. **The molar mass of calcium (Ca) is**
    a) 40 g mol^{-1}
    b) 40 amu
    c) $40 \times N_A \text{ g}$
    d) 40 kg

8. **The molar mass of $MgCl_2$ is**
    a) 95 amu
    b) 95 g mol^{-1}
    c) 95 kg
    d) 95 mol L^{-1}

9. **The molar mass of $CaSO_4 \cdot 2H_2O$ is**
    a) 154 g mol^{-1}
    b) 136 g mol^{-1}
    c) 172 g mol^{-1}
    d) 200 g mol^{-1}

10. **The molar mass of $Fe(NO_3)_3$ is**
    a) 242 g mol^{-1}
    b) 242 amu
    c) 118 g mol^{-1}
    d) 118 amu

11. **The mass of 0.05 mol of $Fe(NO_3)_3$ is**
    a) 111 g
    b) 12.1 g
    c) 242 g
    d) $2.06 \times 10^{-4} \text{ g}$

12. **The number of moles of iron (III) nitrate ($Fe(NO_3)_3$) in 121 g of this substance is**
    a) 2 mol
    b) 1 mol
    c) 0.5 mol
    d) 0 mol

13. **The number of molecules of iron (II) chloride ($FeCl_2$) in 0.1 mol of this substance is**
    a) 0.1 molecule
    b) 1 molecule
    c) 6.022×10^{22} molecules
    d) 6.022×10^{23} molecules

14. **The number of molecules of iron (III) nitrate ($Fe(NO_3)_3$) in 121 g of this substance is**
    a) 3.011×10^{23} molecules
    b) 6.022×10^{23} molecules
    c) 0.5 molecule
    d) 8.3×10^{21} molecules

15. **The number of moles of oxygen (O) in one mole of iron (III) nitrate ($Fe(NO_3)_3$) is**
    a) 1 mol
    b) 3 mol
    c) 6 mol
    d) 9 mol

16. **The number of moles of nitrogen (N) in 121 g of iron (III) nitrate ($Fe(NO_3)_3$) is**
    a) 0 mol
    b) 1 mol
    c) 1.5 mol
    d) 3 mol

17. **The mass of nitrogen (N) in 121 g of iron (III) nitrate ($Fe(NO_3)_3$) is**
    a) 121 g
    b) 14 g

c) 42 g

d) 21g

18. **The number of oxygen (O) atoms in 121 g of iron (III) nitrate $(Fe(NO_3)_3)$ is**

    a) 2.709×10^{24} atoms

    b) 1.806×10^{24} atoms

    c) 6.022×10^{23} atoms

    d) 5.419×10^{23} atoms

19. **The mass percent of oxygen (O) in iron(III) nitrate $(Fe(NO_3)_3)$ is**

    a) 19.8 %

    b) 39.6 %

    c) 59.5 %

    d) 6.61 %

20. **The mass percent of oxygen (O) in 75 g of iron (III) nitrate $(Fe(NO_3)_3)$ is**

    a) 59.5 %

    b) 21.3 %

    c) 19.8 %

    d) 75.0 %

21. **The mass percent of Fe in limonite $(Fe_2O_3 \cdot 3/2H_2O)$ is 59.92%. Calculate the mass of iron in 1 ton of limonite.**

    a) 1000 kg

    b) 56 kg

    c) 112 kg

    d) 599.2 kg

22. **The molar mass of vitamin C is found to be 176 g mol^{-1}. Its simplest formula is $C_3H_4O_3$. What is the molecular formula of vitamin C?**

    a) $C_3H_4O_3$

    b) $2C_3H_4O_3$

    c) $C_6H_8O_6$

    d) $C_{12}H_6O$

23. **A sample of a compound contains 18 g of C, 3 g of H and 24 g of O. The simplest formula of this compound is**

    a) $C_{18}H_3O_{24}$

    b) $C_6H_{12}O$

    c) C_6HO_8

    d) CH_2O

24. **Calculate the molarity of 400 cm^3 of sodium hydroxide (NaOH) solution containing 8.0 g of NaOH.**

    a) 0.05 mol L^{-1}

    b) 0.02 mol L^{-1}

    c) 0.5 mol L^{-1}

    d) 2 mol L^{-1}

25. **Calculate the molality of 400 cm^3 of sodium hydroxide (NaOH) solution prepared by dissolving 8 g of NaOH in water.**

    a) 0.02 kg L^{-1}

    b) 0.5 mol kg^{-1}

    c) 0.5 mol L^{-1}

    d) 8.0 g L^{-1}

26. **Determine the (x, y, z) values which make the following chemical reaction balanced:**

$$x\, MnO_4^- + y\, H_2O_2 + z\, H^+ \rightarrow 2\, Mn^{2+} + 5O_2 + 8H_2O$$

    a) (2, 6, 5)

    b) (2, 5, 6)

    c) (2, 10, 12)

    d) (2, 1, 1)

27. **Which of the following is a balanced chemical equation?**
    a) $2C_2H_6 + 7O_2 \rightarrow 4CO_2 + 6H_2O$
    b) $5H_2 + N_2 \rightarrow 5NH_3$
    c) $2HgO \rightarrow 2Hg + 1/2O_2$
    d) $Ca^{2+} + 2HCO_3^- \rightarrow CaCO_3 + 2CO_2 + H_2O$

28. **Ethane reacts with oxygen to form carbon dioxide and water according to the following reaction:**

$$2C_2H_6 + 7O_2 \rightarrow 4CO_2 + 6H_2O$$

    Calculate the mass in grams of CO_2 formed when 1.35 g of C_2H_6 reacts.
    a) 1.35 g
    b) 3.96 g
    c) 1.98 g
    d) 44.0 g

29. **The following equation is a balanced chemical equation:**

$$2MnO_4^- + 5H_2O_2 + 6H^+ \rightarrow 2Mn^{3+} + 5O_2 + 8H_2O$$

    a) True
    b) False

30. **In one mole of H_2SO_4, the number of moles of oxygen is four and the mass of oxygen equals $4 \times (4 \times 16) = 128$ g.**
    a) True
    b) False

31. **One mole of H_2SO_4 and one mole of $CaCO_3$ have the same number of molecules but different masses.**
    a) True
    b) False

32. **Rubidium has two isotopes: ^{85}Rb which has an atomic mass of 84.9117 amu and ^{87}Rb which has an atomic mass of 86.9085 amu. The average atomic mass of Rb is**

$$(84.9117 + 86.9085)/2 = 85.9101 \text{ amu}$$

    a) True
    b) False

33. **The solubility of KNO_3 in water is 2460 g L^{-1} water at 100°C. When the temperature decreases to 20°C, the solubility remains constant.**
    a) True
    b) False

Calculation: Atomic masses: H = 1, He = 4, O = 16, Fe = 56.

Q4.2   **Copper has two isotopes, ^{63}Cu and ^{65}Cu. The isotope ^{63}Cu has an atomic mass 62.930 amu and an abundance of 69.2% and the isotope ^{65}Cu has an atomic mass 64.928 amu and an abundance of 30.8%. Calculate the average atomic mass of copper.**

Q4.3   **How many moles of limonite $Fe_2O_3 \cdot 3/2H_2O$ are there in 50.0 g of this compound?**

Q4.4   **How many molecules are there in a 5.0×10^{-2} g drop of water H_2O?**

Q4.5   **How many atoms are there in 13 g of He?**

Q4.6.   **How many atoms of oxygen (O) are there in 50.0 g of limonite ($Fe_2O_3 \cdot 3/2H_2O$)?**

Q4.7   **Calculate the mass of iron (Fe) in 150 g of iron oxide (Fe_2O_3).**

Q4.8   **Calculate the mass percent of iron (Fe) and oxygen (O) in iron oxide (Fe_2O_3).**

Q4.9   **Calculate the mass percent of iron (Fe) in 150 g of iron oxide (Fe_2O_3).**

Q4.10 Calculate the mass percent of Na, H, C and O in $NaHCO_3$.

4.11 Complete the following table that gives information contained in the chemical formula of iron(III) carbonate ($Fe_2(CO_3)_3$).
(Atomic masses: C = 12; O = 16; Fe = 56)

Information	Fe	C	O
Number of atoms in one molecule of $Fe_2(CO_3)_3$	Two atoms	_____ atoms	_____ atoms
Number of atoms in one mole of $Fe_2(CO_3)_3$	$2 \times N_A = 1.204 \times 10^{24}$ atoms	_____ atoms	_____ atoms
Mass of atoms in one molecule in amu	_____ amu	$3 \times 12 = 36$ amu	_____ amu
of $Fe_2(CO_3)_3$ in g	_____ g	_____ g	$144/N_A$ g $= 2.392 \times 10^{-24}$ g
Number of moles of the element in one mole of $Fe_2(CO_3)_3$	_____ mol	_____ mol	$3 \times 3 = 9$ mol
Mass of the element per mole of $Fe_2(CO_3)_3$	$56 \times 2 = 112$ g	_____ g	_____ g
Molar mass of $Fe_2(CO_3)_3$	_____ g mol^{-1}		
Mass percent of the element	_____ %	_____ %	_____ %
Simplest formula of $Fe_2(CO_3)_3$	_____		

Molecular and empirical formulas calculations
Atomic masses: H = 1, C = 12, N = 14, O = 16, K = 39, Mn = 55.

Q4.12 Elemental analysis of a compound shows that it contains 15.6 g of potassium (K), 22.0 g of manganese (Mn) and 25.6 g of oxygen (O). Find the empirical and molecular formulas of this compound.

Q4.13 Hexamethylenediamine contains C, H, N. Its molar mass is = 116 g mol^{-1}. When 6.481 g of hexamethylenediamine is burned in oxygen, it gives 14.36 g CO_2 and 7.921 g H_2O. What are the empirical formula and the molecular formula of this compound?

Q4.14 Elemental analysis of lactic acid (M = 90.08 g mol^{-1}) shows that it contains 40.0% (C), 6.7% (H), and 53.3% (O).
Find the empirical formula and the molecular formula of lactic acid.

Stoichiometric calculations
Atomic masses: H = 1, C = 12, N = 14, O = 16, Al = 27, S = 32, Cl = 35.5, Cu = 63.5.

Q4.15 Ammonia (NH_3) is made by reacting nitrogen (N_2) with hydrogen (H_2).
a) Write the balanced chemical equation.
b) Determine the mass in g of NH_3 formed when 1.8 mol H_2 reacts.
c) Calculate the mass in g of N_2 required to forming 1.00 kg NH_3.
d) Calculate the mass in g of H_2 required to react with 6.00 g N_2.

Q4.16
a) Balance the following chemical equation of the combustion of propane:

$$C_3H_8 + O_2 \rightarrow CO_2 + H_2O$$

b) Calculate the mass of CO_2 formed when 11 g of propane reacts.
c) Calculate the mass of H_2O formed when 11 g of propane reacts.
d) Calculate the number of moles of O_2 required to the combustion of 11 g of propane.
e) Calculate the mass of CO_2 formed when 40 g of O_2 reacts.

Q4.17
Calculate the mass of H_2SO_4 required to react with 0.7 mol of copper Cu according to the following balanced chemical equation:

$$2H_2SO_4 + Cu \rightarrow CuSO_4 + 2H_2O + SO_2$$

Q4.18

Aluminum reacts with oxygen to form a protective layer of Al_2O_3 according to the following balanced chemical equation:

$$4Al + 3O_2 \rightarrow 2Al_2O_3$$

Calculate the mass of oxygen (O_2) required to form 250 g of the protective layer.

Solution calculation

4.19

A solution contains 36 % of hydrochloric acid HCl and it has a density of 1.18 g cm^{-3}. Calculate the volume of this solution needed to prepare 100 mL of 0.5 M HCl solution. Atomic masses: H = 1, Cl = 35.5

ANSWERS TO QUESTIONS

Q4.1

1. c
2. b
3. b
4. b
5. c
6. d
7. a
8. b
9. c
10. a
11. b
12. c
13. c
14. a
15. d
16. c
17. d
18. a
19. c
20. a
21. d
22. c
23. d
24. c
25. b
26. b
27. a
28. b
29. b
30. b
31. a
32. b
33. b

Calculation:

Q4.2   The average atomic mass of copper is 63.55 amu.

Q4.3   The number of moles in 50 g of limonite ($Fe_2O_3 \cdot 3/2H_2O$) is 0.26 mol.

Q4.4   The number of molecules in a 5.0×10^{-2} g drop of water (H_2O) is 1.7×10^{21} molecules.

Q4.5   The number of atoms in 13 g of He is 1.96×10^{24} atoms.

Q4.6   The number of atoms of oxygen (O) in 50.0 g of limonite ($Fe_2O_3 \cdot 3/2H_2O$) is 7.22×10^{23} atoms.

Q4.7   The mass of iron Fe in 150 g of iron oxide (Fe_2O_3) is 105 g.

Q4.8   In iron oxide (Fe_2O_3), the mass percent of iron (Fe) is 70% and the mass percent of oxygen (O) is 30%.

Q4.9   The mass percent of iron (Fe) in 150 g of iron oxide (Fe_2O_3) is 70%.

Q4.10   The mass percent of Na in $NaHCO_3$ is %(Na) = 18.7 %
The mass percent of H in $NaHCO_3$ is %(H) = 1.3 %
The mass percent of C in $NaHCO_3$ is %(C) = 16.0 %
The mass percent of O in $NaHCO_3$ is %(O) = 64.0 %

Q4.11   The following table gives information contained in the chemical formula of iron(III) carbonate ($Fe_2(CO_3)_3$):
(Atomic masses: C = 12; O = 16; Fe = 56)

Information	Fe	C	O
Number of atoms in one molecule of $Fe_2(CO_3)_3$	**Two** atoms	**Three** atoms	**Nine** atoms
Number of atoms in one mole of $Fe_2(CO_3)_3$	$2 \times N_A = 1.204 \times 10^{24}$ atoms	$3 \times N_A = 1.807 \times 10^{24}$ atoms	$9 \times N_A = 5.420 \times 10^{24}$ atoms
Mass of atoms in one molecule of $Fe_2(CO_3)_3$ in amu	$2 \times 56 = 112$ amu	$3 \times 12 = 36$ amu	$9 \times 16 = 144$ amu
in g	$112/N_A$ g $= 1.860 \times 10^{-24}$ g	$36/N_A$ g $= 5.978 \times 10^{-24}$ g	$144/N_A$ g $= 2.392 \times 10^{-24}$ g
Number of moles of the element in one mole of $Fe_2(CO_3)_3$	**2** mol	**3** mol	$3 \times 3 = 9$ mol
Mass of the element per mole of $Fe_2(CO_3)_3$	$56 \times 2 = 112$ g	$3 \times 12 = 36$ g	$9 \times 16 = 144$ g
Molar mass of $Fe_2(CO_3)_3$	$M = 2 \times M(Fe) + 3 \times M(C) + 9 \times M(O) = (2 \times 56) + (3 \times 12) + (9 \times 16) = 292$ g mol^{-1}		
Mass percent of the element: %(element) = 100 × m(element)/ m($Fe_2(CO_3)_3$)	**%(Fe) = 100 × 112/292 = 38.4%**	**%(C) = 100 × 36/292 = 12.3%**	**%(O) = 100 × 144/292 = 49.3%**
Simplest formula of $Fe_2(CO_3)_3$	**$Fe_2(CO_3)_3$ (which is the same as the molecular formula)**		

Molecular and empirical formulas calculation

Q4.12   The empirical formula is $KMnO_4$ which is the same as the molecular formula.

Q4.13   The simplest formula is C_3H_8N, and the molecular formula is $C_6H_{16}N_2$.

Q4.14   The empirical formula of lactic acid is CH_2O, and its molecular formula is $C_3H_6O_3$.

Stoichiometric calculation

Q4.15
a) $N_2 + 3H_2 \rightarrow 2NH_3$
b) The mass in g of NH_3 formed is m(NH_3) = 20.4 g
c) The mass in g of N_2 required to form 1.00 kg NH_3 is m(N_2) = 823.5 g
d) The mass in g of H_2 required to react with 6.00 g N_2 is m(H_2) = 1.28 g

Q4.16
a) $C_3H_8 + 5O_2 \rightarrow 3CO_2 + 4H_2O$
It is worth =noting that, when balancing a combustion reaction, it is best to balance the oxygen atoms last.
b) The mass of CO_2 formed when 11 g of propane reacts: m(CO_2) = 33 g
c) The mass of H_2O formed when 11 g of propane reacts: m(H_2O) = 18 g
d) The number of moles of O_2 required to the combustion of 11 g of propane: n(O_2) = 1.25 mol
e) The mass of CO_2 formed when 16 g of O_2 reacts: m(CO_2) = 33 g

Q4.17 Mass of H_2SO_4 required to react with 0.7 mol of copper (Cu) is 137.2 g

Q4.18 Mass of O_2 required to form 250 g of the protective layer Al_2O_3: $m(O_2) = 117.76$ g

Q4.19 The volume to be taken from the stock solution to prepare 100 mL of 0.5 M HCl solution is V = 4.3 mL

KEY EXPLANATIONS

Q4.1

1. 1 amu = $1/N_A$ g = 1.660×10^{-24} g.
2. The atomic mass is commonly expressed in amu. Calcium (Ca) is an element, and its atomic mass is given in the periodic table of elements. The atomic mass of calcium Ca is 40 amu.
3. The atomic mass of chlorine Cl is 35.5 amu; 1 amu = $1/N_A$ g = 1.660×10^{-24} g, so the mass in g of one atom of chlorine (Cl) is $35.5/N_A$ g = 5.895×10^{-23} g.
4. The molecular mass of calcium $CaCl_2$ is the sum of the atomic masses of the constituent elements (Ca and Cl). The atomic masses of Ca and Cl are 40 amu and 35.5 amu, respectively. One molecule of $CaCl_2$ contains 1 atom of Ca and 2 atoms of Cl. The molecular mass of calcium $CaCl_2$ equals $1 \times 40 + 2 \times 35.5 = 111$ amu = $111/N_A$ g = 1.843×10^{-22} g.
5. The molecular mass of sodium carbonate Na_2CO_3 is the sum of the atomic masses of the constituent elements (Na, C and O). One molecule of Na_2CO_3 contains two atoms of Na, 1 atom of C and three atoms of O. The atomic masses are C = 12, O = 16 and Na = 23. The molecular mass of sodium carbonate Na_2CO_3 is $2 \times 23 + 1 \times 12 + 3 \times 16 = 106$ amu. The mass in g of one molecule of sodium carbonate Na_2CO_3 is $106/N_A$ g
6. The mass of one molecule of calcium carbonate $CaCO_3$ is the sum of the atomic masses of the constituent elements (Ca, C and O). The molecular mass of calcium carbonate $CaCO_3$ is $1 \times 40 + 12 + 3 \times 16 = 100$ amu.
   The mass in grams of one molecule of calcium carbonate $CaCO_3$ is $100/N_A$ g
7. The molar mass is expressed in g mol^{-1}. The atomic mass (mass of one atom) of calcium Ca is 40 amu and the molar mass (mass of one mole) of Ca is 40 g mol^{-1}. One mole of Ca contains N_A = 6.022×10^{23} atoms of Ca.
8. The molar mass of $MgCl_2$ is the sum of the molar masses of the constituent elements (Mg and Cl). One mole of $MgCl_2$ contains one mole of Mg and two moles of Cl. The atomic masses are Mg = 24 and Cl = 35.5. The molar mass of $MgCl_2$ is $1 \times 24 + 2 \times 35.5 = 95$ g mol^{-1}.
9. Gypsum ($CaSO_4 \cdot 2H_2O$) is a compound. The molar mass of $CaSO_4 \cdot 2H_2O$ is the sum of the molar masses of the constituent elements (Ca, S, O and H). The atomic masses are H = 1, S = 32, O = 16 and Ca = 40. One mole of $CaSO_4 \cdot 2H_2O$ contains one mole of Ca, one mole of S, four moles of O and two moles of H_2O. The molar mass of $CaSO_4 \cdot 2H_2O$ is equal to $1 \times 40 + 1 \times 32 + 4 \times 16 + 2 \times (1 \times 2 + 1 \times 16) = 172$ g mol^{-1}.
10. The molar mass of $Fe(NO_3)_3$ is the sum of the molar masses of the constituent elements (Fe, N and O). The atomic masses are Fe = 56, O = 16 and N = 14. One mole of $Fe(NO_3)_3$ contains one mole of Fe and three moles of (NO_3). This means that one mole of $Fe(NO_3)_3$ contains one mole of Fe, three moles of N and (3×3 =) nine moles of O. The molar mass of $Fe(NO_3)_3$ is $1 \times 56 + 3 \times 14 + 9 \times 16 = 242$ g mol^{-1}.
11. The number of moles of a substance n = m/M where m is the mass of the substance and M is its molar mass. So, m = n $\times$ M
    The molar mass M of $Fe(NO_3)_3$ equals 242 g mol^{-1}. See question 10 for details.
    The mass of 0.05 mol of $Fe(NO_3)_3$ is m = n $\times$ M = $0.05 \times 242 = 12.1$ g
12. The number of moles of a substance n = m/M where m is the mass of the substance and M is its molar mass.
    The molar mass M of $Fe(NO_3)_3$ equals 242 g mol^{-1} (see question 10 for details).
    The number of moles of iron(III) nitrate ($Fe(NO_3)_3$) in 121 g of this substance is n = m/M = 121/242 = 0.5 mol

13. One mole of a compound contains N_A molecules, (N_A is Avogadro's number). Therefore, n moles contain $n \times N_A$ molecules.
    The number of molecules $N = n \times N_A$ where n is the number of moles of the compound.
    The number of molecules of iron(II) chloride ($FeCl_2$) in 0.1 mol of $FeCl_2$ is
    $N = n \times N_A = 0.1 \times 6.022 \times 10^{23} = 6.022 \times 10^{22}$ molecules.

14. In this question is given the mass of the substance, so first, we should calculate the number of moles (n = m/M) then, by multiplying the number of moles by Avogadro's number, we determine the number of molecules ($N = n \times N_A$).
    The molar mass of iron (III) nitrate ($Fe(NO_3)_3$) is 242 g mol^{-1} (see question 10 for details)
    The number of moles of $Fe(NO_3)_3$ is n = m/M = 121/242 = 0.5 mol
    The number of molecules of $Fe(NO_3)_3$ is
    $N = n \times N_A = 0.5 \times 6.022 \times 10^{23} = 3.011 \times 10^{23}$ molecules.

15. According to the chemical formula of iron (III) nitrate ($Fe(NO_3)_3$), one mole contains one mole of Fe, three moles of N and nine moles of O.

16. This question is about how to determine the number of moles of one constituent element of a compound in m g of the compound.
    First, determine the number of moles of the compound (n = m/M), then determine the number of moles of the constituent element, using the chemical formula of the compound.
    The molar mass M of $Fe(NO_3)_3$ equals 242 g mol^{-1} (see question 10 for details).
    The number of moles of $Fe(NO_3)_3$: n = m/M = 121/242 = 0.5 mol
    According to the chemical formula of iron (III) nitrate ($Fe(NO_3)_3$), one mole contains three moles of nitrogen (N). So, 0.5 mol of $Fe(NO_3)_3$ contains
    $3 \times 0.5 = 1.5$ mol of nitrogen.

17. The number of moles of $Fe(NO_3)_3$:
    n = m/M = 121/242 = 0.5 mol (see question 16 for details).
    One mole of $Fe(NO_3)_3$ contains three moles of N, so 0.5 mol of $Fe(NO_3)_3$ contains
    $3 \times 0.5 = 1.5$ mol of nitrogen (N).
    The mass of nitrogen is obtained by multiplying the number of moles of nitrogen by the molar mass of nitrogen: m(N) = n(N)×M(N) = 1.5 × 14 = 21 g.

18. The number of moles of $Fe(NO_3)_3$
    n = m/M = 121/242 = 0.5 mol
    One mole of $Fe(NO_3)_3$ contains nine moles of O, so 0.5 mol of $Fe(NO_3)_3$ contains
    $9 \times 0.5 = 4.5$ moles of O
    One mole of O contains $N_A = 6.022 \times 10^{23}$ atoms of O
    4.5 moles of O contain $N = n \times N_A = 4.5 \times 6.022 \times 10^{23} = 2.709 \times 10^{24}$ atoms of O.

19. One mole of $Fe(NO_3)_3$ contains $3 \times 3 = 9$ moles of O
    The molar mass of $Fe(NO_3)_3$ equals 242 g mol^{-1} (see question 10 for details) and the mass of oxygen in one mole of $Fe(NO_3)_3$ equals $9 \times 16 = 144$ g
    The mass percent of oxygen (O) is
    %(O) = 100 × (mass of oxygen)/molar mass of $Fe(NO_3)_3$
    = 100 × (144/242) = 59.5%

20. The number of moles of $Fe(NO_3)_3$ in 75 g is
    n = m/M = 75/242 = 0.31 mol
    One mole of $Fe(NO_3)_3$ contains nine moles of O, so 0.31 moles of contains
    $9 \times 0.31 = 2.79$ mol of O.
    The mass of oxygen is m = n × M = 2.8 × 16 = 44.64 g
    The mass percent of oxygen (O) is
    % (O) = 100 × (mass of oxygen/mass of $Fe(NO_3)_3$)
    = 100 × (44.64/75) = 59.5%
    Note that we found the same mass percent for questions 19 and 20:
    The mass percent of an element in a compound is the same when calculated taking one mole of the compound or when considering a definite mass (m) of the compound.

21. The mass percent of Fe in limonite ($Fe_2O_3 \cdot 3/2H_2O$) is 59.92%. The mass of iron in 1 ton (=1,000 kg) of limonite is (59.92 × 1,000)/100 = 599.2 kg.

22. The molar mass of the empirical formula ($C_3H_4O_3$) equals

$3 \times 12 + 4 \times 1 + 3 \times 16 = 88$ g mol^{-1}
The molar mass of vitamin C is 176 g mol^{-1}
The empirical formula is repeated D times to give the molecular formula
D = molar mass of vitamin C/molar mass of the empirical formula ($C_3H_4O_3$)
= 176/88 = 2, so $C_3H_4O_3$ is repeated two times to give vitamin C.
Therefore, the molecular formula of vitamin C is $C_6H_8O_6$.

23. Number of moles of carbon C: n(C) = 18/12 = 1.5 mol
Number of moles of hydrogen (H): n(H) = 3/1 = 3 mol
Number of moles of oxygen (O): n(O) = 24/16 = 1.5 mol
A preliminary formula of this compound is $C_{1.5}H_3O_{1.5}$
Divide by the lowest mol amount (herein, 1.5) to find the relative mol ratios: CH_2O.
Since the subscripts are integers, the simplest (empirical) formula of this compound is CH_2O

24. The molarity (M) is defined as number of moles of solute per liter of solution
M = n/V
The molar mass of NaOH: M(NaOH) = $1 \times 23 + 1 \times 16 + 1 \times 1 = 40$ g mol^{-1}
The number of moles of NaOH is n(NaOH): n(NaOH) = m(NaOH)/M(NaOH)
= 8.0/40 = 0.2 mol
The volume of the NaOH solution is 400 cm^3 = 400 mL = 0.4 L
The molarity of the NaOH solution is M = n(NaOH)/V = 0.2/0.4 = 0.5 mol L^{-1}.

25. The molality is defined as number of moles of solute per kilogram of solvent. It is expressed in SI units as mol kg^{-1}.
The number of moles of solute NaOH, n(NaOH) = 8.0/(23 + 16+1) = 0.2 mol
The density of water is equal to 1. Therefore, the mass of 400 mL of water is 400 g = 0.4 kg.
The molality of the NaOH solution is calculated as:
Molality (m) = (moles of solute)/(kilograms of solvent) = 0.2/0.4 = 0.5 mol kg^{-1}

26. The (x, y, z) values which make the following chemical reaction
x MnO_4^- + y H_2O_2 + z $H^+ \to$ 2 Mn^{2+} + 5 O_2 + $8H_2O$
balanced are (2, 5, 6): 2 MnO_4^- + 5 H_2O_2 + 6 $H^+ \to$ 2 Mn^{2+} + 5 O_2 + $8H_2O$
Check to make sure that the number of atoms of each element is the same on both sides of the equation:

Reactants (left side)	Products (right side)
2 Mn	2 Mn
18 O	18 O
16 H	16 H

27. The balanced chemical equation is: $2C_2H_6 + 7O_2 \to 4CO_2 + 6H_2O$
The number of atoms of each element is the same on both sides of the equation:

Reactants (left side)	Products (right side)
4 C	4 C
14 O	14 O
12 H	12 H

28. Ethane reacts with oxygen to form carbon dioxide and water, according to the following reaction: $2C_2H_6 + 7O_2 \to 4CO_2 + 6H_2O$
From the above chemical reaction, we determine the relationship between the number of moles of ethane, n(C_2H_6), and the number of moles of carbon dioxide n(CO_2):
n(C_2H_6)/2 = n(CO_2)/4 so n(CO_2) = $2 \times$ n(C_2H_6)
Note that this relationship is obtained simply by dividing the number of moles of the substance by its coefficient in the balanced chemical equation.
The number of moles of ethane n(C_2H_6) = 1.35/30 = 0.045 mol
The number of moles of carbon dioxide:

$n(CO_2) = 2 \times n(C_2H_6) = 2 \times 0.045 = 0.09$ mol

The mass of carbon dioxide $m(CO_2) = n(CO_2) \times M(CO_2) = 0.09 \times 44 = 3.96$ g

29. The coefficients provide equal numbers of each type of atom on each side of the equation. However, the sums of charges on each side of the equation are not the same. Indeed, the sum of charges on the left side is $2 \times (-1) + 6 \times (+1) = +4$ and the sum of charges on the right side is equal to $2 \times (+3) = +6$. Therefore, the equation: $2\ MnO_4^- + 5\ H_2O_2 + 6\ H^+ \rightarrow 2\ Mn^{3+} + 5\ O_2 + 8\ H_2O$ is not a balanced chemical equation.

30. One mole of H_2SO_4 contains four moles of oxygen. So, the number of moles of oxygen is four. The mass of oxygen $m = n \times M = 4 \times 16 = 64$ g. Do not use the value "4" twice; i.e., in calculating the number of moles and in calculating the mass of oxygen.

31. One mole of an element or a compound contains $N_A = 6.022 \times 10^{23}$ entities (atoms, molecules or ions). As a result, there is the same number of atoms or molecules in one mole of different elements or compounds. However, the molar masses (defined as the mass of one mole) are not necessarily the same. For example, the molar mass of H_2SO_4 is 98 g mol^{-1} and it differs from the molar mass of $CaCO_3$ which equals 100 g mol^{-1}.

32. The average atomic mass of an element considers the abundance of each isotope. For this reason, the average atomic mass of an element is different from the average of the atomic masses of all isotopes of this element. The average atomic mass of Rb (=85.4678 amu, value given in the periodic table of elements) is different from the average of the atomic masses of its two isotopes, i.e., $(84.9117 + 86.9085)/2 = 85.9101$ amu.

33. Solubility depends strongly on temperature. When temperature increases or decreases, the solubility varies.

Calculations

Q.4.2 The average atomic mass of copper is the sum of the contribution of each isotope to the atomic mass of the element:

$$62.930 \times (69.2 / 100) + 64.928 \times (30.8 / 100) = 63.55 \text{ amu.}$$

Q4.3 Limonite has the formula $Fe_2O_3 \cdot 3/2H_2O$.

One mole of limonite contains two moles of Fe, three moles of O and 3/2 moles of H_2O

The molar mass of $Fe_2O_3 \cdot 3/2H_2O$ is $M = 2 \times 56 + 3 \times 16 + 3/2\ (2 \times 1 + 16) = 187$ g mol^{-1}

The number of moles of $Fe_2O_3 \cdot 3/2H_2O$ in 50 g is $n = m/M = 50/187 = 0.26$ mol.

Q4.4 One mol of H_2O contains two moles of H and one mole of O

The molar mass of water is $M = 2 \times 1 + 1 \times 16 = 18$ g mol^{-1}

Number of moles of H_2O is $n = m/M = (5.0 \times 10^{-2})/18 = 2.8 \times 10^{-3}$ mol

Number of atoms or molecules = number of moles $\times$ Avogadro's number N_A

$$N = n \times N_A = 2.8 \times 10^{-3} \times 6.022 \times 10^{23}$$

$$= 1.7 \times 10^{21} \text{ molecules}$$

Q4.5 Number of moles of He is $n = m/M = 13/4 = 3.25$ mol

Number of atoms of He is N = number of moles $\times$ Avogadro's number N_A

$$N = n \times N_A = 3.25 \times 6.022 \times 10^{23}$$

$$= 1.96 \times 10^{24} \text{ atoms}$$

Q4.6 The number of moles of $Fe_2O_3 \cdot 3/2H_2O$ in 50.0 g is $n = m/M = 50.0/187 = 0.267$ mol (see question Q4.2 for details).

One mole of limonite contains 4.5 mol of oxygen (3 mol of O (in FeO_3) and 3/2 mol of O (in $3/2H_2O$)).

The number of moles of oxygen in 0.267 mol of limonite is $n = 4.5 \times 0.267 = 1.20$ mol

The number of atoms of oxygen N = number of moles of oxygen × Avogadro's number N_A

$$N = n \times N_A = 1.20 \times 6.022 \times 10^{23}$$

$$= 7.22 \times 10^{23} \text{ atoms}$$

Q4.7   One mol of Fe_2O_3 contains two mol of Fe and three mol of O

The molar mass of Fe_2O_3 is $M = 2 \times 56 + 3 \times 16 = 160$ g mol^{-1}

Number of moles of Fe_2O_3 is $n = m/M = 150/160 = 0.94$ mol

One mol of Fe_2O_3 contains two mol of Fe, so the number of moles of Fe in 0.938 mol of Fe_2O_3 is $n(Fe) = 2 \times 0.94 = 1.88$ mol

The mass of Fe in 150 g of Fe_2O_3 is $m(Fe) = n(Fe) \times M(Fe)$

$$m(Fe) = 1.88 \times 56 = 105 \text{ g}$$

Q4.8   One mol of Fe_2O_3 contains two mol of Fe and three mol of O

The mass percent of an element in a compound m(%) is

$$m(\%) = \frac{n \times \text{atomic mass of the element}}{\text{molecular mass of the compound}} \times 100$$

$$\%(Fe) = \frac{2 \times 56}{160} \times 100 = 70\%$$

$$\%(O) = \frac{3 \times 16}{160} \times 100 = 30\%$$

Note that % (Fe) + % (O) = 100%.

The mass percent of iron (or oxygen) could be used to determine the mass of iron (or oxygen) in a given mass of Fe_2O_3. For example, the mass of Fe in 150 g of Fe_2O_3 is m(Fe) = %(Fe)×150/100 = 70 × 150/100 = 105 g (the same result was found in Q4.6)

Q4.9   The molar mass of Fe_2O_3 is $M = 2 \times 56 + 3 \times 16 = 160$ g mol^{-1}

Number of moles of Fe_2O_3 is $n = m/M = 150/160 = 0.94$ mol

One mol of Fe_2O_3 contains two mol of Fe, so the number of moles of Fe in 0.938 mol of Fe_2O_3 is $n(Fe) = 2 \times 0.94 = 1.88$ mol

The mass of Fe is m(Fe) = n(Fe) × M(Fe) = $1.88 \times 56 = 105$ g

The mass percent of Fe in 150 g of Fe_2O_3 is % (Fe) = 100 × m(Fe)/m(Fe_2O_3)

% (Fe) = 100 × (105/150) = 70%

Note that the mass percent of iron (Fe) in Fe_2O_3 calculated with respect to one molecule (Q4.8) or with respect to a given mass of Fe_2O_3 (in 150g of Fe_2O_3 in this case), is exactly the same.

Q4.10   Calculation of the mass percent of Na, H, C and O in $NaHCO_3$:

The molar mass of $NaHCO_3$ is

$$M = 1 \times 14 + 1 \times 1 + 1 \times 12 + 3 \times 16 = 75 \text{ g } mol^{-1}$$

One mole of $NaHCO_3$ contains one mole of Na, which means that 75 g of $NaHCO_3$ contains 14 g of Na. The mass percent of Na in $NaHCO_3$ is:

$$\%(Na) = 100 \times \text{mass}(Na)/\text{mass}(NaHCO_3) = 100 \times 14/75 = 18.7\%$$

One mole of $NaHCO_3$ contains one mole of H, which means that 75 g of $NaHCO_3$ contains 1 g of H. The mass percent of H in $NaHCO_3$ is:

$$\%(H) = 100 \times \text{mass}(H)/\text{mass}(NaHCO_3) = 100 \times 1/75 = 1.3\%$$

One mole of $NaHCO_3$ contains one mole of C, which means that 75 g of $NaHCO_3$ contains 12 g of C. The mass percent of C in $NaHCO_3$ is:

$$\%(C) = 100 \times mass(C)/mass(NaHCO_3) = 100 \times 12/75 = 16.0\%$$

One mole of $NaHCO_3$ contains three moles of O, which means that 75 g of $NaHCO_3$ contains $3 \times 16 = 48$ g of O. The mass percent of O in $NaHCO_3$ is:

$$\%(O) = 100 \times mass(O)/mass(NaHCO_3) = 100 \times 48/75 = 64.0\%$$

Check calculation: If you add up all the mass percent compositions, you should get 100%.

$$\%(Na) + \%(H) + \%(C) + \%(O) = 18.7 + 1.3 + 16.0 + 64.0 = 100\%$$

Molecular and empirical formula calculations

Q4.12 Elemental analysis of a compound shows that it contains 15.6 g of potassium (K), 22.0 g of manganese (Mn) and 25.6 g of oxygen (O). (Atomic masses: K = 39, Mn = 55, O = 16)

Step 1: To determine the empirical and molecular formulas of this compound, we determine first the number of moles of each constituent element from its mass:
- Number of moles of K: $n(K) = m(K)/M(K) = 15.6/39 = 0.4$ mol
- Number of moles of Cr: $n(Cr) = m(Mn)/M(Mn) = 22.0/55 = 0.4$ mol
- Number of moles of K: $n(K) = m(O)/M(O) = 25.6/16 = 1.6$ mol

Step 2: Divide each mole number by the smallest number of moles (0.4):
- Relative mole ratio of K = 0.4/0.4 = 1.0
- Relative mole ratio of Mn = 0.4/0.4 = 1.0
- Relative mole ratio of O = 1.6/0.4 = 4

Step 3: Use the relative mol ratios as subscripts to obtain a preliminary formula: $KMnO_4$

Step 4: Since the relative mol ratios are integers, the empirical formula is $KMnO_4$, which is the same as the molecular formula.

Q4.13 The molar mass of CO_2 is $M(CO_2) = 1 \times 12 + 2 \times 16 = 44$ g mol^{-1}.

The number of moles of CO_2 is

$$n(CO_2) = m(CO_2)/M(CO_2) = 14.36/44 = 0.33 \text{ mol}$$

One mole of CO_2 contains one mol of C so, the number of moles of C is

$$n(C) = n(CO_2) = 0.33 \text{ mol}$$

The mass of C is $m(C) = n(C) \times M(C) = 0.33 \times 12 = 4.0$ g

Number of moles of H_2O is $n(HO_2) = m(HO_2)/M(HO_2) = 7.921/(2 \times 1.0 + 16) = 0.44$ mol

One mole of H_2O contains two mol of H then, the number of H moles is

$$n(H) = 2 \times n(H_2O) = 2 \times 0.44 = 0.88 \text{ mol}$$

The mass of H is $m(H) = n(H) \times M(H) = 0.88 \times 1.0 = 0.88$ g

A mass of 6.481 g of hexamethylenediamine contain 4.0 g of C, 0.88 g of H and m(N) g of N, so the mass of N is $m(N) = 6.481 - (4.0 + 0.88) = 1.6$ g

The number of moles of nitrogen N is $n(N) = m(N)/M(N) = 1.6/14 = 0.11$ mol

A preliminary formula of hexamethylenediamine is $C_{0.33}H_{0.88}N_{0.11}$

Divide by smallest number of moles (0.11) to get the relative mole ratio and the number of moles in the empirical formula:

The relative mol ratio of C = 0.33/0.11 = 3

The relative mol ratio of H = 0.88/0.11 = 8

The relative mol ratio of N = 0.11/0.11 = 1

The simplest (empirical) formula is C_3H_8N

The molar mass of the simplest formula is $M(C_3H_8N) = 3 \times 12 + 8 \times 1.0 + 1 \times 14 = 58 \text{ g mol}^{-1}$

Or the molar mass of the molecular formula is 116 g mol^{-1}, so the greatest common denominator

$$D = \frac{\text{molar mass of the molecular formula}}{\text{molar mass of the empirical formula}}$$

$$= 116/58 = 2$$

The molecular formula of hexamethylenediamine is obtained by multiplying the subscripts of the simplest formula C_3H_8N by D = 2.

The molecular formula of hexamethylenediamine is $C_6H_{16}N_2$.

Q4.14

Step 1: determine the number of moles of each constituent element:

Assuming there are 100 g of lactic acid, the constituent elements are 40 g of C, 6.71 g of H and 53.3 g of O.

- The number of moles of C: $n(C) = m(C)/M(C) = 40/12 = 3.33 \text{ mol}$
- The number of moles of H: $n(H) = m(H)/M(H) = 6.7/1 = 6.7 \text{ mol}$
- The number of moles of O: $n(O) = m(O)/M(O) = 53.3/16 = 3.33 \text{ mol}$

Step 2: Divide each mole number by the smallest number of moles (3.33):

- Relative mole ratio of C = 3.33/3.33 = 1
- Relative mole ratio of H = 6.7/3.33 = 2
- Relative mole ratio of O = 3.33/3.33 = 1

Step 3: Use the relative mol ratios as subscripts to obtain a preliminary formula: CH_2O

Step 4: Since the relative mol ratios are integers, the empirical formula is CH_2O.

The molar mass of the empirical formula CH_2O is

$$M(CH_2O) = 12 + 2 \times 1 + 16 = 30 \text{ g mol}^{-1}$$

Or the molar mass of the molecular formula is M(lactic acid) = 90 g mol^{-1}

So, $D = M(\text{lactic acid})/M(CH_2O) = 90/30 = 3$.

The molecular formula is obtained by multiplying the subscripts of the empirical formula by 3. Therefore, the molecular formula of lactic acid is $C_2H_6O_2$.

Q4.15

a) Ammonia (NH_3) is made by reacting nitrogen (N_2) with hydrogen (H_2). The balanced chemical equation is: $N_2 + 3H_2 \rightarrow 2NH_3$

From this balanced chemical equation, we have the following relationships between the number of moles of reactants and products:

$$\frac{n(N_2)}{1} = \frac{n(H_2)}{3} = \frac{n(NH_3)}{2}$$

Note that these relationships are obtained simply by dividing the number of moles of the substance by its coefficient (the number in front of the chemical formula of the substance) in the balanced chemical equation.

b) This question relates NH_3 to H_2. According to the balanced chemical equation, the relationship between the number of moles of H_2 and the number of moles of NH_3 is $2 \times n(H_2) = 3 \times n(NH_3)$. Or 1.8 mol H_2 reacts therefore

$$n(NH_3) = 2/3 \times n(H_2) = (2/3) \times 1.8 = 1.2 \text{ mol}$$

The molar mass of NH_3 is $M(NH_3) = 1 \times 14 + 3 \times 1 = 17$ g mol the number of moles of H_2.

The mass in g of NH_3 formed is $m(NH_3) = n(NH_3) \times M(NH_3)$

$= 1.2 \times 17 = 20.4 \text{ g}$

You can also use the cross-multiplication method:
3 mol H_2 → 2 mol NH_3
1.8 mol → n (NH_3) =?
n (NH_3) = (1.8 × 2)/3 = 1.2 mol
The mass in g of NH_3 formed is m(NH_3) = n(NH_3) × M(NH_3)
= 1.2 × (14 + 3 × 1) = 20.4 g

c) This question relates N_2 to NH_3.
Use the cross-multiplication method:
1 mol N_2 → 2 mol NH_3
2 × 14 = 28 g N_2 → 2 × (14 + 3 × 1) = 34 g NH_3
m(N_2) =? → 1 kg = 1000 g
The mass in g of N_2 required to form 1.00 kg NH_3 is
m(N_2) = (28 × 1000)/34 = 823.5 g

d) This question relates N_2 to H_2.
Use the cross-multiplication method:
1 mol N_2 → 3 mol H_2
2 × 14 = 28 g N_2 → 3 × (2 × 1) = 6 g H_2
6 g → m(H_2) =?
The mass in g of H_2 required to react with 6.00 g N_2 is
m(H_2) =(6 × 6)/28 = 1.28 g

Q4.16

a) To balance the chemical equation of the combustion of propane (C_3H_8), start by balancing C or H since they appear in one reactant or one product. Thereafter, balance the oxygen. The balanced chemical equation of the combustion of propane is:

$$C_3H_8 + 5O_2 \rightarrow 3CO_2 + 4H_2O$$

From this equation, we have the following relationships between the number of moles of reactants and products:

$$\frac{n(C_3H_8)}{1} = \frac{n(O_2)}{5} = \frac{n(CO_2)}{3} = \frac{n(H_2O)}{4}$$

Note that these relationships are obtained simply by dividing the number of moles of the substance by its coefficient in the balanced chemical equation.

b) The molar mass of propane M(C_3H_8) = 3 × 12 + 8 × 1 = 44 g mol^{-1}.
Number of moles of propane n(C_3H_8) = m(C_3H_8)/M(C_3H_8) = 11/44 = 0.25 mol

$$\frac{n(C_3H_8)}{1} = \frac{n(CO_2)}{3}$$

Number of moles of CO_2: n(CO_2) = 3 × n(C_3H_8) = 3 × 0.25 = 0.75 mol
The molar mass of CO_2: M(CO_2) = 1 × 12 + 2 × 16 = 44 g mol^{-1}.
Mass of CO_2 formed: m(CO_2) = n(CO_2) × M(CO_2) = 0.75 × 44 = 33 g

c)

$$\frac{n(C_3H_8)}{1} = \frac{n(H_2O)}{4}$$

Number of moles of H_2O: n(H_2O) = 4 × n(C_3H_8) = 4 × 0.25 = 1 mol
The molar mass of H_2O: M(H_2O) = 2 × 1 + 1 × 16 = 18 g mol^{-1}.
Mass of H_2O formed: m(H_2O) = n(H_2O) × M(H_2O) = 1 × 18 = 18 g

d)

$$\frac{n(C_3H_8)}{1} = \frac{n(O_2)}{5}$$

Number of moles of O_2: n(O_2) = 5 × n(C_3H_8) = 5 × 0.25 = 1.25 mol

e) The molar mass of oxygen $M(O_2) = 2 \times 16 = 32$ g mol^{-1}.
Number of moles of oxygen $n(O_2) = m(O_2)/M(O_2) = 40/32 = 1.25$ mol

$$\frac{n(O_2)}{5} = \frac{n(CO_2)}{3}$$

Number of moles of CO_2: $n(CO_2) = (3/5) \times n(O_2) = (3/5) \times 1.25 = 0.75$ mol
Mass of CO_2 formed: $m(CO_2) = n(CO_2) \times M(CO_2) = 0.75 \times 44 = 33$ g
Note that, when 11 g of propane are burned, 33 g of CO_2 (question b) and 1.25 mol of oxygen (question c) are obtained. The mass of oxygen required is $m(O_2) = n(O_2) \times M(O_2) = 1.25 \times 32 = 40$ g which is the same mass as in question e). So, it is expected that 33 g of CO_2 should be obtained when 40 g (containing 1.25 mol) of oxygen react.

Q4.17

From the following balanced chemical equation:

$$2H_2SO_4 + Cu \rightarrow CuSO_4 + 2H_2O + SO_2$$

we deduce the following relationship between the number of moles of reactants and products:

$$\frac{n(H_2SO_4)}{2} = \frac{n(Cu)}{1}$$

Note that this relationship is obtained simply by dividing the number of moles of the substance by its coefficient in the balanced chemical equation.
Number of moles of H_2SO_4 is $n(H_2SO_4) = 2 \times n(Cu) = 2 \times 0.7 = 1.4$ mol
The molar mass of H_2SO_4 is $M(H_2SO_4) = (2 \times 1) + (1 \times 32) + (4 \times 16) = 98$ g/mol
Mass of H_2SO_4 required for reacting with 0.7 mol of cupper (Cu):

$$m(H_2SO_4) = n(H_2SO_4) \times M(H_2SO_4) = 1.4 \times 98 = 137.2 \text{ g}$$

Q4.18

From the following balanced chemical equation:

$$4Al + 3O_2 \rightarrow 2Al_2O_3$$

we deduce the following relationship between the number of moles of reactants and products:

$$\frac{n(O_2)}{3} = \frac{n(Al_2O_3)}{2}$$

Note that this relationship is obtained simply by dividing the number of moles of the substance by its coefficient in the balanced chemical equation.

$$n(O_2) = (3/2) \times n(Al_2O_3)$$

The molar mass of Al_2O_3 is $M(Al_2O_3) = (2 \times 27) + (3 \times 16) = 102$ g mol^{-1}
Number of moles of Al_2O_3 is $n(Al_2O_3) = m(Al_2O_3)/M(Al_2O_3) = 250/102 = 2.45$ mol
Number of moles of O_2 is $n(O_2) = (3/2) \times n(Al_2O_3) = 1.5 \times 2.45 = 3.68$ mol
Mass of O_2 required to form 250 g of the protective layer, Al_2O_3:

$$m(O_2) = n(O_2) \times M(O_2) = 3.68 \times (2 \times 16) = 117.76 \text{ g}.$$

4.19

A solution contains 36% by mass of hydrochloric acid (HCl). This means that the mass of HCl in a volume taken from this solution $V_{solution}$ is

$$mass\%\left(HCl\right)=\frac{m_{HCl}}{m_{solution}}\times100$$

Or $m_{solution}=d\times V_{solution}$

Therefore, we have $mass\%\left(HCl\right)=\dfrac{m_{HCl}}{d\times V_{solution}}\times100$

Rearranging gives $m_{HCl}=\dfrac{mass\%\left(HCl\right)\times d\times V_{solution}}{100}$

On the other hand, the number of moles of HCl is given by $n_{HCl}=\dfrac{m_{HCl}}{M_{HCl}}$

where M_{HCl} (=35.5 + 1 = 36.5 g mol^{-1}) is the molar mass of HCl and m_{HCl} is the mass of HCl.

Combining the last two equations, we obtain:

$$n_{HCl}=\frac{mass\%\left(HCl\right)\times d\times V_{solution}}{100\times M_{HCl}}$$

We consider that the concentration of the solution to be prepared is M and its volume is V. We have $M=\dfrac{n_{HCl}}{V}$ where M is expressed in mol L^{-1}.

Rearranging gives $n_{HCl}=M\times V=\dfrac{mass\%\left(HCl\right)\times d\times V_{solution}}{100\times M_{HCl}}$

and $V_{solution}=\dfrac{100\times M_{HCl}\times\dfrac{M}{1000}\times V}{mass\%\left(HCl\right)\times d}$ where V is expressed in mL and d in g mL^{-1}.

Resolving gives $V_{solution}=\dfrac{100\times36.5\times0.5\times100}{36\times1000\times1.18}=4.3$ mL.

Thermochemistry

5

5.1 OBJECTIVES

At the end of the present chapter, the student will be able to:

1. Define energy, work and heat.
2. Distinguish between the system and the surroundings.
3. Define exothermic and endothermic processes.
4. Define the internal energy and the enthalpy of a system.
5. Perform calculations involving change of energy.
6. Define calorimetry and distinguish between specific heat and heat capacity.
7. Recognize the different phase changes of a substance and calculate the heat absorbed or released during phase changes.
8. Define the standard enthalpy of formation and the bond enthalpy.
9. Recognize Hess's law and calculate the enthalpy of reaction.
10. Define the Born-Haber cycle and calculate the lattice enthalpy change of ionic compounds.

5.2 ENERGY, HEAT AND WORK

Energy is the capacity to do work. Energy cannot be created or destroyed; it can only be transformed. The science that concerns the study of energy and its transformation is thermodynamics and the branch of thermodynamics which concerns the study of heat energy changes associated with chemical reactions or physical changes is called thermochemistry. In thermochemistry, we often concern ourselves with energy transfer as work (w) or as heat (q). Heat is the transfer of thermal energy (defined as the kinetic energy related to the random motion of atoms or molecules) between two matters at different temperatures. Matter can absorb or release heat during a chemical reaction or a physical process. Work is the energy used to cause an amount of matter to move. Work (w) can be calculated by the following equation:

$$w = F \times d$$

where F is the applied force and d is the distance traveled by the matter due to this force.

5.3 SYSTEM AND SURROUNDINGS

The system is a portion of space which includes the particles of interest that we are studying. The surroundings are everything else. For example, if we are interested in the combustion reaction of methane (CH_4) with oxygen (O_2) in a cylinder, then CH_4 and O_2 in the cylinder form the system, and everything else is the surroundings. The entire system and surroundings constitute the universe. The system is bounded by a real or fictitious surface through which the exchanges of energy and matter are made with the surroundings (Figure 5.1).

DOI: 10.1201/9781003257059-5

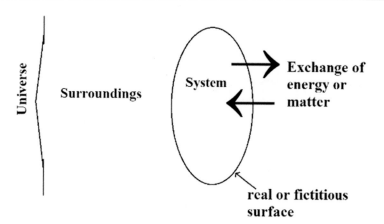

FIGURE 5.1   System, surroundings and universe. An open system can exchange energy and matter with the surroundings

Depending on the nature of the exchange with the surroundings, a system can be open, closed or isolated. An open system can exchange both matter and energy with the surroundings, whereas an isolated system can exchange neither energy nor matter with the surroundings. A closed system can exchange energy but not matter with the surroundings.

5.4 ENERGY EXCHANGE

The energy conservation law (or first law of thermodynamics) states that energy is neither created nor destroyed. Energy can only be transformed from one form to another. This implies that the energy of the universe (system and surroundings) is constant.

The exchange of energy between the system and the surroundings occurs as heat or as work. The work can be done by the system on the surroundings or done on the system by the surroundings. The signs of heat (q) and work (w) exchanged between the system and the surroundings are attributed to the system (Figure 5.2). When heat is absorbed by the system, the process is endothermic, and the heat amount is considered to be a positive quantity (q > 0). When heat is released by the system, the process is exothermic, and the heat amount is considered to be a negative quantity (q < 0). The work done on the system by the surroundings is considered to be a positive quantity (w > 0), whereas the work done by the system on the surroundings is considered to be a negative quantity (w < 0). In other words, energy is transferred from the surroundings to the system when it absorbs heat from the surroundings or when the surroundings do work on the system. Conversely, energy is transferred from the system to the surroundings when heat is lost from the system, or when the system does work on the surroundings.

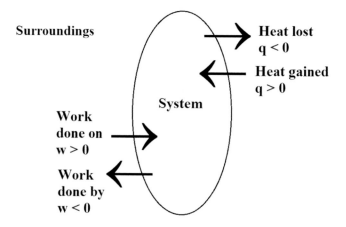

FIGURE 5.2   Signs of the heat and the work exchanged between the system and the surroundings

Practice 5.1 Calculate the work done by a gas in a cylindrical container as it expands from 30 L to 50 L against a constant pressure of 3.2 atm.

Answer:

During expansion of the gas (Figure 5.3.), the gas moves by Δh and the volume increases ($\Delta V > 0$):

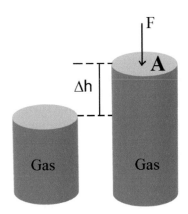

FIGURE 5.3   Increase of the gas volume during expansion

The work w is given by $w = F \times d$ where F is the applied force and d is the distance traveled by the matter. In our case, the gas is moved by Δh, so $d = \Delta h$.

The pressure (P) is the force applied perpendicular to the surface of an object per unit area over which that force is applied.

We have the pressure $P = \dfrac{F}{A}$ where F is the applied force and A is the area over which the force is applied.

And $\Delta V = A \times \Delta h$

So, we have $w = F \times \Delta h$ and $F = P \times A$

Rearranging gives $w = P \times A \times \Delta h = P\Delta V$

During the expansion of the gas, the volume increases ($\Delta V > 0$) and work is done by the system on the surroundings ($w < 0$). Therefore, we should add a negative sign to obtain a negative value for the work:

$$w = -P\Delta V$$

Resolving gives $w = -3.2\,\text{atm} \times (50 - 30)\text{L} = -64\,\text{L.atm}$

We need to express the work in joule. We know that 1 L atm = 101.3 J

Converting gives $w = -64\ \text{L.atm} \times \dfrac{101.3\,\text{J}}{1\ \text{L.atm}} = -6423.2\,\text{J} = -6.4232\,\text{kJ}$

GET SMART

INTERPRETATION AND USE OF THE EQUATION $w = -P\Delta V$

- During compression of the gas, the volume decreases ($\Delta V < 0$) and work is done on the system by the surroundings $\left(w = -P\Delta V > 0 \right)$.
- During expansion of the gas, the volume increases ($\Delta V > 0$) and work is done by the system on the surroundings ($w < 0$).
- For transformations that occur at constant volume, called isochore transformations, $\Delta V = 0$ and $w = -P\Delta V = 0$
- P is the pressure exerted by the surroundings on the system, i.e., the external pressure. This pressure is different from the internal pressure of the gas which

gradually changes to reach the external pressure during compression or expansion of the gas.

- Use the conversion factor $\left(\dfrac{101.3\,J}{1\,L.atm} \right)$ to convert work unit from L atm to J.

5.5 INTERNAL ENERGY

The internal energy, denoted by U, characterizes the energy content of the constituent particles of a substance, such as atoms, molecules, ions, nuclei and electrons. It is equal to the sum of all possible types of energy present in the substance. There are two main types of energy in a substance, namely the kinetic energy and the potential energy. The kinetic energy is associated with the movement of particles of the substance and the potential energy is related to the interactions between these particles. Energy is stored in a substance by increasing the motion (vibration, rotation or straight-line motion) of the constituent particles and therefore by increasing the kinetic energy. Conversely, when the system loses thermal energy, the intensities of these motions decrease and the kinetic energy decreases.

In practice, we cannot measure the internal energy of a substance since we cannot determine the sum of all energies, but we can measure a change in the internal energy, ΔU. During a chemical reaction or a physical change, the internal energy of the system can change and an exchange of energy, as work (w) or as heat (q), between the system and the surroundings can be observed. The change in the internal energy ΔU is given by:

$$\Delta U = U_F - U_I = q + w$$

where U_I and U_F are the initial and final internal energies of the system, respectively.

The internal energy is a state function which means that it depends only on the initial and final states of the system. The internal energy does not depend on either the path by which the system achieved these states, or on how it is used.

Example 1. The internal energy of (n) moles of H_2O at 50°C does not depend on whether we heat (n) moles of H_2O at 25°C to 50°C or we cool (n) moles of H_2O from 100°C to 50°C.

Example 2. A battery can be discharged by producing heat and light in a flashlight or by producing heat and work when running a fan. The change in internal energy of the battery is the same in both cases.

GET SMART

ARE THE WORK AND THE HEAT STATE FUNCTIONS?

The internal energy is a state function; however, the work w and the heat q are not state functions. The work exchanged during a transformation depends on how this transformation is performed. Consequently, the heat $q = \Delta U - w$ depends also on how this transformation is performed. However,

- when q = 0 (for adiabatic transformation), $\Delta U = w$,
- when w = 0 (for isochore transformation, $\Delta V = 0$), $\Delta U = q$

In both cases, the work or the heat does not depend on how the transformation is performed.

Practice 5.2 Calculate the change in the internal energy ΔU of a system if 250 J of heat are released by the system and if 400 J of work are done on the system.

Answer:
250 J of heat are released by the system. This means that the system loses heat, so the amount of heat is considered to be negative. Therefore, we have q = –250 J.

400 J of work are done on the system. So, the amount of work is considered to be positive. Therefore, we have w = +400 J.

The change in the internal energy is given by: $\Delta U = q + w$

Replacing gives $\Delta U = -250 + 400 = 150$ J

This means that the system gains 150 J, but the surroundings lose 150 J.

Practice 5.3 Calculate the change in the internal energy ΔU of a system if 300 J of heat are released by the surroundings and if 450 J of work are performed on the surroundings.

Answer:

Heat of 300 J are released by the surroundings which means that 300 J of heat are absorbed by the system. As a consequence, the amount of heat is considered to be positive. Therefore, we have q = 300 J.

Work (450 J) is performed on the surroundings, meaning that 450 J of work are done by the system. Consequently, the amount of work is considered to be negative. Therefore, we have w = −450 J.

The change in the internal energy is given by: $\Delta U = q + w$

Replacing gives $\Delta U = 300 + (-450) = -150$ J

This means that the system loses 150 J, but the surroundings gain 150 J.

5.6 CALORIMETRY

5.6.1 PRINCIPLES OF CALORIMETRY

Calorimetry is defined as the measurement of heat released or absorbed during a chemical reaction or a physical process. The measurement of heat can be performed using a calorimeter (Figure 5.4). A calorimeter consists of a metallic vessel filled with water and suspended above a reaction chamber. A stirrer and a thermometer are immersed in the metallic vessel. The thermometer allows measurement of any change in thermal properties inside the metallic vessel. The vessel and the reaction chamber are kept in insulating jackets to prevent heat loss into the surroundings. The heat release or absorption causes a change in the water temperature. For example, when an exothermic reaction occurs in the reaction chamber, the heat produced by the reaction is absorbed by the water, the temperature of which increases. Conversely, when an endothermic reaction occurs, the heat required is absorbed from the thermal energy of the water, which decreases in temperature. The temperature change, along with the specific heat and mass of water, allows determination of the amount of heat involved in the exothermic and endothermic processes.

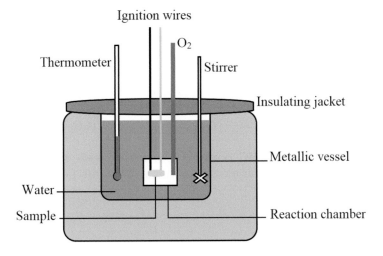

FIGURE 5.4   Schematic illustration of a calorimeter

5.6.2 HEAT CAPACITY AND SPECIFIC HEAT

The heat capacity is the amount of energy required to increase the temperature of a substance by 1 K (or 1°C). The heat capacity, expressed in J K^{-1}, is given by the following equation:

$$C = \frac{q}{\Delta T}$$

where q is the heat and ΔT is the temperature change. Rearranging gives $q = C \times \Delta T$.

The specific heat capacity (or simply, specific heat) Cs is the amount of energy required to raise the temperature of 1 g of a substance by 1°C (or 1 K). If the amount of the substance is one mole, the variable is the molar heat capacity. The specific heat Cs, expressed in J g^{-1}°C, is given by the following equation:

$$Cs = \frac{q}{m \times \Delta T}$$

Rearranging gives $q = Cs \times m \times \Delta T$, where m and ΔT are the mass and the change in temperature of the substance, respectively. Note that heat is an extensive property (that depends on the amount of substance).

In a calorimeter, the heat gained by the substance is equal to the heat lost by the water, and, conversely, the heat lost by the substance is equal to the heat gained by the water. We have:

$$q_{substance} = -q_{water} = -Cs \times m \times \Delta T$$

where Cs, m and ΔT are the specific heat, the mass and the temperature change of the water, respectively. For an endothermic process, the water releases heat which is gained by the substance. In this case, $\Delta T < 0$ since water loses energy and consequently, $q_{substance} > 0$. For an exothermic process, the substance releases heat which is gained by the water. In that case, $\Delta T > 0$ since water gains energy, and consequently, $q_{substance} < 0$.

GET SMART

CAN WE USE THE EQUATION $q = Cs \times m \times \Delta T$ FOR PHASE CHANGES?

The relationship does not apply for phase changes because the heat gained or lost during a phase change does not change the temperature. Indeed, the temperature remains constant during a phase change, and energy is used to overcome attractive forces between molecules. So, how is the heat absorbed or released during a phase change calculated?

We use the following equation: $q = m \times L$, where L is the latent heat which is defined as the heat or energy that is absorbed or released during a phase change of a substance.

Practice 5.4 Calculate the energy required to heat 100 g of water from 25°C to 60°C. The specific heat of water is 4.184 J g^{-1}°C.

Answer:
We have $q = Cs \times m \times \Delta T$

$$\text{Resolving gives } q = 4.184 \frac{J}{g \cdot °C} \times 100 \text{ g} \times (60 - 25)°C = 14644 \text{ J} = 14.644 \text{ kJ}$$

Practice 5.5 Water (30 g) at 25°C (sample 1) is mixed with 47 g of water at 45°C (sample 2). Calculate the final temperature obtained after mixing. The specific heat of water is 4.184 J g^{-1}°C.

Answer:
We will use the equation $q = Cs \times m \times \Delta T$

Heat transfers from the hotter substance to the colder one. Let us consider q_1 to be the heat gained by sample 1 when its temperature increases to reach T_F, and q_2 is the heat lost by sample 2 when its temperature decreases to reach T_F. We have:

$$q_1 = Cs \times m_1 \times \Delta T = Cs \times m_1 \times (T_F - T_1)$$

where Cs is the specific heat of water, m_1 is the mass of sample 1, and T_F and T_1 are the initial and final temperatures of sample 1, respectively;

and $q_2 = Cs \times m_2 \times \Delta T = Cs \times m_2 \times (T_F - T_2)$

where Cs is the specific heat of water, m_2 is the mass of sample 2, and T_F and T_2 are the initial and final temperatures of sample 2, respectively.

The heat (q_1) gained by sample 1 is equal to the heat (q_2) lost by sample 2. Therefore, we have $q_1 = -q_2$

Replacing gives $Cs \times m_1 \times (T_F - T_1) = -Cs \times m_2 \times (T_F - T_2)$

Rearranging gives $(m_1 + m_2) \times T_F = m_1 \times T_1 + m_2 \times T_2$

and $T_F = \dfrac{m_1 \times T_1 + m_2 \times T_2}{(m_1 + m_2)}$

Resolving gives Resolving gives $T_F = \dfrac{\left[30 \text{ g} \times 25°C\right] + \left[47 \text{ g} \times 45°C\right]}{(30+45)\text{g}} = 38.2°C.$

Note that, if we mixed two different substances with different specific heat capacities, at different temperatures, the final temperature is given by $T_F = \dfrac{[C_1 \times m_1 \times T_1] + [C_2 \times m_2 \times T_2]}{([C_1 \times m_1] + [C_2 \times m_2])}$

5.7 ENTHALPY

5.7.1 DEFINITION

Enthalpy (H) is the heat transferred between the system and surroundings under constant pressure. Enthalpy is given by:

$$H = U + PV$$

In practice, we cannot measure the enthalpy of a substance, but we can measure a change in enthalpy, ΔH. The change in enthalpy, ΔH, is given by:

$$\Delta H = H_F - H_I = \Delta(U + PV)$$

where H_I and H_F are the initial and final enthalpies of the system, respectively.

Rearranging gives $\Delta H = \Delta U + P\Delta V + V\Delta P$

and, since $\Delta U = w + q = -P\Delta V + q$ and $\Delta P = 0 \, (P = \text{constant})$

substituting gives $\Delta H = q$

where q is the heat transferred at constant pressure P. Note that most transformations are performed at constant pressure, which is often the atmospheric pressure. The enthalpy is a state function which means that it depends only on the initial and final states of the system. The enthalpy does not depend either on the path by which the system achieved these states, or on how it is used.

5.7.2 ENTHALPY CHANGES DURING PHASE CHANGES – LATENT HEAT

Solids are highly ordered and have strongly attractive intermolecular forces. Gases are very disordered and have weakly attractive intermolecular forces. Liquids are more ordered than gases but less ordered than solids. They show weaker attractive forces than solids but stronger attractive forces than gases. Phase changes are transformations from one phase to another that are accompanied by a change in the system's energy, that is, the enthalpy change required to go from one state to another state (Figure 5.5). The transformation from solid phase to liquid phase is called melting or fusion, the transformation from liquid phase to gas phase is called vaporization and the transformation from solid phase to gas phase is called sublimation. These phase changes, that occur from a more-ordered to a less-ordered state, are endothermic. They require energy (heat) to provide the molecules of the substance with enough kinetic energy to allow them to overcome the intermolecular attractive forces (forces between molecules). The enthalpy changes associated with the phase changes fusion, vaporization and sublimation are depicted by ΔH_{fus}, ΔH_{vap} and ΔH_{sub}, respectively. Conversely, the transformation from liquid phase to solid phase is called freezing, the transformation from gas phase to liquid phase is called condensation and the transformation from gas phase to solid phase is called deposition. All these processes, that occur from a less-ordered to a more-ordered state, are exothermic. They release energy. The heat absorbed or released during a phase change can be calculated by the following equation:

$$q = m \times L$$

where m is the mass of the substance and L is the latent heat. The latent heat of fusion, L_{fus}, is the heat required for a substance to go from the solid phase to the liquid phase, or *vice versa*. The latent heat of vaporization, L_{vap}, is the heat required for a substance to go from the liquid phase to the gas phase, or *vice versa*, and the latent heat of sublimation, L_{sub}, is the heat required for a substance to go from the solid phase to the gas phase, or *vice versa*. We have $L_{sub} = L_{fus} + L_{vap}$. When a phase change occurs at constant pressure, the corresponding enthalpy change $\Delta H = q = m \times L$.

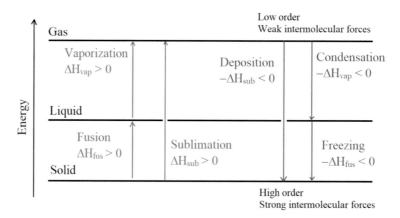

FIGURE 5.5   Enthalpy changes which accompany phase transformations. Fusion, vaporization and sublimation (in red) are the endothermic processes, whereas condensation, freezing and deposition (in blue) are the exothermic processes

The different states of a substance under different conditions of temperature and pressure can be represented by a phase diagram which is a graphical way that has pressure on the y-axis and temperature on the x-axis (Figure 5.6). A phase change occurs when we cross the lines or the

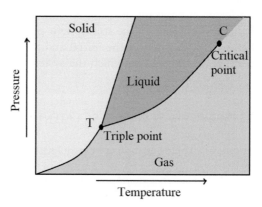

FIGURE 5.6   Typical phase diagram showing the critical and triple points

curves on the phase diagram. During a phase change, the temperature remains constant, and energy is used to overcome attractive forces between molecules. A typical phase diagram shows the critical point (C), beyond which the liquid and gas phases become indistinguishable, and the triple point (T), where liquid, solid and gas phases are in equilibrium.

Practice 5.6 How many joules are required to heat 50 g of iced water from −25°C to 120°C. The specific heat of water is Cs = 4.184 J g^{-1}°C. For water, the latent heat of fusion L$_{fus}$ = 3.33 × 10^5 J kg^{-1} and the latent heat of vaporization L$_{vap}$ = 22.6 × 10^5 J kg^{-1}.

Answer:
Let as consider the following heating curve (Figure 5.7), showing the phase changes over time as water absorbs heat:

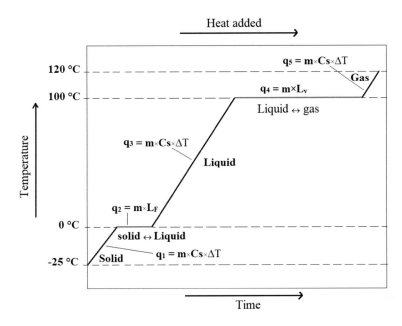

FIGURE 5.7   Heat curve showing the phase changes of water with time

o When the temperature increases from −25°C to 0°C, the water is solid, and it absorbs heat. The heat absorbed = $q_1 = m \times Cs \times \Delta T$
where m is the mass of water, Cs is the specific heat of water and $\Delta T (= T_F - T_I)$ is the temperature change of water.

Resolving gives $q_1 = 50g \times 4.184 \dfrac{J}{g \cdot °C} \times \left(0 - (-25)\right)°C = 5230 \, J$

o At 0°C, the iced water melts, and the heat absorbed = $q_2 = m \times L_F$

Resolving gives $q_2 = 50 \times 10^{-3} \, kg \times 3.33 \times \dfrac{10^5 J}{kg} = 16650 \, J$.

o When temperature increases from 0°C to 100°C, the water is liquid, and it absorbs heat. The heat absorbed $q_3 = m \times Cs \times \Delta T$

Resolving gives $q_3 = 50 \times 4.184 \dfrac{J}{g \cdot °C} \times (100 - 0)°C = 20920 \, J$.

o At 100°C, the liquid water vaporizes, and the heat absorbed = $q_4 = m \times L_V$

Resolving gives $q_4 = 50 \times 10^{-3} \, kg \times 22.6 \times \dfrac{10^5 \, J}{kg} = 113000 \, J$.

Note that the heat absorbed by water during vaporization (q_4) is significantly higher than the heat absorbed during fusion (q_2) which means that the same quantity of water needs more energy to vaporize than to melt. This is because the latent heat of vaporization is significantly higher than the latent heat of fusion. This explains also why, in the heating curve, the segment corresponding to vaporization of water is longer than that corresponding to fusion.

○ When temperature increases from 100°C to 120°C, the water is liquid, and it absorbs heat. The heat absorbed = $q_5 = m \times C_s \times \Delta T$

Resolving gives $q_5 = 50 \times 4.184 \dfrac{J}{g \cdot °C} \times (120 - 100)°C = 4184$ J.

○ The energy Q required to heat 50 g of iced water from –25°C to 120°C is given by:

$$Q = q_1 + q_2 + q_3 + q_4 + q_5$$

Resolving gives $Q = 5230 + 16650 + 20920 + 113000 + 4184 = 159984$ J $= 159.984$ kJ

Note that the energy released when 50 g of water is cooled from 120°C to –25°C is –159.984 kJ.

5.7.3 ENTHALPY OF REACTION

For a chemical reaction (reactants → products), the change of enthalpy, ΔH_{rxn}, called the enthalpy of reaction (or heat of reaction), is given by:

$$\Delta H_{rxn} = \sum \Delta H_{products} - \sum \Delta H_{reactants}$$

When $\Delta H_{rxn} < 0$, the reaction is exothermic; the system loses heat. When $\Delta H_{rxn} > 0$, the reaction is endothermic and the system gains heat. When $\Delta H_{rxn} = 0$, no heat is lost or gained, and the reaction is athermic.

Examples

○ Water is produced according to the following reaction:

$$2H_{2(g)} + O_{2(g)} \rightarrow 2H_2O_{(lq)} \qquad \Delta H_{rxn} = -571.2 \text{ kJ}$$

The production of water is an exothermic process ($\Delta H_{rxn} < 0$). The reaction would feel hot and 571.2 kJ of heat are released to the surroundings when two moles of water are formed.

○ Hydrogen iodide HI can be dissociated according to the following reaction:

$$2HI_{(g)} \rightarrow H_{2(g)} + I_{2(g)} \qquad \Delta H_{rxn} = 51.9 \text{ kJ}$$

The dissociation of HI is an endothermic process ($\Delta H_{rxn} > 0$), and 51.9 kJ of heat are gained by the system when two moles of HI are dissociated.

○ In the following reaction, no heat is released or gained by the system:

$$C_2H_5OH + CH_3COOH \leftrightarrow CH_3COOC_2H_5 + H_2O \qquad \Delta H_{rxn} \approx 0 \text{ kJ/mol}$$

Enthalpy is an extensive property that means it depends on the starting amounts of the reactants.
Example: Let us consider the combustion reaction of methane (CH_4) with oxygen (O_2):

$$CH_{4(g)} + 2O_{2(g)} \rightarrow CO_{2(g)} + 2H_2O_{(g)} \qquad \Delta H_{rxn} = -802 \text{ kJ}$$

This equation indicates that when one mole of CH_4 reacts with two moles of O_2 at a given temperature and pressure to form one mole of CO_2 and two moles of H_2O at the same temperature and pressure, 802 kJ of heat are released into the surroundings. When the value of ΔH is given in kJ rather than kJ mol^{-1}, it is the value of ΔH which corresponds to the reaction of the molar quantities of reactants and products given in the balanced chemical equation. If we multiply the coefficients by 2, for example, the enthalpy of reaction should be multiplied by 2:

$$2CH_{4(g)} + 4O_{2(g)} \rightarrow 2CO_{2(g)} + 4H_2O_{(g)} \qquad \Delta H_{rxn} = -802 \times 2 = -1604 \text{ kJ}$$

If we reverse the reaction, the sign of the enthalpy change should be reversed.

Example: $CO_{2(g)} + 2H_2O_{(g)} \rightarrow CH_{4(g)} + 2O_{2(g)}$ $\Delta H_{rxn} = +802 \text{ kJ}$

Enthalpy change depends also on the states of the reactants and products.

Example: $CH_{4(g)} + 2O_{2(g)} \rightarrow CO_{2(g)} + 2H_2O_{(g)}$ $\Delta H_{rxn} = -802 \text{ kJ}$

$\qquad\qquad CH_{4(g)} + 2O_{2(g)} \rightarrow CO_{2(g)} + 2H_2O_{(lq)}$ $\Delta H_{rxn} = -890 \text{ kJ}$

The difference in the enthalpy change between the above two reactions, where water is in the gas phase in the first reaction and in the liquid phase in the second reaction, is the enthalpy change ΔH_{cond} corresponding to the condensation of water vapor:

$$2H_2O_{(g)} \rightarrow 2H_2O_{(lq)} \qquad \Delta H_{cond} = -88 \text{ kJ} = -890 \text{ kJ} - (+802) \text{ kJ}$$

Practice 5.7 From the following combustion reaction of methane (CH_4):

$$CH_{4(g)} + 2O_{2(g)} \rightarrow CO_{2(g)} + 2H_2O_{(lq)} \qquad \Delta H_{rxn} = -890 \text{ kJ}$$

Calculate the heat released by the reaction when 102.4 g of O_2 reacts.

Atomic mass: O = 16.

Answer:

We have the mass of O_2 that should be converted to number of moles then to kJ.

We will use the equation $n = \dfrac{m}{M}$, where n is the number of moles of O_2

Resolving gives $n = \dfrac{102.4 \text{ g}}{2 \times 16 \dfrac{\text{g}}{\text{mol}}} = 3.2 \text{ mol}$

The equation $CH_{4(g)} + 2O_{2(g)} \rightarrow CO_{2(g)} + 2H_2O_{(lq)}$ tells us that 2 moles of O_2 release 890 kJ.

Therefore, 3.2 mol O_2 release $3.2 \text{ mol} \times \dfrac{890 \text{ kJ}}{2 \text{ mol}} = 1442 \text{ kJ}$

5.7.4 STANDARD ENTHALPY OF FORMATION

Enthalpy change depends on the temperature and pressure at which the reaction is carried out. Typically, the values of enthalpy changes are tabulated for reactions in which both the reactants and products are at the same conditions. A standard state refers to a substance under a pressure of 1 atm and a solution at 1 M. However, it does not specify the temperature even though the values of enthalpy changes are usually tabulated for a standard state of 1 atm and 25°C. The symbol ($\Delta H°$) is used to designate an enthalpy change under standard conditions.

The standard enthalpy of formation, designated by $\Delta H_f^°$, is the change in enthalpy when one mole of a substance is formed under standard conditions (P = 1 atm and T = 25°C) from its pure elements in the same standard conditions. It is given in kJ mol^{-1}. Conventionally, the standard enthalpy of formation of a pure element in its most stable form is zero.

Examples:

$$C(\text{graphite}) + O_2(g) \rightarrow CO_2(g), \quad \Delta H_f^° = -393.1 \text{ kJ mol}^{-1}.$$

$$2C(\text{graphite}) + 3H_2(g) + \frac{1}{2}O_2(g) \rightarrow C_2H_5OH(lq), \quad \Delta H_f^° = -277.7 \text{ kJ mol}^{-1}$$

$$O_2(g) \rightarrow O_2(g), \quad \Delta H_f^° = 0 \text{ kJ mol}^{-1}$$

$$C(\text{graphite}) \rightarrow C(\text{graphite}), \quad \Delta H_f^° = 0 \text{ kJ mol}^{-1}$$

Note that

- The standard state of carbon is graphite, whatever the temperature.
- The product should be one mole of a substance.

The standard enthalpy of reaction ΔH_{rxn}° can be calculated from the standard enthalpy of formation of products and reactants by the following equation:

$$\Delta H_{rxn}^{\circ} = \sum_i \gamma_i \Delta H_{f,i}^{\circ}(products) - \sum_j \gamma_j \Delta H_{f,j}^{\circ}(reactants)$$

where γ_i and γ_j are the stoichiometric coefficients of the products and reactants, respectively.

Example: when taking the reaction:

$$3H_2(g) + N_2(g) \rightarrow 2NH_3(g)$$

we have $\Delta H_{rxn}^{\circ} = 2 \times \Delta H_f^{\circ}(NH_3(g)) - \left[(3 \times \Delta H_f^{\circ}(H_2(g))) + \Delta H_f^{\circ}(N_2(g))\right]$

H_2 and N_2 are pure elements, so $\Delta H_f^{\circ}(H_2(g)) = \Delta H_f^{\circ}(N_2(g)) = 0\,kJmol^{-1}$

Replacing gives $\Delta H_{rxn}^{\circ} = 2 \times \Delta H_f^{\circ}(NH_3(g))$

Practice 5.8 Calculate the standard enthalpy of combustion of ethane (C₂H₆) at 298 K:

$$2C_2H_6(g) + 7O_2(g) \rightarrow 4CO_2(g) + 6H_2O(lq), \quad \Delta H_{comb}^{\circ} = ?$$

Standard enthalpy of formation at 298 K:

$$\Delta H_f^{\circ}(C_2H_6) = -84.7\ kJ\,mol^{-1}.$$

$$\Delta H_f^{\circ}(CO_2) = -393.1\ kJ\,mol^{-1}.$$

$$\Delta H_f^{\circ}(H_2O)lq = -285.6\ kJ\,mol^{-1}.$$

Answer:

The standard enthalpy of combustion of ethane (C₂H₆), ΔH_{comb}°, is given by:

$$\Delta H_{comb}^{\circ} = \sum_i \gamma_i \Delta H_{f,i}^{\circ}(products) - \sum_j \gamma_j \Delta H_{f,j}^{\circ}(reactants)$$

We have
$$\Delta H_{comb}^{\circ} = \left[4 \times \Delta H_f^{\circ}(CO_2(g)) + 6 \times \Delta H_f^{\circ}(H_2O(lq))\right]$$
$$- \left[(2 \times \Delta H_f^{\circ}(C_2H_6)) + 7 \times \Delta H_f^{\circ}(O_2(g))\right]$$

Note that since O_2 is a pure element, $\Delta H_f^{\circ}(O_2(g)) = 0\,kJmol^{-1}$

Resolving gives
$$\Delta H_{comb}^{\circ} = \left[4 \times (-393.1) + 6 \times (-285.6)\right] - \left[(2 \times (-84.7))\right]$$
$$= -3116.6\ kJmol^{-1}.$$

5.7.5 HESS'S LAW

Hess's law states that, if a chemical reaction is carried out in a series of steps, the enthalpy change, ΔH, for the overall reaction is equal to the sum of the enthalpy changes for the individual steps.

Example:

$$C \ (graphite) + O_2 \ (g) \xrightarrow{\Delta H} CO_2 \ (g)$$

$$\Delta H_1 \searrow \qquad \nearrow \Delta H_2$$

$$CO \ (g) + \frac{1}{2} O_2 \ (g)$$

According to Hess's law, $\Delta H = \Delta H_1 + \Delta H_2$

Practice 5.9 Using the following equations:

$$\mathbf{Ca}(sd) + \mathbf{O_2}(g) \rightarrow \mathbf{CaO}(sd), \quad \Delta H_1^\circ = -635.7 \text{ kJ mol}^{-1}.$$

$$\mathbf{CaO}(sd) + \mathbf{H_2O}(lq) \rightarrow \mathbf{Ca(OH)_2}(sd), \quad \Delta H_2^\circ = -65.2 \text{ kJ mol}^{-1}.$$

$$\mathbf{H_2}(g) + \frac{1}{2}\mathbf{O_2}(g) \rightarrow \mathbf{H_2O}(lq), \quad \Delta H_3^\circ = -285.6 \text{ kJ mol}^{-1}.$$

Calculate the standard enthalpy of formation of Ca(OH)$_2$ solid, $\Delta H_f^\circ \left(Ca(OH)_2 \right)$.

Answer:
The formation reaction of Ca(OH)$_2$ is

$$Ca(sd) + H_2(g) + O_2(g) \rightarrow Ca(OH)_2(sd), \quad \Delta H_f^\circ \left(Ca(OH)_2 \right) = ?$$

which can be carried out in the following individual reactions:

$$
\begin{aligned}
Ca \ (sd) + \tfrac{1}{2}O_2 \ (g) &\rightarrow CaO \ (sd), & \Delta H_1^\circ &= -635.7 \ kJmol^{-1}. \\
CaO \ (sd) + H_2O \ (lq) &\rightarrow Ca(OH)_2 \ (sd), & \Delta H_2^\circ &= -65.2 \ kJmol^{-1}. \\
H_2 \ (g) + \tfrac{1}{2}O_2 \ (g) &\rightarrow H_2O \ (lq), & \Delta H_3^\circ &= -285.6 \ kJmol^{-1} \\
\hline
Ca \ (sd) + O_2 \ (g) + H_2 \ (g) &\rightarrow Ca(OH)_2 \ (sd), & \Delta H_f^\circ \ (Ca(OH)_2) &= ?
\end{aligned}
$$

Applying Hess's law gives

$$\Delta H_f^\circ \left(Ca(OH)_2 \right) = \Delta H_1^\circ + \Delta H_2^\circ + \Delta H_3^\circ$$

Resolving gives $\Delta H_f^\circ \left(Ca(OH)_2 \right) = (-635.7) + (-65.2) + (-285.6) = -986.5 \text{kJmol}^{-1}$

5.8 BOND ENTHALPY

A substance can be formed from its pure elements in the gaseous state. During the formation reaction, chemical bonds are formed between the gaseous atoms which lead to the formation of the corresponding substance. The bond making is an exothermic process (it releases energy). The stronger the bond between atoms, the more energy is released during the bond making. However, the bond breaking requires energy. The energy required to break one mole of that chemical bond is called the bond enthalpy or bond dissociation enthalpy. It is also a measure of the bond strength. Bond enthalpy values are always positive since bond breaking is an endothermic process.

Example: The bond enthalpy of the H–H bond, ΔH(H–H), is equal to 436 kJ mol^{-1} which means that we need an energy of 436 kJ to break the H–H bonds in one mole of H–H bonds. The bond enthalpy of the C=C bond, ΔH(C=C), is equal to 610 kJmol^{-1}, which means that we need an energy of 610 kJ to break all C=C bonds in one mole of C=C bonds. By comparing the values of the bond enthalpy of H–H and C=C, we can say that the bond C=C is stronger than the bond H–H.

Practice 5.10 The dissociation reaction of gaseous CH₃OH is

$$CH_3OH(g) \rightarrow C(g) + 4H(g) + O(g), \quad \Delta H^{\circ}_{rxn}(CH_3OH) = 2041\,kJ$$

Calculate the bond enthalpy of the bond C–O in CH₃OH.

$$\Delta H(C-H) = 416\,kJmol^{-1}, \quad \Delta H(O-H) = 463\,kJmol^{-1}$$

Answer:
The structural formula of CH₃OH is given in Figure 5.8.

FIGURE 5.8   Structural formula of CH₃OH

In a CH₃OH molecule, there are three C–H bonds, one C–O bond and one O–H bond. The energy of the dissociation reaction of CH₃OH, $\Delta H^{\circ}_{rxn}(CH_3OH)$, is the energy needed to break all these bonds in one mole of CH₃OH. We have

$$\Delta H^{\circ}_{rxn}(CH_3OH) = 3 \times \Delta H(C-H) + \Delta H(C-O) + \Delta H(O-H)$$

Rearranging gives $\Delta H(C-O) = \Delta H^{\circ}_{rxn}(CH_3OH) - 3 \times \Delta H(C-H) - \Delta H(O-H)$
Resolving gives $\Delta H(C-O) = 2041 - 3 \times 416 - 463 = 330\,kJmol^{-1}$

A chemical reaction can be described as involving the breaking of bonds in the reactants and the making of bonds in the products. Thus, if the bond enthalpies of the reactants and products are known, we can calculate the standard enthalpy of the reaction, ΔH°_{rxn}:

$$\Delta H^{\circ}_{rxn} = \sum \Delta H(broken\ bonds) - \sum \Delta H(formed\ bonds)$$

Practice 5.11 Calculate the standard enthalpy of the hydrogenation reaction of ethene (C₂H₄):

$$C_2H_4(g) + H_2(g) \rightarrow C_2H_6(g), \quad \Delta H^{\circ}_{rxn} = ?$$

$$\Delta H(C-H) = 416\,kJmol^{-1}, \Delta H(H-H) = 436\,kJmol^{-1}, \Delta H(C-C) = 348\,kJmol^{-1}\ and$$

$$\Delta H(C=C) = 612\,kJmol^{-1}$$

Answer:
For the hydrogenation of ethene reaction, the broken bonds are one C=C bond and four C–H bonds in C₂H₄ and one H–H bond in H₂. The bonds formed are one C–C bond and six C–H bonds in C₂H₆. The enthalpy of the reaction is given by:

$$\Delta H^{\circ}_{rxn} = \left[\Delta H(C=C) + 4 \times \Delta H(C-H) + \Delta H(H-H)\right]$$
$$- \left[\Delta H(C-C) + 6 \times \Delta H(C-H)\right]$$

Resolving gives

$$\Delta H^{\circ}_{rxn} = [612 + 4 \times 416 + 436] - [348 + 6 \times 416] = -132\ kJmol^{-1}$$

GET SMART

FACTORS AFFECTING THE BOND ENTHALPY

- The atomic environment: In HCF_3 and C_2H_2, $\Delta H(C–H)$ equals −446 kJ/mol and −556 kJ/mol, respectively.
- Delocalized electrons: In CH_4 and C_6H_6, $\Delta H(C–H)$ equals −438 kJ/mol and −464 kJ/mol, respectively.
- Bond multiplicity (single, double, or triple bonds): In C_2H_6, $\Delta H(C–C) = -376\,kJ/mol$ and in C_2H_2, $\Delta H(C\equiv C) = -965$ kJ/mol.

Generally, an average value of the bond enthalpy is tabulated in literature and used for energy calculations.

5.9 LATTICE ENERGY AND THE BORN–HABER CYCLE

Lattice energy can be defined in two opposite ways. In one definition, it is defined as the amount of energy that is spent to separate an ionic compound into its constituent gaseous ions. In the second definition, it is the energy released when gaseous ions bind to form an ionic compound. Accordingly, this process is exothermic, and the values for lattice energy, expressed in kJ mol^{-1}, are negative. The lattice energy cannot be directly determined by experiments. However, it can be calculated by applying Hess's law on a series of individual reactions forming a cycle. This cycle is called the Born–Haber cycle. It can be obtained by the following steps:

1- Sublimation of the metal
2- Dissociation of the gaseous non-metal
3- Ionization of the gaseous metal
4- Formation of the gaseous non-metal anion
5- Formation of the ionic compound

    Example: Let us consider the formation of NaCl ionic compound:

$$Na(sd) + \frac{1}{2}Cl_2(g) \rightarrow NaCl(sd), \qquad \Delta H_1 = -441.2\,kJmol^{-1}$$

This formation reaction can be carried out by series of individual reactions:

$$\Delta H_1^\circ = -411\,kJmol^{-1}$$

$$Na(sd) + \frac{1}{2}Cl_2(g) \longrightarrow NaCl(sd)$$

$$\Delta H_2^\circ = \Delta H_{sub}(Na) = 107\,kJmol^{-1}$$

$$Na(g) + \frac{1}{2}Cl_2(g)$$

$$\Delta H_3^\circ = \frac{1}{2}\Delta H_{diss}(Cl_2) = 122\,kJmol^{-1} \qquad \Delta H_6^\circ = \Delta H_{Lattice} = ?$$

$$Na(g) + Cl(g)$$

$$\Delta H_4^\circ = I(Na) = 496\,kJmol^{-1}$$

$$Na^+(g) + Cl(g) \longrightarrow Na^+(g) + Cl^-(g)$$

$$\Delta H_5^\circ = Ae(Cl) = -348\,kJmol^{-1}$$

Born–Haber cycle for NaCl

Applying Hess's law, we have:

$$\Delta H_1^\circ = \Delta H_2^\circ + \Delta H_3^\circ + \Delta H_4^\circ + \Delta H_5^\circ + \Delta H_6^\circ$$

Replacing gives $\Delta H_1^\circ = \Delta H_{sub}(Na) + \dfrac{1}{2} \times \Delta H_{diss}(Cl_2) + I(Na) + Ae(Cl) + \Delta H_{Lattice}$

where $\Delta H_{sub}(Na)$ is the enthalpy of sublimation of sodium (Na(sd) → Na(g)),
$\Delta H_{diss}(Cl_2)$ is the enthalpy bond of the Cl–Cl bond,
I(Na) is the ionization energy of Na. The ionization energy is defined as the amount of energy it takes to detach one electron from a gaseous neutral atom.
Ae(Cl) is the electron affinity of Cl. The electron affinity is defined as the amount of energy released when an electron is added to a gaseous neutral atom.

Rearranging gives

$$\Delta H_{Lattice} = \Delta H_1^\circ - \left[\Delta H_{sub}(Na) + \frac{1}{2} \Delta H_{diss}(Cl_2) + I(Na) + Ae(Cl) \right]$$

Resolving gives

$$\Delta H_{Lattice} = -411 - \left[107 + \frac{1}{2} \times 244 + 496 - 348 \right] = -788 \text{ kJmol}^{-1}$$

CHECK YOUR READING

What are the different types of energy?
What are the system and the surroundings?
How can the work and heat exchanged between a system and the surroundings be calculated?
What is the difference between the specific heat and the heat capacity?
What happens in a calorimeter?
What is the difference between the internal energy and the enthalpy?
What is the standard enthalpy of formation?
How is the heat absorbed or released during a phase change calculated?
How is the enthalpy change of a chemical reaction calculated?
What is the bond enthalpy?
How is the lattice enthalpy of an ionic compound calculated?
How can you tell whether a chemical reaction or a physical process is endothermic or exothermic?

SUMMARY OF CHAPTER 5

Energy is the capacity to do work or transfer energy. Energy can only be transformed from one form to another. The energy of the universe is constant.

The **system** is a portion of space which includes the particles of interest that we study. The **surroundings** are everything else. The entire system and surroundings constitute the **universe**. The system is bounded by a real or fictitious surface through which the exchanges of energy and matter are made with the surroundings.

Energy can transfer as **work (w)** or as **heat (q)**. Heat is the transfer of thermal energy (defined as the kinetic energy related to the random motion of atoms or molecules) between two matters at different temperatures. Work is the energy used to cause an amount of matter to move. At constant pressure, **w = −PΔV**.

When the heat is absorbed by the system, the process is endothermic (q > 0). When the heat is released by the system, the process is exothermic (q < 0). The work done on the system by the surroundings is considered to be a positive quantity (w > 0) and the work done by the system on the surroundings is considered to be a negative quantity (w < 0).

A **state function** depends only on the initial and final states of the system. The internal energy does not depend on either the path by which the system achieved these states, or on how it is used.

The **internal energy**, U, is a state function. It is equal to the sum of all possible types of energy present in a substance, mainly the kinetic energy and the potential energy. The kinetic energy is associated with the movement of particles of a substance and the potential energy is related to the interactions between these particles. Energy is stored in a substance by increasing the motion of the constituent particles (the kinetic energy increases). When the system loses thermal energy, the intensities of these motions decrease (the kinetic energy decreases).

In practice, we cannot measure the internal energy of a substance, but we can measure a change in the internal energy **ΔU**, which is given by:

$$\Delta U = U_F - U_I = q + w$$

where U_I and U_F are the initial and final internal energies of the system, respectively.

Calorimetry is defined as the measurement of heat released or absorbed during a chemical reaction or a physical process. The measurement of heat can be performed using a **calorimeter** which consists of a metallic vessel filled with water and suspended above a reaction chamber. The heat release or absorption causes a change in the water temperature.

The heat gained by the substance is equal to the heat lost by water and, conversely, the heat lost by the substance is equal to the heat gained by water. We have:

$$q_{substance} = -q_{water} = -Cs \times m \times \Delta T$$

where Cs, m and ΔT are the specific heat, the mass, and the temperature change of water, respectively.

Enthalpy (H) is the heat transferred between the system and the surroundings under constant pressure. Enthalpy is a state function. The enthalpy change is $\Delta H = q$ where P is constant.

Phase changes are transformations from one phase to another that are accompanied by a change in the enthalpy.

During a phase change, **the temperature remains constant,** and energy is used to overcome attractive forces between molecules. The heat absorbed or released during a phase change is given by the following equation:

$$\Delta H = q = m \times L$$

where m is the mass of the substance and L is the latent heat.

For a chemical reaction (reactants → products), the change of enthalpy ΔH_{rxn}, called the enthalpy of reaction (or heat of reaction), is given by:

$$\Delta H_{rxn} = \sum \Delta H_{products} - \sum \Delta H_{reactants}$$

When $\Delta H_{rxn} < 0$, the reaction is **exothermic**; the system loses heat. When $\Delta H_{rxn} > 0$, the reaction is **endothermic**; the system gains heat. When $\Delta H_{rxn} = 0$, no heat is lost or gained. The reaction is athermic.

Enthalpy is an **extensive property** which means it depends on the starting amounts of the reactants.

Example: $CH_{4(g)} + 2O_{2(g)} \rightarrow CO_{2(g)} + 2H_2O_{(g)}$ $\qquad \Delta H_{rxn} = -802$ kJ

$\qquad\qquad 2CH_{4(g)} + 4O_{2(g)} \rightarrow 2CO_{2(g)} + 4H_2O_{(g)}$ $\qquad \Delta H_{rxn} = -802 \times 2 = -1604$ kJ

If we reverse the reaction, the signs of the enthalpy change should be reversed.

Example: $CH_{4(g)} + 2O_{2(g)} \rightarrow CO_{2(g)} + 2H_2O_{(g)}$ $\qquad \Delta H_{rxn} = -802$ kJ

$\qquad\qquad CO_{2(g)} + 2H_2O_{(g)} \rightarrow CH_{4(g)} + 2O_{2(g)}$ $\qquad \Delta H_{rxn} = +802$ kJ

Enthalpy change depends also on the reactants and the product states.

Example: $CH_{4(g)} + 2O_{2(g)} \rightarrow CO_{2(g)} + 2H_2O_{(g)}$ $\qquad \Delta H_{rxn} = -802$ kJ

$\qquad\qquad CH_{4(g)} + 2O_{2(g)} \rightarrow CO_{2(g)} + 2H_2O_{(lq)}$ $\qquad \Delta H_{rxn} = -890$ kJ

The difference in the enthalpy change between the above two reactions, where water is in the gas phase in the first reaction and in the liquid phase in the second reaction, is the enthalpy change ΔH_{cond} corresponding to the condensation of water vapor:

$$2H_2O_{(g)} \rightarrow 2H_2O_{(lq)} \qquad \Delta H_{cond} = -88 \text{ kJ} = -890 \text{kJ} - (+802) \text{ kJ}$$

The standard enthalpy of formation, designated by ΔH_f°, is the change in enthalpy when **one mole** of a substance is formed under standard conditions (P = 1 atm and T = 25°C) from its pure elements under the same standard conditions. It is given in kJ mol^{-1}.

Example: $C(\text{graphite}) + O_2(g) \rightarrow CO_2(g),$ $\qquad \Delta H_f^\circ = -393.1$ kJ mol^{-1}.

The standard enthalpy of formation of a pure element in its most stable form is zero.

Example: $C(\text{graphite}) \rightarrow C(\text{graphite}),$ $\qquad \Delta H_f^\circ = 0$ kJmol^{-1}

The standard enthalpy of a reaction, ΔH_{rxn}°, can be calculated from the standard enthalpy of formation of the products and reactants, using the following equation:

$$\Delta H_{rxn}^\circ = \sum_i \gamma_i \Delta H_{f,i}^\circ (\textbf{\textit{products}}) - \sum_j \gamma_j \Delta H_{f,j}^\circ (\textbf{\textit{reactants}})$$

Example: $2C_2H_6(g) + 7O_2(g) \rightarrow 4CO_2(g) + 6H_2O(lq),$ $\qquad \Delta H_{comb}^\circ = ?$

$$\Delta H_{comb}^\circ = \left[4 \times \Delta H_f^\circ (CO_2(g)) + 6 \times \Delta H_f^\circ (H_2O(lq)) \right]$$

$$- \left[(2 \times \Delta H_f^\circ (C_2H_6)) + 7 \times \Delta H_f^\circ (O_2(g)) \right]$$

Hess's law states that if a chemical reaction is carried out in a series of steps, the enthalpy change, ΔH, for the overall reaction is equal to the sum of the enthalpy changes for the individual steps.

Example:

According to Hess's law, $\Delta H = \Delta H_1 + \Delta H_2$

Bond enthalpy is the energy required to break one mole of that chemical bond. It is also a measure of the bond strength. Bond enthalpy values are always positive since bond breaking is an endothermic process.

Example: The bond enthalpy of the H–H bond, $\Delta H(H–H)$, is equal to 436 kJ mol^{-1}, which means that we need an energy of 436 kJ to break the H–H bonds in one mole of H–H bonds.

A chemical reaction can be described as the breaking of bonds in the reactants and the making of bonds in the products. Thus, we can calculate the standard enthalpy of the reaction, ΔH°_{rxn}:

$$\Delta H^{\circ}_{rxn} = \sum \Delta H \left(\textbf{\textit{broken bonds}} \right) - \sum \Delta H \left(\textbf{\textit{formed bonds}} \right)$$

Example: For the ethene hydrogenation reaction:

$$C_2H_4(g) + H_2(g) \rightarrow C_2H_6(g), \qquad \Delta H^{\circ}_{rxn} = ?$$

$$\Delta H^{\circ}_{rxn} = \left[\Delta H(C=C) + 4 \times \Delta H(C-H) + \Delta H(H-H) \right]$$
$$- \left[\Delta H(C-C) + 6 \times \Delta H(C-H) \right]$$

Lattice energy can be defined in two opposite ways:

- ○ It is the amount of energy that is spent to separate an ionic crystal into its constituent gaseous ions.
- ○ It is the energy released when gaseous ions bind to form an ionic compound. This process is exothermic, and the values for lattice energy, expressed in kJ mol^{-1}, are negative.

The lattice energy can be calculated by applying Hess's law on a series of individual reactions, forming a cycle. This cycle is called the Born–Haber cycle.

The **Born–Haber cycle** is used to calculate the lattice enthalpy of an ionic compound, formed by a metal and a non-metal. It can be obtained by the following steps:

1. Sublimation of the metal
2. Dissociation of the gaseous non-metal
3. Ionization of the gaseous metal
4. Formation of the gaseous non-metal anion
5. Formation of the ionic compound

   Example: The Born–Haber cycle for NaCl

$$\Delta H^{\circ}_1 = -411 \; kJmol^{-1}$$

$$Na(sd) + \tfrac{1}{2}Cl_2(g) \longrightarrow NaCl(sd)$$

$$\Delta H^{\circ}_2 = \Delta H_{sub}(Na) = 107 kJmol^{-1}$$

$$Na(g) + \tfrac{1}{2}Cl_2(g)$$

$$\Delta H^{\circ}_3 = \tfrac{1}{2}\Delta H_{diss}(Cl_2) = 122 \; kJmol^{-1}$$

$$\Delta H^{\circ}_6 = \Delta H_{Lattice} = ?$$

$$Na(g) + Cl(g)$$

$$\Delta H^{\circ}_4 = I(Na) = 496 \; kJmol^{-1}$$

$$Na^+(g) + Cl(g) \longrightarrow Na^+(g) + Cl^-(g)$$

$$\Delta H^{\circ}_5 = Ae(Cl) = -348 \; kJmol^{-1}$$

$$\Delta H^{\circ}_1 = \Delta H_{sub}(Na) + \frac{1}{2} \times \Delta H_{diss}(Cl_2) + I(Na) + Ae(Cl) + \Delta H_{Lattice}$$

$$\Delta H_{Lattice} = \Delta H^{\circ}_1 - \left[\Delta H_{sub}(Na) + \frac{1}{2}\Delta H_{diss}(Cl_2) + I(Na) + Ae(Cl) \right] = -788 \; kJmol^{-1}$$

$$\Delta H^{\circ}_1 = \Delta H_{sub}(Na) + \frac{1}{2} \times \Delta H_{diss}(Cl_2) + I(Na) + Ae(Cl) + \Delta H_{Lattice}$$

$$\Delta H_{Lattice} = \Delta H^{\circ}_1 - \left[\Delta H_{sub}(Na) + \frac{1}{2}\Delta H_{diss}(Cl_2) + I(Na) + Ae(Cl) \right] = -788 \; kJ \; mol^{-1}$$

PRACTICE ON CHAPTER 5

Q.5.1 Choose the correct answer

1. A system
   a) is the capacity to do work or transfer energy
   b) is a portion of space bounded by a real or fictitious surface and which can exchange energy and matter with its surroundings
   c) and the universe constitute the surroundings
   d) can only absorb energy from the surroundings

2. Which of the following observations about energy is correct?
   a) energy can be destroyed
   b) energy of the universe increases
   c) energy can transfer as work or as heat
   d) a system cannot exchange energy with surroundings

3. When heat is absorbed by the system, the process is
   a) athermic
   b) exothermic
   c) endothermic
   d) all of these

4. When work is a positive quantity (w > 0), it is done
   a) on the system by the surroundings
   b) by the system on the surroundings
   c) by the system on the system
   d) by the surroundings on the surroundings

5. At constant pressure, the work w =
   a) $V\Delta P$
   b) $-P\Delta V$
   c) $P\Delta V$
   d) $C\Delta T$

6. A state function
   a) depends only on the initial and final states of the system
   b) depends on the path by which the chemical reaction is achieved
   c) depends on how it is used
   d) all the above are correct

7. The internal energy is
   a) equal to the sum of all possible types of energy present in a substance
   b) the kinetic energy of the particles of a substance
   c) the potential energy of the particles of a substance
   d) the heat transfer at constant pressure

8. When a substance stores energy,
   a) its kinetic energy decreases
   b) its kinetic energy increases
   c) its kinetic energy remains constant
   d) its internal energy remains constant

9. Which of the followings describes what happens in a calorimeter?
   a) the heat gained by the substance is equal to the heat lost by water
   b) the heat lost by the substance is equal to the heat gained by water
   c) $q_{substance} = -q_{water} = -C_s \times m \times \Delta T$
   d) all the above are correct

10. Enthalpy is defined as
   a) the transfer of energy as work
   b) The transformation of a substance from one phase to another
   c) the sum of all possible types of energy present in a substance
   d) the heat transferred between the system and the surroundings under constant pressure

11. **The transformation of a substance from a solid to a gas is called**
    a) sublimation
    b) vaporization
    c) fusion
    d) condensation

12. **Deposition is**
    a) an endothermic process
    b) an exothermic process
    c) an athermic process
    d) the transformation of a substance from liquid to solid.

13. **During a phase change, the temperature**
    a) increases
    b) decreases
    c) remains constant
    d) increases or decreases depending on the enthalpy change.

14. **When $\Delta H_{rxn} < 0$, the reaction is**
    a) exothermic
    b) endothermic
    c) athermic
    d) endothermic or exothermic depending on the products

15. **The standard enthalpy of formation**
    a) is designated by ΔH_f°
    b) is the change in enthalpy when one mole of a substance is formed under standard conditions
    c) of a pure element in its most stable form is zero
    d) all the above are correct

16. **Bond enthalpy is**
    a) the energy required to break one mole of a chemical bond
    b) is a measure of the bond strength
    c) positive because bond breaking is an endothermic process
    d) all the above are correct

17. **Hess's law states that**
    a) if a chemical reaction is carried out in a series of steps, the enthalpy change ΔH for the overall reaction is equal to the sum of the enthalpy changes for the individual steps
    b) the standard enthalpy of a pure element in its most stable form is zero
    c) the heat gained by the substance is equal to the heat lost by surroundings
    d) the energy of the universe (system and surroundings) is constant

Calculations

Q5.2 **Calculate the work done by a gas in a cylindrical container as it compresses from 40 to 20 L against a constant pressure of 2.5 atm.**

Q.5.3 **Calculate the change in the internal energy ΔU of a system if 500 J of heat is released by the system and if 700 J of work is done by the system.**

Q.5.4 **Calculate the change in the internal energy ΔU of a system if 700 J of heat is absorbed by the system and if 850 J of work is done on the system.**

Q5.5 **Calculate the energy required to heat 150 g of water from −18°C to 65°C.**
    $C_s = 4.184$ J g^{-1}°C and $L_{fus} = 3.33 \times 10^5$ J kg^{-1}.

Q.5.6 **A granite stone of 0.5 kg at 100°C is added to 7 kg water at 25°C. Calculate the final temperature of the water. The specific heat of water is 4.184 J g^{-1}°C and the specific heat of granite stone is 0.790 J g^{-1}°C.**

Q.5.7 **From the following chemical reaction:**

$$CH_{4(g)} + 2O_{2(g)} \rightarrow CO_{2(g)} + 2H_2O_{(lq)} \qquad \Delta H = -890 \text{ kJ}$$

calculate the mass of CH_4 needed to release a heat of 728 kJ.
Atomic masses: $H = 1$, $C = 12$.

Q.5.8 N_2 reacts with H_2 to form NH_3 according to the following reaction:

$$N_{2(g)} + 3H_{2(g)} \rightarrow 2NH_{3(g)}$$

When 0.03 mol of N_2 reacts with 0.09 mol of H_2 to form 0.06 mol of NH_3, 2.778 kJ of heat are produced. Calculate the enthalpy change of this reaction when one mole of N_2 reacts.

Q.5.9 HNO_3 can be formed according to the following reaction:

(1) $3NO_2(g) + H_2O(lq) \rightarrow 2HNO_3(aq) + NO(g)$

Use the reactions here to determine the $\Delta H°$ for reaction (1):

(2) $3NO_2(g) \rightarrow \dfrac{3}{2}N_2(g) + 3O_2(g) \qquad \Delta H_2° = -99.6 \text{ kJ}$

(3) $H_2(g) + \dfrac{1}{2}O_2(g) \rightarrow H_2O(lq) \qquad \Delta H_3° = -285.8 \text{ kJ}$

(4) $\dfrac{1}{2}N_2(g) + \dfrac{1}{2}O_2(g) \rightarrow NO(g) \qquad \Delta H_4° = +90.2 \text{ kJ}$

(5) $\dfrac{1}{2}H_2(g) + \dfrac{1}{2}N_2(g) + \dfrac{3}{2}O_2(g) \rightarrow HNO_3(aq) \qquad \Delta H_5° = -207.4 \text{ kJ}$

Q.5.10 Calculate the bond enthalpy of the C–H bond in CH_4 using the following reactions:

$$CH_4(g) \rightarrow C(g) + 4H(g), \qquad \Delta H_1 = ?$$

$$C(graphite) + 2H_2(g) \rightarrow CH_4(g), \qquad \Delta H_f°(CH_4) = -47.8 \text{kJmol}^{-1}$$

$$C(graphite) \rightarrow C(g), \qquad \Delta H_{sub}(graphite) = 718.6 \text{kJmol}^{-1}$$

$$H_2(g) \rightarrow 2H(g), \quad \Delta H(\text{ H–H }) = 436 \text{kJmol}^{-1}$$

Q.5.11 Determine the Born–Haber cycle for $MgCl_2$ and calculate its lattice enthalpy change, using the following reactions and their corresponding enthalpy changes.

$$Mg(sd) \rightarrow Mg(g), \qquad \Delta H_{sub} = 148 \text{kJmol}^{-1}$$

$$Cl_2(g) \rightarrow 2Cl(g), \qquad \Delta H_{diss} = 244 \text{kJmol}^{-1},$$

$$Mg(g) \rightarrow Mg^+(g) \qquad I_1 = 738 \text{kJmol}^{-1}$$

$$Mg^+(g) \rightarrow Mg^{2+}(g), \qquad I_2 = 1451 \text{kJmol}^{-1}$$

$$Cl(g) \rightarrow Cl^-(g), \qquad Ae(Cl) = -348 \text{kJmol}^{-1}$$

ANSWERS TO QUESTIONS

Q5.1
1. a
2. c

3. c
4. a
5. b
6. a
7. a
8. b
9. d
10. d
11. a
12. b
13. c
14. a
15. d
16. d
17. a

Calculations

Q5.2

$$w = 5.065 \, kJ$$

Q5.3

$$\Delta U = -1.2 \, kJ$$

Q5.4

$$\Delta U = 1.55 \, kJ$$

Q5.5

$$Q = 102.0408 \, kJ$$

Q5.6

$$T_F = 26°C$$

Q5.7

$$m = 13.2 \, g$$

Q5.8

$$\Delta H = -92.6 \, kJ$$

Q5.9

$$\Delta H° = -138.4 \, kJ$$

Q5.10

$$\Delta H(C-H) = 409.6 \, kJmol^{-1}$$

Q5.11

$$\Delta H_{Lattice} = -2526 \, kJmol^{-1}$$

KEY EXPLANATIONS

Q5.1

1. The system is a portion of space which includes the particles of interest that we study. The surroundings are everything else. The entire system and surroundings constitute the universe. The system is bounded by a real or fictitious surface through which the exchanges of energy and matter with the surroundings are made.

2. The energy conservation law (or first law of thermodynamics) states that energy is neither created nor destroyed, which implies that the energy of the universe is constant. A system can exchange energy and matter with the surroundings. Energy can transfer as work or as heat.

3. When heat is absorbed by the system, the process is endothermic, and the heat amount is considered to be a positive quantity ($q > 0$). When the heat is released by the system, the process is exothermic, and the heat amount is considered to be a negative quantity ($q < 0$).

4. The work done on the system by the surroundings is considered to be a positive quantity ($w > 0$) and the work done by the system on the surroundings is considered to be a negative quantity ($w < 0$).

5. $w = -P\Delta V$ where P is the pressure exerted by the surroundings on the system (external pressure). During compression of the gas, the volume decreases ($\Delta V < 0$) and work is carried out on the system by the surroundings $\left(w = -P\Delta V > 0\right)$. During expansion of the gas, the volume increases ($\Delta V > 0$) and the work is carried out by the system on the surroundings ($w < 0$).

6. A state function depends only on the initial and final states of the system. The internal energy depends on neither the path by which the system achieved these states, nor on how it is used.

7. The internal energy, U, is a state function. It is equal to the sum of all possible types of energy present in a substance, mainly the kinetic energy and the potential energy. The kinetic energy is associated with the movement of particles of a substance, whereas the potential energy is related to the interactions between these particles.

8. Energy is stored in a substance by increasing the motion of the constituent particles (the kinetic energy increases). When the system loses thermal energy, the intensities of these motions decrease (the kinetic energy decreases).

9. The measurement of heat can be performed using a calorimeter which consists of a metallic vessel filled with water and suspended above a reaction chamber. The heat gained by the substance is equal to the heat lost by water and conversely, the heat lost by the substance is equal to the heat gained by water:

$$q_{substance} = -q_{water} = -Cs \times m \times \Delta T$$

where Cs, m and ΔT are the specific heat, the mass and the temperature change of water, respectively.

10. Enthalpy H is the heat transferred between the system and surroundings under constant pressure. Internal energy U is the sum of all possible types of energy present in a substance. We have $H = U + PV$. H and U are both state functions.

11. and 12. The phase changes: melting or fusion (transformation from solid to liquid), vaporization (the transformation from liquid to gas) and sublimation (the transformation from solid to gas) which occur from a more-ordered to a less-ordered state, are endothermic. They require energy (heat). Conversely, freezing (the transformation from liquid to solid), condensation (the transformation from gas to liquid) and deposition (the transformation from gas to solid), that occur from a less-ordered to a more-ordered state, are exothermic. They release energy.

13. A phase change occurs when we cross the lines or the curves on the phase diagram. The temperature remains constant during a phase change (Figure 5.9) and energy is used to overcome attractive forces between molecules.

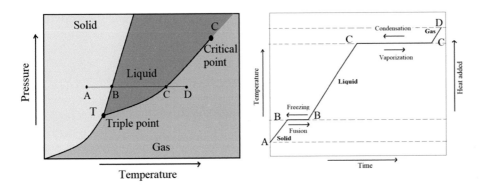

FIGURE 5.9 Phase changes and temperature variations when moving from the point A to the point D in the phase diagram.

14. For a chemical reaction, when $\Delta H_{rxn} < 0$ the reaction is exothermic, when $\Delta H_{rxn} > 0$ the reaction is endothermic and when $\Delta H_{rxn} = 0$ the reaction is athermic (no heat is gained or lost).

15. The standard enthalpy of formation, designated by ΔH_f°, is the change in enthalpy when one mole of a substance is formed under standard conditions (P = 1 atm and T = 25°C) from its pure elements under the same standard conditions. Conventionally, the standard enthalpy of formation of a pure element in its most stable form is zero.

16. Bond breaking requires energy. The energy required to break one mole of that chemical bond is called the bond enthalpy or bond dissociation enthalpy. It is also a measure of the bond strength. Bond enthalpy values are always positive since bond breaking is an endothermic process. However, bond making is an exothermic process (it releases energy). Note that a chemical reaction can be described as the breaking of bonds in the reactants and the making of bonds in the products. Thus, if the bond enthalpies of the reactants and products are known, we can calculate the standard enthalpy of the reaction, ΔH_{rxn}°:

$$\Delta H_{rxn}^\circ = \Sigma \Delta H \left(\text{broken bonds} \right) - \Sigma \Delta H \left(\text{formed bonds} \right)$$

17. Hess's law states that, if a chemical reaction is carried out in a series of steps, the enthalpy change ΔH for the overall reaction is equal to the sum of the enthalpy changes for the individual steps. Hess's law is used to determine the enthalpy of chemical reactions and it can be applied using the Born–Haber cycle to determine the lattice energy for ionic compounds.

Calculations

Q5.2

We will calculate the work done by a gas in a cylindrical container as it compresses from 40 to 20 L against a constant pressure of 2.5 atm (Figure 5.10).

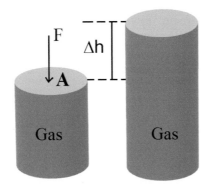

FIGURE 5.10 Decrease of the gas volume during compression.

During compression of the gas, the volume decreases ($\Delta V < 0$) and the work is done on the system by the surroundings:

$$w = -P\Delta V > 0$$

Resolving gives $w = -2.5\,\text{atm} \times (20 - 40)\text{L} = 50\,\text{L} \cdot \text{atm}$

We need to express the work in joule. We know that 1 L atm = 101.3 J

Converting gives $w = 50\,\text{L} \cdot \text{atm} \times \dfrac{101.3\,\text{J}}{1\,\text{L} \cdot \text{atm}} = 5065\,\text{J} = 5.065\,\text{kJ}$

Q.5.3

We will calculate the change in the internal energy ΔU of a system if 500 J of heat is released by the system and if 700 J of work is done by the system.
500 J of heat is released by the system. This means that the system loses heat, so the amount of heat is considered to be negative. Therefore, we have $q = -500$ J.
700 J of work is done by the system. So, the amount of work is considered to be negative. Therefore, we have $w = -700$ J.
The change in the internal energy is given by: $\Delta U = q + w$

Replacing gives $\Delta U = -500 + (-700) = -1200\,\text{J} = -1.2\,\text{kJ}$
This means that the system loses 1200 J but the surroundings gain 1200 J.

Q.5.4

We will calculate the change in the internal energy ΔU of a system if 700 J of heat is absorbed by the system and if 850 J of work is done on the system.
700 J of heat is absorbed by the system. This means that the system gains heat, so the amount of heat is considered to be positive. Therefore, we have $q = 700$ J.
850 J of work is done on the system. The amount of work is therefore considered to be positive. Therefore, we have $w = 850$ J.
The change in the internal energy is given by: $\Delta U = q + w$

Replacing gives $\Delta U = 700 + 850 = 1550\,\text{J} = 1.55\,\text{kJ}$

This means that the system gains 1550 J, but the surroundings lose 1550 J.

Q5.5

We will calculate the energy required to heat 150 g of water from –18°C to 65°C.
$Cs = 4.184\,\text{J g}^{-1}°\text{C}$ and $L_{fus} = 3.33 \times 10^5\,\text{J kg}^{-1}$.

- When the temperature increases from –18°C to 0°C, the water is solid, and it absorbs heat. The heat absorbed $= q_1 = m \times Cs \times \Delta T$
  where m is the mass of water, Cs is the specific heat of water and $\Delta T\ (= T_F - T_I)$ is the temperature change of water.

  Resolving gives $q_1 = 150\,\text{g} \times 4.184\,\dfrac{\text{J}}{\text{g} \cdot °\text{C}} \times (0 - (-18))°\text{C} = 11296.8\,\text{J}$.

- At 0°C, the iced water melts, and the heat absorbed $= q_2 = m \times L_F$

  Resolving gives $q_2 = 150 \times 10^{-3}\,\text{kg} \times 3.33 \times \dfrac{10^5\,\text{J}}{\text{kg}} = 49950\,\text{J}$.

- When the temperature increases from 0°C to 65°C, the water is liquid and it absorbs heat. The heat absorbed $= q_3 = m \times Cs \times \Delta T$

  Resolving gives $q_3 = 150\,\text{g} \times 4.184\,\dfrac{\text{J}}{\text{g} \cdot °\text{C}} \times (65 - 0)°\text{C} = 40794\,\text{J}$.

- The energy Q required to heat 150 g of iced water from –18°C to 65°C is given by:

$$Q = q_1 + q_2 + q_3$$

Resolving gives $Q = 11296.8 + 49950 + 40794 = 102040.8\,\text{J} = 102.0408\,\text{kJ}$

Q.5.6

We will calculate the final temperature of the water if a granite stone of 0.5 kg at 100°C is added to 7 kg water at 25°C.

The specific heat of water is 4.184 J g^{-1}°C and the specific heat of granite stone is 0.790 J g^{-1}°C.

We will use the equation $q = Cs \times m \times \Delta T$

Heat transfers from the hotter substance to the colder one. Let us consider q_1 is the heat gained by water when its temperature increases from 25°C to reach T_F and q_2 is the heat lost by the granite stone when its temperature decreases from 100°C to reach T_F. We have:

$$q_1 = Cs \times m_1 \times \Delta T = Cs \times m_1 \times (T_F - T_1)$$

where Cs is the specific heat of water, m_1 is the mass of water and T_F and T_1 are the initial and final temperatures of this water, respectively,

and $q_2 = Cs \times m_2 \times \Delta T = Cs \times m_2 \times (T_F - T_2)$

where Cs is the specific heat of the granite stone, m_2 is the mass of the granite stone and T_F and T_2 are the initial and final temperatures of the granite stone, respectively. Note that T_F is the same for water and the granite stone.

The heat (q_1) gained by water is equal to the heat (q_2) lost by the granite stone. Therefore, we have $q_1 = -q_2$

Replacing gives $C_1 \times m_1 \times (T_F - T_1) = -C_2 \times m_2 \times (T_F - T_2)$

Rearranging gives $(C_1 \times m_1 + C_2 \times m_2) \times T_F = C_1 \times m_1 \times T_1 + C_2 \times m_2 \times T_2$

and $T_F = \dfrac{C_1 \times m_1 \times T_1 + C_2 \times m_2 \times T_2}{(C_1 \times m_1 + C_2 \times m_2)}$

Resolving gives $T_F = \dfrac{4.184\dfrac{J}{g°C} \times 7000\,g \times 25°C + 0.790\dfrac{J}{g°C} \times 500\,g \times 100°C}{\left(4.184\dfrac{J}{g°C} \times 7000\,g + 0.790\dfrac{J}{g°C} \times 500\,g\right)} = 26°C$

Q.5.7

$$CH_{4(g)} + 2O_{2(g)} \rightarrow CO_{2(g)} + 2H_2O_{(lq)} \qquad \Delta H = -890 \text{ kJ}$$

The heat released (728 kJ) should be converted to number of moles and then to mass.

The equation $CH_{4(g)} + 2O_{2(g)} \rightarrow CO_{2(g)} + 2H_2O_{(lq)}$ tells us that one mole of CH_4 releases 890 kJ. Therefore, the heat of 728 kJ is released by $\dfrac{728\,kJ}{890\,kJ} \times 1\,mol = 0.82\,mol$

Using the equation $n = \dfrac{m}{M}$, we determine the mass (m) of CH_4

Rearranging gives $m = n \times M$

Resolving gives $m = 0.82 \times (12 + 4 \times 1) = 13.2\,g$ of CH_4.

Q.5.8

N_2 reacts with H_2 to form NH_3 according to the following reaction:

$$N_{2(g)} + 3H_{2(g)} \rightarrow 2NH_{3(g)}$$

When 0.03 mol of N_2 reacts with 0.09 mol of H_2 to form 0.06 mol of NH_3, 2.778 kJ of heat are produced.

The heat is produced by this reaction, which means that it is an exothermic reaction and the heat q = –2.778 kJ
We have the following relations between the number of moles of the reactants and products:

$$n_{N_2} = \frac{n_{H_2}}{3} = \frac{n_{NH_3}}{2} = 0.03\,mol$$

The reactants are provided in stoichiometric amounts. The amount of N_2 can be used to calculate the enthalpy change, ΔH. Since ΔH is an extensive property, it is proportional to the amount of N_2 that reacts:

$$\Delta H = 1\,mol\,N_2 \times \frac{\left(-2.778\,kJ\right)}{0.03\,mol\,N_2} = -92.6\,kJ$$

Q.5.9

HNO_3 can be formed according to the following reaction:

(1) $3NO_2(g) + H_2O(lq) \rightarrow 2HNO_3(aq) + NO(g)$

The reactions here can be used to determine the $\Delta H°$ for reaction (1):

(2) $3NO_2(g) \rightarrow \frac{3}{2}N_2(g) + 3O_2(g)$ $\Delta H_2° = -99.6$ kJ

(3) $H_2(g) + \frac{1}{2}O_2(g) \rightarrow H_2O(lq)$ $\Delta H_3° = -285.8$ kJ

(4) $\frac{1}{2}N_2(g) + \frac{1}{2}O_2(g) \rightarrow NO(g)$ $\Delta H_4° = +90.2$ kJ

(5) $\frac{1}{2}H_2(g) + \frac{1}{2}N_2(g) + \frac{3}{2}O_2(g) \rightarrow HNO_3(aq)$ $\Delta H_5° = -207.4$ kJ

We should find the right combination of reactions (2–5) such that they add up to reaction (1).
Reaction 1 indicates that

- 3 moles of $NO_2(g)$ are needed as a reactant. This can be obtained by reaction (2).

$$3NO_2(g) \rightarrow \frac{3}{2}N_2(g) + 3O_2(g)   \Delta H_2° = -99.6kJ$$

- 1 mole of $H_2O(lq)$ is needed as a reactant. This can be obtained by reversing reaction (3) which means that the sign of $\Delta H_3°$ is also reversed.

$$H_2O(lq) \rightarrow H_2(g) + \frac{1}{2}O_2(g)   - \Delta H_3° = +285.8kJ$$

2 moles of $HNO_3(aq)$ are needed as a product. This can be obtained by multiplying reaction (5) by 2 which means that the $\Delta H_5°$ is also multiplied by 2:

$$H_2(g) + N_2(g) + 3O_2(g) \rightarrow 2HNO_3(aq)   2 \times \Delta H_5° = -414.8kJ$$

- 1 mole of $NO(g)$ is needed as a product. This can be obtained by reaction (4).

$$\frac{1}{2}N_2(g) + \frac{1}{2}O_2(g) \rightarrow NO(g)   \Delta H_4° = +90.2kJ$$

The reaction of HNO_3 formation can be carried out by the following individual reactions:

$$3NO_2(g) \rightarrow \frac{3}{2}N_2(g) + 3O_2(g) \qquad \Delta H_2^{\circ} = -99.6 \text{ kJ}$$

$$H_2O(lq) \rightarrow H_2(g) + \frac{1}{2}O_2(g) \qquad -\Delta H_3^{\circ} = +285.8 \text{ kJ}$$

$$H_2(g) + N_2(g) + 3O_2(g) \rightarrow 2HNO_3(aq) \qquad 2 \times \Delta H_5^{\circ} = -414.8 \text{ kJ}$$

$$\frac{1}{2}N_2(g) + \frac{1}{2}O_2(g) \rightarrow NO(g) \qquad \Delta H_4^{\circ} = +90.2 \text{ kJ}$$

$$\overline{3NO_2(g) + H_2O(lq) \rightarrow 2HNO_3(aq) + NO(g) \quad \Delta H^{\circ} = ?}$$

We have $\Delta H^{\circ} = \Delta H_2^{\circ} - \Delta H_3^{\circ} + 2 \times \Delta H_5^{\circ} + \Delta H_4^{\circ}$

Resolving gives $\Delta H^{\circ} = -99.6 \text{kJ} + 285.8 \text{kJ} - 414.8 \text{kJ} + 90.2 \text{kJ} = -138.4 \text{kJ}$

Q.5.10

We can calculate the bond enthalpy of the C–H bond in CH_4, using Hess's law:

$$\Delta H_1$$
$$CH_4(g) \quad \rightarrow \quad C(g) + 4H(g)$$

$$\Delta H_2 \downarrow \qquad\qquad \uparrow \Delta H_4$$

$$C(graphite) + 2H_2(g) \quad \rightarrow \quad C(g) + 2H_2(g)$$
$$\Delta H_3$$

ΔH_1 is the enthalpy of dissociation reaction of CH_4: $\Delta H_1 = 4 \times \Delta H(\text{C–H})$

ΔH_2 is the inverse of the standard enthalpy of CH_4 formation: $\Delta H_2 = -\Delta H_f^{\circ}(CH_4)$

ΔH_3 is the enthalpy of the sublimation reaction of carbon (C (graphite) $\rightarrow$ C (g)): $\Delta H_1 = \Delta H_{sub}$

ΔH_4 is the enthalpy of the dissociation reaction of H_2: $\Delta H_4 = 2 \times \Delta H(\text{H–H})$

Applying Hess's law, we have

$$\Delta H_1 = \Delta H_2 + \Delta H_3 + \Delta H_4$$

$$4 \times \Delta H(\text{C–H}) = -\Delta H_f^{\circ}(CH_4) + \Delta H_{sub} + 2 \times \Delta H(\text{H–H})$$

Rearranging gives $\Delta H(\text{C–H}) = \frac{1}{4} \times \left[-\Delta H_f^{\circ}(CH_4) + \Delta H_{sub} + 2 \times \Delta H(\text{H–H}) \right]$

Resolving gives $\Delta H(\text{C–H}) = \frac{1}{4} \times [47.8 + 718.6 + 2 \times 436] = 409.6 \text{kJmol}^{-1}$

Q.5.11

The ionic compound $MgCl_2$ can be formed according to the following reaction:

$$Mg(sd) + Cl_2(g) \rightarrow MgCl_2(sd), \qquad \Delta H_1 = -641 \text{kJmol}^{-1}$$

This formation reaction can take place in the following steps:

$$Mg(sd) + Cl_2(g) \rightarrow Mg(g) + Cl_2(g), \qquad \Delta H_2 = 148 \text{kJmol}^{-1},$$

$$Mg(g) + Cl_2(g) \rightarrow Mg(g) + 2Cl(g), \qquad \Delta H_3 = 244 \text{kJmol}^{-1},$$

$$Mg(g) + 2Cl(g) \rightarrow Mg^+(g) + 2Cl(g), \qquad \Delta H_4 = I_1 = 738 \text{kJmol}^{-1}$$

$$Mg^+(g) + 2Cl(g) \rightarrow Mg^{2+}(g) + 2Cl(g), \qquad \Delta H_5 = I_2 = 1451\,kJmol^{-1}$$

$$Mg^{2+}(g) + 2Cl(g) \rightarrow Mg^{2+}(g) + 2Cl^-(g),$$

$$\Delta H_6 = Ae(Cl) = 2 \times (-348) = -696\,kJ\,mol^{-1}$$

$$Mg^{2+}(g) + 2Cl^-(g) \rightarrow MgCl_2(sd), \qquad \Delta H_{Lattice} = ?$$

The Born–Haber cycle for $MgCl_2$ is as follows:

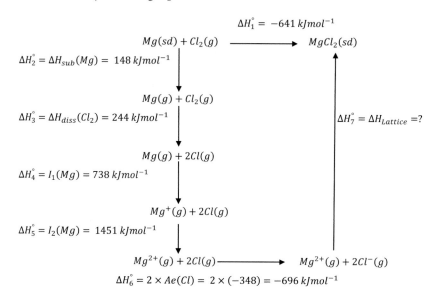

Applying Hess's law, we have

$$\Delta H_1^\circ = \Delta H_2^\circ + \Delta H_3^\circ + \Delta H_4^\circ + \Delta H_5^\circ + \Delta H_6^\circ + \Delta H_7^\circ$$

Replacing gives

$$\Delta H_1^\circ = \Delta H_{sub}(Mg) + \Delta H_{diss}(Cl_2) + I_1(Mg) + I_2(Mg) + 2 \times Ae(Cl) + \Delta H_{Lattice}$$

where $\Delta H_{sub}(Mg)$ is the enthalpy of sublimation of sodium ($Mg(sd) \rightarrow Mg(g)$)

$\Delta H_{diss}(Cl_2)$ is the enthalpy bond of the Cl–Cl bond

$I_1(Mg)$ is the first ionization energy of Mg $\left(Mg(g) \rightarrow Mg^+(g)\right)$.

$I_2(Mg)$ is the second ionization energy of Mg $\left(Mg^+(g) \rightarrow Mg^{2+}(g)\right)$.

$Ae(Cl)$ is the electron affinity of Cl.

Rearranging gives

$$\Delta H_{Lattice} = \Delta H_1^\circ - \left[\Delta H_{sub}(Mg) + \Delta H_{diss}(Cl_2) + I_1(Mg) + I_2(Mg) + 2 \times Ae(Cl) \right]$$

Resolving gives

$$\Delta H_{Lattice} = -641 - \left[148 + 244 + 738 + 1451 - 696\right] = -2526\,kJmol^{-1}$$

Introduction to Quantum Theory

6

6.1 OBJECTIVES

At the end of this chapter, the student will be able to:

1. Recognize the wave particle duality of light and the wave particle duality of matter;
2. Explain the atomic line spectrum of a hydrogen atom;
3. Explain the photoelectric effect; and
4. Define the different rules for electronic configurations and write the electronic configuration of any atom.

6.2 ELECTROMAGNETIC RADIATION – LIGHT

Electromagnetic radiation is a form of energy that is propagated through space and takes many forms, such as radio waves, microwaves, infrared, ultraviolet, light and X-rays. Each electromagnetic radiation can be characterized by its wavelength and/or frequency. The **wavelength (λ)** is the distance from one crest to the next in the wave (Figure 6.1). It is measured in units of distance. **The frequency (ν)** is the number of complete cycles per second, expressed in **s^{-1}** or Hz. The frequency ν of a wave is inversely proportional to its wavelength, λ:

$$\nu = \frac{C}{\lambda}$$

where the **speed of light, C**, is the same for all electromagnetic radiation (**$C = 3 \times 10^8$ m s^{-1}** in a vacuum).

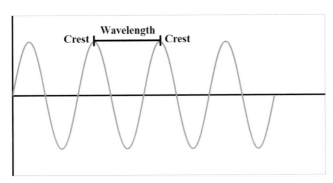

FIGURE 6.1 The wavelength λ

Practice 6.1 The yellow light given off by a sodium lamp has $\lambda = 589$ nm. What is the frequency of this radiation?

Answer:
$C = 3.00 \times 10^8$ m s^{-1} and $\lambda = 589$ nm $= 589 \times 10^{-9}$ m for the yellow light from a sodium lamp.
The frequency of the yellow light is $\nu = C/\lambda = 3.00 \times 10^8/589 \times 10^{-9} = 5.09 \times 10^{14}$ $s^{-1} = 5.09 \times 10^{14}$ Hz

DOI: 10.1201/9781003257059-6

6.3 WAVE-PARTICLE DUALITY OF LIGHT

Planck proposed in 1900 that atoms can gain or lose only certain quantities of energy by absorbing or emitting radiation:

$$\Delta E_{atom} = nh\nu$$

where n is a positive integer (n = 1, 2, 3, etc.) and h is Planck's constant (h = 6.62 × 10^{-34} J s). This means that energy is quantized. Einstein in 1905 proposed the wave-particle duality of light, which means that light consists of particle-like packets of energy called photons, also have wave-like properties. The energy of a photon is given by the following equation:

$$E_{photon} = h\nu = h\frac{c}{\lambda}$$

Furthermore, Einstein proposed that atoms could gain or lose certain quantities of energy by absorbing or emitting photons. This means that

$$\Delta E_{atom} = E_{photon} = h\nu = h\frac{C}{\lambda}$$

This conclusion was of a great importance since it allowed explanation of the photoelectric effect and the discontinuous line of the hydrogen spectrum.

Practice 6.2 The wavelength of the radiation in a microwave oven is 1.20 × 10^{-2} m. Calculate the energy of one photon of this microwave radiation?

C = 3.00 × 10^8 m s^{-1} and h = 6.63 × 10^{-34} J s.

Answer:
The energy of one photon of this microwave radiation is:

$$E_{photon} = h\frac{c}{\lambda}$$

$$E_{photon} = 6.63 \times 10^{-34}\frac{3.00 \times 10^8}{1.20 \times 10^{-2}} = 1.66 \times 10^{-23} J$$

6.4 PHOTOELECTRIC EFFECT

The photoelectric effect is the emission of a flow of electrons when light shines on plates of certain metals (Figure 6.2). Although bright light caused more electron flow than weak light, electron flow started immediately with both strong and weak light.

The view that light behaves like a wave could not explain this phenomenon. Indeed, according to the wave theory:

1. a significant delay should be detected between light incidence and electron ejection;
2. the intensity of light should affect the kinetic energy of the electrons emitted; and
3. electrons will be emitted regardless of the light frequency.

However, experiments show that:

1. no delay is detected between light incidence and electron ejection;
2. the intensity of light increases the number of emitted electrons, but it does not increase the velocity of the electrons (there was a maximum kinetic energy);
3. there was a threshold light frequency (ν_0) for electrons to be emitted. This means that the photoelectric effect is not observed for all electromagnetic radiations. For example, red light (λ = 700 nm) cannot cause electron emission.

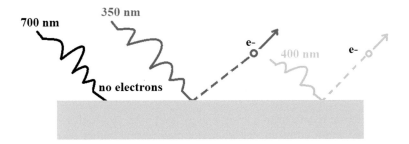

FIGURE 6.2   The photoelectric effect

The photoelectric effect was explained by the wave-particle duality of light. Indeed, a photon may penetrate the metal plate and transfer its energy to an electron. If the energy of the photon is at least as high as the energy required for ejecting the electron from the metal plate (known as the work function, $\Phi, = h\nu_0$, which depends on the metal), the electron will have enough energy to emerge from the metal. By conservation of energy, the kinetic energy K_E of the electron is given by the following equation:

$$K_E = E_{photon} - \Phi = h\frac{c}{\lambda} - \Phi$$

or $K_E = \frac{1}{2}\,mV^2$ (where m is the mass of the electron, and V is its velocity), therefore:

$$\frac{1}{2}mV^2 = h\frac{c}{\lambda} - \Phi$$

If the energy of the photon is less than the work function Φ, this photon will never be able to eject electrons.

6.5 LINE SPECTRA OF THE HYDROGEN ATOM

6.5.1 BOHR'S MODEL

The *emission spectrum of the hydrogen atom is discontinuous* and has been divided into several spectral line series. Niels Bohr in 1913 proposed a model that explained the line spectrum of the hydrogen atom. The Bohr model was based on the following assumptions:

1. The electron in a hydrogen atom travels around the nucleus in a circular orbit. The orbit radius is given by the following equation:

$$r_n = n^2 \times (0.5)\,\text{Å}$$

   where n is an integer (n = 1, 2, 3,...∞).
2. The hydrogen atom has only certain allowable energy levels for the electron orbits (orbits are quantized). The energy of the electron in an orbit is inversely proportional to its distance from the nucleus; the further the electron is from the nucleus, the more energy it has. The energy level for each orbit (E_n) is given by the following equation:

$$E_n = \frac{-2.18 \times 10^{-18}}{n^2}\text{J}$$

   where n = 1, 2, 3,...∞
   The lowest possible energy level at which an electron can be is called the ground state (n = 1). An excited state is an energy level higher than the ground state (n > 1). When the electron is very far from the nucleus (n = ∞), it is free (not attracted by the nucleus) and its energy is the highest (E = 0). The energy of the electron decreases when it approaches the nucleus. This explains the sign (−) in the formula of E_n.

3. When the electron moves from one orbit ($n_{initial}$) to another orbit (n_{final}), it absorbs or emits a photon (light), the energy of which equals the difference in energy between the two orbits:

$$\Delta E_{atom} = h\frac{C}{\lambda} = -2.18 \times 10^{-18} \times \left(\frac{1}{n_{final}^2} - \frac{1}{n_{initial}^2} \right) J$$

When the electron moves from a lower energy level orbit to a higher energy level one, it absorbs energy, and it emits energy when it moves from a higher energy level orbit to a lower energy level one. The lines of the hydrogen spectrum correspond to the transition of the electron from a higher energy level to a lower energy level. The wavelength of each line in the hydrogen atom spectrum can be determined using the following equation:

$$\lambda = \frac{hC}{\Delta E_{atom}}$$

or using the Rydberg formula:

$$\frac{1}{\lambda} = R \times \left(\frac{1}{n_1^2} - \frac{1}{n_2^2} \right)$$

where R is the Rydberg constant ($R = 1.09677 \times 10^7$ m^{-1}) and $n_2 > n_1$.

Practice 6.3 Calculate the wavelength of the line that corresponds to the transition of an electron from n = 4 to n = 2.

$C = 3.00 \times 10^8$ m s^{-1} and $h = 6.63 \times 10^{-34}$ J s

Answer:
The electron moves from a higher energy orbit ($n_{initial} = 4$) to a lower energy orbit ($n_{final} = 2$). It emits a photon (light) (Figure 6.3).

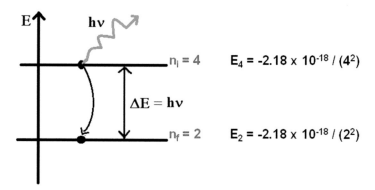

FIGURE 6.3   When an electron moves from one energy orbit ($n_{initial} = 4$) to a lower- energy orbit ($n_{final} = 2$), it emits light

$$\Delta E_{atom} = h\frac{C}{\lambda} = -2.18 \times 10^{-18} \times \left(\frac{1}{n_{final}^2} - \frac{1}{n_{initial}^2} \right)$$

$$\Delta E_{atom} = h\frac{C}{\lambda} = -2.18 \times 10^{-18} \times \left(\frac{1}{2^2} - \frac{1}{4^2} \right)$$

$$\Delta E_{atom} = -4.09 \times 10^{-19} J$$

Note that there is no negative energy. Here, the negative sign (–) means that the electron loses energy and the sign (–) should not be considered when calculating the wavelength of the emitted light (there is no negative wavelength value):

$$\lambda = \frac{hC}{\Delta E_{atom}} = \frac{6.63 \times 10^{-34} \times 3.00 \times 10^{8}}{4.09 \times 10^{-19}} = 4.86 \times 10^{-7} \text{ m} = 486 \text{ nm } \left(\text{green light}\right)$$

Therefore, the green line in the hydrogen atom spectrum is explained by the transition of an electron from $n = 4$ to $n = 2$. In the same manner, we can explain all the spectral line series of the hydrogen atom (Figure 6.4).

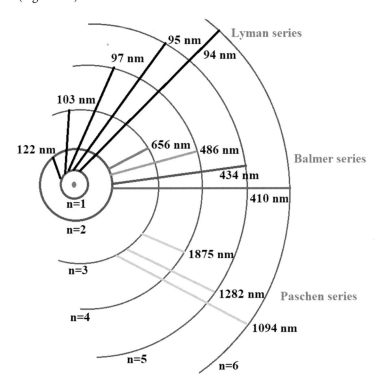

FIGURE 6.4   Spectral line series of the hydrogen atom

We can use the Rydberg formula to determine the wavelength λ:

$$\frac{1}{\lambda} = R \times \left(\frac{1}{n_1^2} - \frac{1}{n_2^2}\right)$$

$$\frac{1}{\lambda} = R \times \left(\frac{1}{n_1^2} - \frac{1}{n_2^2}\right) = 1.09677 \times 10^{7} \times \left(\frac{1}{2^2} - \frac{1}{4^2}\right) = 2.0564 \times 10^{6} \text{ m}^{-1}$$

and $\lambda = 4.86278 \times 10^{-7} \text{ m} = 486 \text{ nm}$.

6.5.2 SUCCESSES AND LIMITATIONS OF THE BOHR MODEL

The Bohr model has had many successes. For example:

- ○ It succeeded in explaining the emission spectrum of the hydrogen atom. Furthermore, it succeeded in explaining the spectral line series of species that, like hydrogen atom, have only one electron in the valence shell ("hydrogenic species"), such as He^+ and Li^{2+}. For the hydrogenic species:

$$\Delta E_{atom} = h\frac{C}{\lambda} = -2.18 \times 10^{-18} \times Z^2 \left(\frac{1}{n_{final}^2} - \frac{1}{n_{initial}^2} \right)$$

where Z is the atomic number of the hydrogenic species.

- It introduced the concept of "allowed orbits" which was developed further with Schrödinger's theory.
- It was an important contribution to the model of the atom.

However, the Bohr model also has shown many limitations:

- It could not explain why orbits were allowed.
- It explained the emission spectrum of the hydrogen atom and atoms or ions that have only one electron in the valence shell (hydrogenic species) but did not always explain those of other elements.
- It could not explain the splitting of the spectral lines in the presence of a magnetic field (the so-called Zeeman effect) (Figure 6.5).

Magnetic field off   Magnetic field on

FIGURE 6.5   Zeeman effect

GET SMART

HOW TO CALCULATE THE ENERGY AND THE WAVELENGTH OF A LINE IN THE SPECTRAL LINE SERIES OF ATOMS OF HYDROGEN AND HYDROGENIC SPECIES?

The emission spectrum of a hydrogen atom is discontinuous and is divided into spectral line series. The wavelength of each line is given by the Rydberg formula:

$$\frac{1}{\lambda} = R \times \left(\frac{1}{n_1^2} - \frac{1}{n_2^2} \right)$$

or $\Delta E = \dfrac{h \times C}{\lambda} = h \times C \times R \times \left(\dfrac{1}{n_1^2} - \dfrac{1}{n_2^2} \right)$

where h = 6.63 × 10⁻³⁴ J s, R = 1.09677 × 10⁷ m⁻¹ and C = 3.00 × 10⁸ m s⁻¹

$$\Delta E = 6.63 \times 10^{-34} \times 3.00 \times 10^8 \times 1.09677 \times 10^7 \times \left(\frac{1}{n_1^2} - \frac{1}{n_2^2} \right)$$

$$= 2.18 \times 10^{-18} \left(\frac{1}{n_1^2} - \frac{1}{n_2^2} \right) \text{ and } n_2 > n_1.$$

When $n_2 < n_1$, $\dfrac{1}{\lambda} = R \times \left(\dfrac{1}{n_2^2} - \dfrac{1}{n_1^2} \right)$ and $\Delta E = 2.18 \times 10^{-18} \left(\dfrac{1}{n_2^2} - \dfrac{1}{n_1^2} \right)$

For the hydrogenic species:

$$\frac{1}{\lambda} = R \times Z^2 \left(\frac{1}{n_1^2} - \frac{1}{n_2^2} \right) \text{ and } \Delta E = 2.18 \times 10^{-18} \times Z^2 \left(\frac{1}{n_1^2} - \frac{1}{n_2^2} \right)$$

where Z is the atomic number of the hydrogenic species and $n_2 > n_1$.

6.6 De Broglie HYPOTHESIS – THE WAVE-PARTICLE DUALITY OF MATTER

De Broglie in 1923 hypothesized that every particle also has a wave nature. Since light behaves like a particle, as shown by Einstein to explain the photoelectric effect, then particles could behave like a wave. However, it is only truly evident when a particle is very light, such as an electron ($m = 9.11 \times 10^{-31}$ kg). The wavelength of the corresponding wave is given by the de Broglie equation:

$$\lambda = \frac{h}{mV}$$

where h is Planck's constant, m is the mass of the particle and V is its velocity.

Practice 6.4 Determine the wavelength of an electron with a velocity of 6.00×10^6 m s^{-1} and the wavelength of a tennis ball with a speed of 20.0 ms^{-1} and a mass of 10.0 g.

Mass of the electron: $m = 9.11 \times 10^{-31}$ kg

Plank's constant: $h = 6.63 \times 10^{-34}$ J s

Answer:
The wavelength of the electron is

$$\lambda_{\text{electron}} = \frac{h}{mV} = \frac{6.63 \times 10^{-34}}{9.11 \times 10^{-31} \times 6.00 \times 10^6} = 1.21 \times 10^{-10} \text{ m} = 1.21 \text{Å}$$

With a wavelength of 1.21 Å, the electron could have a wave behavior.
   The wavelength of the tennis ball is

$$\lambda_{\text{ball}} = \frac{h}{mV} = \frac{6.63 \times 10^{-34}}{10.0 \times 10^{-3} \times 20.0} = 3.31 \times 10^{-31} \text{ m} = 3.31 \times 10^{-21} \text{Å}$$

The wavelength of the tennis ball $\left(3.31 \times 10^{-21} \text{Å}\right)$ is so incredibly small that such a wave cannot be detected.

6.7 HEISENBERG UNCERTAINTY PRINCIPLE

The Heisenberg uncertainty principle expresses a limitation on the accuracy of simultaneous measurements of both the position and the momentum of a particle. We cannot know both the position and the speed of a particle at a given time with the same accuracy. We cannot observe both the wave nature and the particle nature of the electron at the same time. This lack of accuracy is expressed mathematically by the following expression:

$$\Delta x \cdot m\Delta V \geq \frac{h}{4\pi}$$

where Δx and ΔV are the uncertainty on the position and the speed of the particle, respectively, and m is its mass.

Practice 6.5 An electron moves near an atomic nucleus with a speed of $6 \times 10^6 \pm 1\%$ ms^{-1}. Calculate the uncertainty in its position (Δ x).

Mass of the electron: $m = 9.11 \times 10^{-31}$ kg

Planck's constant: $h = 6.63 \times 10^{-34}$ J s

Answer:

The uncertainty in speed of the electron is 1% = 0.01, so

$$\Delta V = 6 \times 10^6 \times 0.01 = 6 \times 10^4 \, \text{ms}^{-1}.$$

From the uncertainty equation $\left(\Delta x \cdot m\Delta V \geq \dfrac{h}{4\pi} \right)$, we deduce the uncertainty in the position of the electron Δx:

$$\Delta x \geq \frac{h}{4\pi m \Delta V}$$

$$\Delta x \geq \frac{6.63 \times 10^{-34}}{4\pi \times 9.11 \times 10^{-31} \times 6 \times 10^4}$$

$$\Delta x \geq 10^{-9} \, \text{m}$$

6.8 INTRODUCTION TO QUANTUM THEORY

6.8.1 ATOMIC ORBITAL

Schrödinger in 1935 used mathematical equations to describe the electron in an atom. This atomic model is known as the quantum model of the atom. In this model, the nucleus is surrounded by an electron cloud. The probability of finding the electron is greatest when the cloud is most dense. Conversely, the electron is less likely to be found where the cloud is less dense. The atomic orbital is defined as a region in space in which there is a **high probability of finding an electron**.

6.8.2 QUANTUM NUMBERS

The atomic orbital is specified by three quantum numbers (n, ℓ and m_ℓ). The electron is specified by the quantum number, m_s (or s).

6.8.2.1 THE PRINCIPAL QUANTUM NUMBER

The principal quantum number (n) indicates the main energy levels allowed for an electron in an atom (Figure 6.6).

n is an integer : n = 1, 2, 3, 4...

As n increases, the main levels become closer to each other. Each main energy level has sub-levels.
For $n \leq 4$, the maximum number of electrons in a principal energy level is $2n^2$. For n higher than 4, this equation is no longer valid. For example, for $n = 5$ and $n = 6$, the maximum number of electrons is 18 electrons and for $n = 7$, the maximum number of electrons is two electrons.

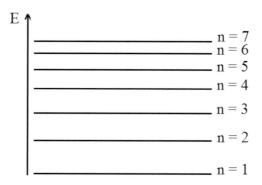

FIGURE 6.6   The main energy levels allowed for an electron in an atom

Practice 6.6 Determine the maximum number of electrons in the principal energy level n = 4?

Answer: The maximum number of electrons is $2n^2 = 2 \times (4)^2 = 32$ electrons.

6.8.2.2 THE ANGULAR MOMENTUM QUANTUM NUMBER

The angular momentum quantum number (also called the orbital quantum number or the azimuthal quantum number) (ℓ) indicates the number and shape of the orbital sublevels (also called subshells) (Figure 6.7). The angular momentum quantum number (ℓ) takes values from zero to (n – 1):

$$\ell = 0, 1, ...(n-1)$$

When $\ell = 0$, the sublevel is designated by the letter (s). The sublevels (s) have a spherical shape.

When $\ell = 1$, the sublevel is designated by the letter (p). The sublevels (p) have a dumbbell-like shape.

When $\ell = 2$, the sublevel is designated by the letter (d). The sublevels (d) look like a dumbbell but with a donut around the middle or like eggs stacked on one another.

When $\ell = 3$, the sublevel is designated by the letter (f). The sublevels (f) look like a dumbbell but with two donuts between the ends of the barbell or like balloons tied together in the center.

Practice 6.7 How many sublevels are there for n = 1 and for n = 4?

Answer:

$$\ell = 0, 1,..., (n-1)$$

o n = 1 then $\ell = 0$. There is one sublevel for n = 1
o n = 4 then $\ell = 0, 1, 2$ and 3. There are four sublevels for n = 4

In the same manner, we could determine the number of sublevels in each main level.

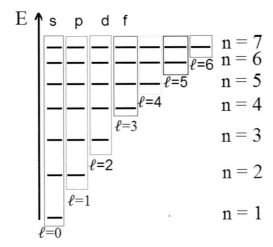

FIGURE 6.7   Sublevels in each main level

6.8.2.3 THE MAGNETIC QUANTUM NUMBER

The magnetic quantum number designated by (m_ℓ) distinguishes the orbitals available within a sublevel. It gives the exact orbital and indicates the orientation of the orbital in space. The magnetic quantum number (m_ℓ) represents the number of orbitals in a sublevel and takes the integer values, including zero ranging from $-\ell$ to $+\ell$.

Practice 6.8 How many orbitals are there for n = 1 and for n = 4?

Answer:

$$\ell = 0, 1, \ldots, (n-1) \text{ and } -\ell \leq m_\ell \leq +\ell.$$

- o n = 1 then $\ell = 0$ and $m_\ell = 0$. There is one orbital in sublevel s ($\ell = 0$) for n = 1.
- o n = 4 then $\ell = 0, 1, 2$ and 3. There are four sublevels for n = 4.
  $\ell = 0$ then $m_\ell = 0$. There is one orbital in sublevel s ($\ell = 0$).
  $\ell = 1$ then $m_\ell = -1, 0$ and 1. There are three orbitals in sublevel p ($\ell = 1$).
  $\ell = 2$ then $m_\ell = -2, -1, 0, 1$ and 2. There are five orbitals in sublevel d ($\ell = 2$).
  $\ell = 3$ then $m_\ell = -3, -2, -1, 0, 1, 2$ and 3. There are seven orbitals in sublevel f ($\ell = 3$).

In total, there are (1 + 3 + 5 + 7 =) 16 orbitals for n = 4 (Figure 6.8). In the same manner, we could determine the number of orbitals in each main level.

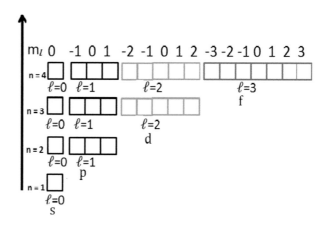

FIGURE 6.8   Number of orbitals in each main level for n = 0 to n = 4

6.8.2.4 THE ELECTRON SPIN QUANTUM NUMBER

The electron spin quantum number designated by m_s (or s) specifies the electron. An electron in an atom spins around an axis and has both angular momentum and orbital angular momentum. The electron spin quantum number indicates the magnitude and direction in which an electron spins in an orbital (clockwise or counterclockwise direction). The electron spin quantum number (s) takes the values −1/2 when it spins in the clockwise direction and +1/2 when it spins in the counterclockwise direction (Figure 6.9). An electron in an orbital is represented by an arrow in a square.

FIGURE 6.9   An electron, in an orbital, spins in the counterclockwise direction (s = 1/2) (on the left) or in the clockwise direction (s = −1/2) (on the right)

GET SMART

HOW CAN YOU DIFFERENTIATE BETWEEN TWO SUBLEVELS (S)?

Two different sublevels (s) are situated in different main energy levels (n). To distinguish between these two sublevels, we add the n quantum number in front of the name of the sublevel, i.e. 1s, 2s, 3s etc. When (n) increases, and the distance from the electron to the nucleus increases, the probability of finding the electron increases and the sublevel orbital volume increases. Thus, in order of **increasing volume**, we have 1s < 2s < 3s < 4s, etc. This is also true for sublevels p, d and f.

The number of electrons in a subshell is denoted as an exponent of the subshell name. **The number of electrons in an atom should be the same as the sum of the number of electrons in the subshells.**

Two electrons in the sublevel s of the main energy level n = 1 are represented as follows (Figure 6.10):

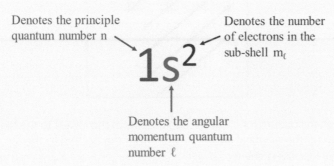

FIGURE 6.10   Representation of two electrons in the sublevel s of the main energy level n = 1.

6.8.3 ELECTRONIC CONFIGURATION

The electronic configuration is how the electrons are distributed among the various atomic orbitals in an atom. There are three rules to building up electronic configurations: the Aufbau (building-up) Principle, Hund's Rule and Pauli's Exclusion Principle.

6.8.3.1 THE AUFBAU PRINCIPLE

The Aufbau ((building-up) principle states that electrons enter the orbitals in order of ascending energy, following the (n+ℓ) rule. This means that electrons fill the atomic orbitals of the lowest available energy level before occupying higher energy levels.

The (n+ℓ) rule means that the orbital that has increased (n+ℓ) has an increased energy. When two orbitals have the same value of (n+ℓ), the orbital that has the lowest energy is the orbital with the lowest energy level (n).

Practice 6.9 Arrange the following orbitals, 1s, 2s, 2p, 3s, 3d and 4s, in order of increasing energy.

Answer:
For 1s, (n+ ℓ) =1 + 0 = 1 and for 2s, (n+ ℓ) = 2 + 0 = 2. Therefore, 1s has a lower energy than 2s.

For 3s, (n+ ℓ) = 3 + 0 = 3 and for 2p, (n+ ℓ) = 2 + 1 = 3. The orbital sublevels 3s and 2p have the same value of (n+ℓ), and 2p has the lowest energy level (n = 2). Therefore, orbital 2p has a lower energy than the orbital 3s.

For 4s, (n+ℓ) = 4 + 0 = 4 and for 3d, (n+ ℓ) = 3 + 2 = 5. Therefore, orbital 4s has a lower energy than orbital 3d, even though 3d has a lower n than 4s.

The correct order of increasing energy of these orbitals is 1s < 2s < 2p < 3s < 4s < 3d

The general pattern for filling the sublevels from the lowest to the highest energy is:

1s < 2s < 2p < 3s < 3p < 4s < 3d < 4p < 5s < 4d < 5p < 6s < 4f < 5d < 6p < 7s …

and it could be obtained easily from the following scheme (Figure 6.11):

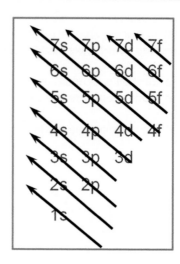

FIGURE 6.11   General pattern for filling the sublevels from the lowest to the highest energy. Start from (1s) and move up the diagonal arrows in order from the bottom to the top. This will give the sublevels in order of increasing energy.

Note that:

- Sublevel s exists in all the main levels.
- Sublevel p exists from $n = 2$ (there is no sublevel 1p), sublevel d exists from $n = 3$ and sublevel f exists from $n = 4$.
- Sublevel 4s has a lower energy than sublevel 3d and sublevel 5s has a lower energy than sublevel 4f.

6.8.3.2 HUND'S RULE

Hund's rule states that, when in orbitals of equal energy, electrons will try to remain unpaired as far as is possible. This means that the electrons occupy degenerate orbitals singly to the maximum extent possible and with their spins parallel. When two electrons are placed in one orbital there will be some electrostatic repulsion between them since they are both negatively charged. Placing each electron in a separate orbital reduces the repulsion and the atom becomes more stable (Figure 6.12):

HOW TO PLACE TWO OR THREE ELECTRONS IN THE SUBLEVEL 2P?

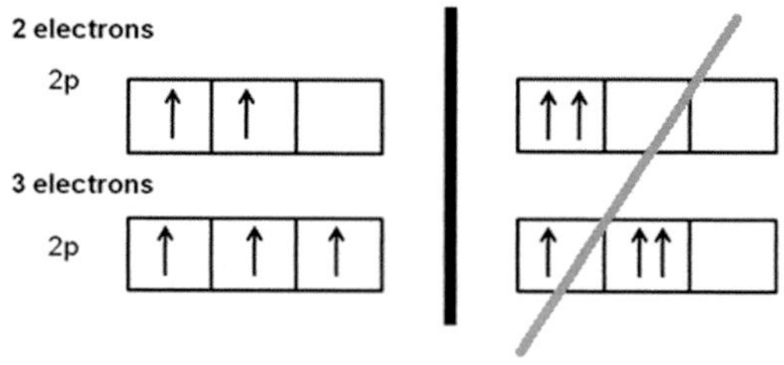

2 electrons

2p | ↑ | ↑ | | | ↑↑ | | |

3 electrons

2p | ↑ | ↑ | ↑ | | ↑ | ↑↑ | |

FIGURE 6.12

YES

The placement of the electrons in the orbitals follows Hund's rule. Placing each electron in a separate orbital reduces the repulsion and the atom becomes more stable.

NO

The placement of the electrons in the orbitals is incorrect and does not follow Hund's rule. Placing two electrons in the same orbital increases the repulsion since both electrons are negatively charged and the atom becomes less stable.

6.8.3.3 PAULI'S EXCLUSION PRINCIPLE

Pauli's exclusion principle states that no two electrons in the same atom have the same set of the four quantum numbers (n, ℓ, m_ℓ and m_s). This means that:

- ○ Orbitals of the same energy must be occupied singly and with the same spin before pairing up of electrons occurs (Hund's rule).
- ○ Electrons occupying the same orbital must have opposite spins. In that manner, the two electrons will have the same quantum numbers (n, ℓ and m_ℓ) but different spin quantum numbers m_s.
- ○ An orbital could be occupied at most by two electrons with opposite spins (Figures 6.13 and 6.14):

FIGURE 6.13   Schematic illustration of an orbital containing paired electrons with opposite spins

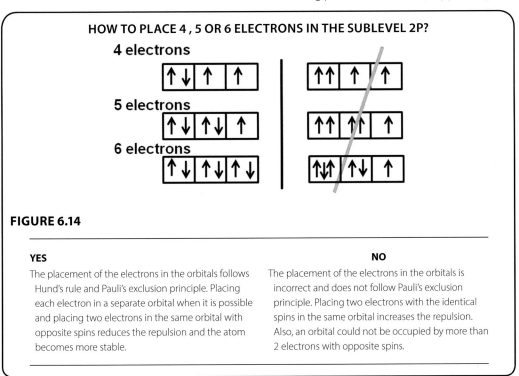

FIGURE 6.14

YES	NO
The placement of the electrons in the orbitals follows Hund's rule and Pauli's exclusion principle. Placing each electron in a separate orbital when it is possible and placing two electrons in the same orbital with opposite spins reduces the repulsion and the atom becomes more stable.	The placement of the electrons in the orbitals is incorrect and does not follow Pauli's exclusion principle. Placing two electrons with the identical spins in the same orbital increases the repulsion. Also, an orbital could not be occupied by more than 2 electrons with opposite spins.

6.8.3.4 ELECTRONIC CONFIGURATION OF ATOMS AND IONS

The electronic configuration is the distribution of electrons of an atom or an ion in the atomic orbitals. The electronic configuration is of great importance for describing the chemical bonds that hold atoms together. The three rules for building up electronic configurations (the Aufbau principle, Hund's rule and Pauli's exclusion principle) should be respected. It is important, before determining the electronic configuration of an atom or an ion, to know the number of electrons in this atom or ion. For example, the electronic configuration of the hydrogen ($_1$H) atom is $1s^1$, the electronic configuration of the sodium ($_{11}$Na) atom is $1s^2 2s^2 2p^6 3s^1$ and the electronic configuration of the sodium ion ($_{11}$Na$^+$) is $1s^2 2s^2 2p^6$.

Practice 6.10 Determine the electronic configuration of the neon atom ($_{10}$Ne), chloride atom ($_{17}$Cl), iron atom ($_{26}$Fe) and bromide ion ($_{35}$Br⁻).

Answer:

The $_{10}$Ne atom contains 10 electrons and its electronic configuration is $1s^22s^22p^6$.

Note that for the neon atom, all orbitals are completely filled.

The $_{17}$Cl atom contains 17 electrons and its electronic configuration is $1s^22s^22p^63s^23p^5$.

The $_{26}$Fe atom contains 26 electrons and its electronic configuration is $1s^22s^22p^63s^23p^64s^23d^6$.

Note that for the iron atom, sublevel 3d is not completely filled.

The $_{35}$Br atom contains 35 electrons and the bromide anion $_{35}$Br⁻ contains 36 electrons. The electronic configuration of $_{35}$Br⁻ is $1s^22s^22p^63s^23p^64s^23d^{10}4p^6$.

Practice 6.11 Determine the number of unpaired electrons in the neon ($_{10}$Ne) and phosphorus atoms ($_{15}$P).

Answer:

$_{10}$Ne: $1s^22s^22p^6$

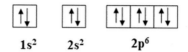

There are no unpaired electrons in the Ne atom.

$_{15}$P: $1s^22s^22p^63s^23p^3$

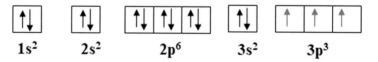

There are three unpaired electrons in the P atom.

GET SMART

PLACING ELECTRONS IN ORBITALS – AN ANALOGY

Let's pretend we are moving visitors (electrons) into a new hotel (atom). The hotel building contains floors (1st, 2nd, 3rd, 4th floor, etc.) (main energy level). There are houses in these floors. The houses (sublevels) are called s, p, d, and f houses. The s house has one bedroom, the p house has three bedrooms, the d house has five bedrooms, and the f house has seven bedrooms. In each bedroom (orbital), there is only one bed, so that, at most, two visitors can sleep in one bedroom.

6.8.3.5 ELECTRONIC CONFIGURATION OF TRANSITION METALS
          AND TRANSITION METAL CATIONS

A transition metal is an element whose atom has an **incomplete d subshell** or which can give rise to cations with an incomplete d subshell. For example, Ti and Fe are transition metals:

$_{22}$Ti: $1s^22s^22p^63s^23p^64s^23d^2$ (subshell d is incomplete)
$_{26}$Fe: $1s^22s^22p^63s^23p^64s^23d^6$ (subshell d is incomplete)

To obtain cations from transition metals, electrons in the **ns** orbital are removed **before** any electrons in the **(n – 1)d** orbitals. For example:

$_{26}$Fe: $1s^22s^22p^63s^23p^64s^23d^6$
$_{26}$Fe^{2+}: $1s^22s^22p^63s^23p^63d^6$ (two electrons are removed from 4s).
$_{26}$Fe^{3+}: $1s^22s^22p^63s^23p^64s^23d^5$ (the third electron is removed from 3d).

Note that, for the zinc atom, for example, the electronic configuration is:

$_{30}$Zn: $1s^22s^22p^63s^23p^64s^23d^{10}$

The d subshell is completely filled. When Zn atom loses two electrons, it becomes Zn^{2+} ion with the electronic configuration:

$_{30}$Zn^{2+}: $1s^22s^22p^63s^23p^63d^{10}$

Zn^{2+} has a complete d subshell and does not meet the definition of a transition metal. Therefore, Zn is not a transition metal.

6.8.3.6 EXCEPTIONS IN THE ELECTRONIC CONFIGURATION OF SOME TRANSITION METALS

According to the Aufbau principle, electrons should always fill orbitals according to increasing energy levels. However, a few atoms are more stable when their electrons fill or half fill a d (or f) sublevel than a partially filled d sublevel. There are two main *exceptions* to *electron configuration,* which are chromium and copper:

$_{24}$Cr: $1s^22s^22p^63s^23p^64s^13d^5$ (the d sublevel is half filled)
$_{29}$Cu: $1s^22s^22p^63s^23p^64s^13d^{10}$ (the d sublevel is completely filled)

Note that the expected electronic configuration of chromium is $1s^22s^22p^63s^23p^64s^23d^4$
    and the expected electronic configuration of copper is $1s^22s^22p^63s^23p^64s^23d^9$, so an *electron* from the 4s orbital moves to a 3d orbital, leading to more stable atoms.

6.8.3.7 ISOELECTRONIC CONFIGURATION

Atoms and ions that have the same number of electrons show an isoelectronic configuration (i.e., have the same electronic configuration). For example, $_8$O^{2-}, $_{11}$Na$^+$ and $_9$F$^-$ show an isoelectronic configuration:

$_8$O^{2-}: $1s^2 2s^2 2p^6$
$_{11}$Na$^+$: $1s^2 2s^2 2p^6$
$_9$F$^-$: $1s^2 2s^2 2p^6$

6.8.3.8 VALENCE SHELL ELECTRONS

The most outer sublevels (for the highest n) are called valence shells. Electrons in the most outer sublevels are called the valence electrons.
    For transition metals, the valence shell is **ns (n – 1)d.**

Practice 6.12 Determine the number of valence electrons of neon ($_{10}$Ne), sodium (Na), sodium ion (Na$^+$), iron ($_{26}$Fe) and bromine ($_{35}$Br).

Answer:

$_{10}$Ne: $1s^22s^22p^6$
    The highest n is 2. The valence shell is 2s2p and the number of electrons in this valence shell is eight.

$_{11}$Na: $1s^2 2s^2 2p^6 3s^1$

The highest n is 3. The valence shell is 3s and the number of electrons in this valence shell is one.

$_{11}$Na$^+$: $1s^2 2s^2 2p^6$

Note that the cation Na$^+$ is obtained by removing one electron from the valence shell 3s of Na. The number of valence electrons is ten.

$_{26}$Fe: $1s^2 2s^2 2p^6 3s^2 3p^6 4s^2 3d^6$

Iron is a transition metal. For transition metals, the valence shell is ns (n − 1)d. Therefore, the valence shell of iron is 4s 3d and the number of electrons in this valence shell is eight.

$_{35}$Br: $1s^2 2s^2 2p^6 3s^2 3p^6 4s^2 3d^{10} 4p^5$

It is important to note that bromine is not a transition metal since it has a completely filled d sub-shell and its valence shell is not 4s3d. Indeed, the highest n in the electronic configuration of bromine is four. The valence shell is 4s4p and the number of electrons in this valence shell is seven.

6.8.3.9 SHORT CUT FOR WRITING ELECTRONIC CONFIGURATIONS

The electronic configuration of an element can be written in a short-cut manner, using the noble gas that comes just before this element in the periodic table of elements. To write the short-cut configuration, follow the following steps:

○ Find the element in the periodic table or identify its atomic number.
○ Go back to the last noble gas that was passed (or find the noble gas that has an atomic number just lower than that of the element in question).
○ Write the symbol of the noble gas in brackets which means the electronic configuration of the noble gas and add the valence shell of the element in the short-cut electronic configuration.

Examples:
$_{10}$**Ne: 1s² 2s² 2p⁶** (noble gas). This electronic configuration can be represented by **[Ne].**
$_{17}$**Cl: 1s² 2s² 2p⁶ 3s² 3p⁵**

The short-cut presentation of the electronic configuration of chlorine is $_{17}$**Cl: [Ne] 3s²3p⁵**

$_{18}$**Ar: 1s² 2s² 2p⁶ 3s² 3p⁶** (noble gas). This electronic configuration can be represented by **[Ar]**
$_{28}$**Ni: 1s² 2s² 2p⁶ 3s² 3p⁶ 4s² 3d⁸**

The short-cut presentation of the electronic configuration of nickel is $_{28}$**Ni: [Ar] 3d⁸ 4s².**

Note that in the shortcut electronic configuration of an element, we write the noble gas in brackets then the valence shell of the element in order of increasing n.

GET SMART

HOW TO PRESENT THE SHORT-CUT ELECTRONIC CONFIGURATION OF AN ELEMENT?

Find the noble gas that has an atomic number just lower than that of the element and the short-cut electronic configuration presentation is based on this noble gas.

Practice: Considering the following noble gases $_2$He, $_{10}$Ne and $_{18}$Ar, find the short-cut electronic configuration of sodium ($_{11}$Na).

Answer: The atomic number of sodium ($_{11}$Na) is 11. The noble gas that has an atomic number just lower than sodium is neon ($_{10}$Ne). Therefore, the short-cut electronic configuration of sodium is based on neon (Ne):

$_{10}$**Ne: 1s² 2s² 2p⁶**. This electronic configuration can be represented by **[Ne]**.

$_{11}$**Na: 1s² 2s² 2p⁶ 3s¹**

The short-cut electronic configuration of sodium is $_{11}$**Na: [Ne]3s¹**

Practice 6.13 Considering the following noble gases $_2$He, $_{10}$Ne, $_{18}$Ar, $_{36}$Kr, $_{54}$Xe, $_{86}$Rn and $_{118}$Og,

find the short-cut electronic configurations of iron ($_{26}$Fe) and strontium ($_{38}$Sr).

Answer:

- The atomic number of iron ($_{26}$Fe) is 26. The noble gas that has an atomic number just lower than that of iron is argon ($_{18}$Ar). Therefore, the short-cut electronic configuration of iron is based on argon (Ar):

    $_{18}$**Ar: 1s² 2s² 2p⁶ 3s² 3p⁶**. This electronic configuration can be represented by **[Ar]**.

    $_{26}$**Fe: 1s² 2s² 2p⁶ 3s² 3p⁶ 4s²3d⁶**

    $_{26}$**Fe: [Ar] 3d⁶ 4s²** (short-cut electronic configuration of iron). Note that **4s3d** is the valence shell of iron and the valence shell electrons (VSE) = eight electrons.

- The atomic number of strontium ($_{38}$Sr) is 38. The noble gas that has an atomic number just lower than that of strontium is krypton ($_{36}$Kr). Therefore, the short-cut electronic configuration of strontium is based on krypton (Kr):

    $_{36}$**Kr: 1s² 2s² 2p⁶ 3s² 3p⁶ 4s² 3d¹⁰ 4p⁶**. This electronic configuration can be represented by **[Kr]**.

    $_{38}$**Sr: 1s² 2s² 2p⁶ 3s² 3p⁶ 4s² 3d¹⁰ 4p⁶ 5s²**

    $_{38}$**Sr: [Kr]5s²** (short-cut electronic configuration of strontium). Note that **5s** is the valence shell of strontium and VSE = two electrons.

CHECK YOUR READING

How was the photoelectric effect explained?

What does the wave-particle duality of light mean? Which particle shows a wave-like behavior?

How were the line spectra of a hydrogen atom explained?

How is the wavelength of a line in the spectrum of the hydrogen atom calculated?

What are the quantum numbers?

What are the rules for placing electrons in the atomic orbitals?

Which main elements show an exception to the electronic configuration?

What is an isoelectronic configuration?

How is the electronic configuration of a transition metal ion written?

How is the valence shell of an element determined?

How to write the shortcut electronic configuration of an element?

SUMMARY OF CHAPTER 6

The **wavelength (λ)** is the distance from one crest to the next in the wave. **The frequency (ν)** is the number of complete cycles per second, expressed in **s⁻¹** or Hz. The frequency ν of a wave is inversely proportional to its wavelength λ:

$$\nu = \frac{C}{\lambda}$$

where C is the speed of light (C = 3 × 10⁸ m s⁻¹ in a vacuum).

The wave-particle duality of light means that light consists of particle-like packets of energy called **photons** that also have wave-like properties.

Atoms can gain or lose certain quantities of energy by absorbing or emitting photons:

$$\Delta E_{atom} = E_{photon} = h\nu = h\frac{c}{\lambda}$$

where h is Planck's constant ($h = 6.63 \times 10^{-34}$ J s).

The photoelectric effect is the emission of a flow of electrons when light shines on plates of certain metals. The kinetic energy, K_E, of the electron is given by the following equation:

$$K_E = \frac{1}{2}mV^2 = E_{photon} - \Phi = h\frac{c}{\lambda} - \Phi$$

where Φ is the work function which depends on the matter, and m and V are the mass and the speed of the electron, respectively.

The *emission spectrum* of the *hydrogen atom is* **discontinuous** and can be divided into a number of spectral line series.

The hydrogen atom has only certain allowable energy levels for the electron orbits. When the electron moves from one orbit ($n_{initial}$) to another orbit (n_{final}), it absorbs or emits a photon (light) whose energy equals the difference in energy between the two orbits:

$$\Delta E_{atom} = h\frac{c}{\lambda} = -2.18 \times 10^{-18} \times \left(\frac{1}{n_{final}^2} - \frac{1}{n_{initial}^2}\right) J$$

Every particle has a wave nature as well. However, it is only truly evident when a particle is extremely light, such as the electron ($m = 9.11 \times 10^{-31}$ kg). The wavelength of the corresponding wave is given by the de Broglie equation:

$$\lambda = \frac{h}{mV}$$

However, we cannot observe both the wave nature and the particle nature of the electron at the same time, which is the Heisenberg uncertainty principle.

The atomic orbital is a region in space in which there is a **high probability of finding an electron**. The atomic orbital is specified by three quantum numbers (n, ℓ and m_ℓ). The electron is specified by the quantum number m_s (or s) (Table 6.1).

TABLE 6.1   Summary of quantum numbers

Symbol	Name	Permitted values	Property
n	Principle quantum number	n integer (n = 1, 2, 3,...)	Indicates the main energy level
ℓ	Angular momentum quantum number	$\ell = 0, 1, ...(n-1)$ $\ell = 0$ corresponds to the s orbital $\ell = 1$ corresponds to the p orbital $\ell = 2$ corresponds to the d orbital $\ell = 3$ corresponds to the f orbital	Gives the number and shape of the sublevels
m_ℓ	Magnetic quantum number	$-\ell \leq m_\ell \leq +\ell$	Indicates the orbital orientations
s or m_s	Spin quantum number	$m_s = 1/2$ and $-1/2$	Indicates the magnitude and direction in which an electron spins in an orbital

---

The **electronic configuration** is how the electrons are distributed among the various atomic orbitals in an atom. There are three rules to building up electronic configurations:

o The Aufbau (building-up) principle states that electrons enter the orbitals in order of ascending energy following the $(n+\ell)$ rule.

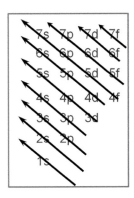

The general pattern for filling the sublevels is from the lowest to the highest energy.

○ Hund's rule states that when in orbitals of equal energy, electrons will try to remain unpaired as far as is possible.
○ Pauli's exclusion principle states that no two electrons in the same atom have the same set of the four quantum numbers (n, ℓ, m_ℓ and m_s).

Placing each electron in a separate orbital when it is possible and placing two electrons in the same orbital with opposite spins reduces the repulsion and the atom becomes more stable.

The **electronic configuration** is the distribution of electrons of an atom or an ion in the atomic orbitals. Before determining the electronic configuration of an atom or an ion, you should know the number of electrons in this atom or ion.

Examples of electronic configuration:
$_{17}$Cl: $1s^2\ 2s^2\ 2p^6\ 3s^2\ 3p^5$
$_{17}$Cl$^-$: $1s^2\ 2s^2\ 2p^6\ 3s^2\ 3p^6$

Atoms and ions that have the same number of electrons show an **isoelectronic configuration**. For example, $_8$O^{2-}, $_{11}$Na$^+$ and $_9$F$^-$ show the isoelectronic configuration $1s^2\ 2s^2\ 2p^6$.

The electronic configuration of an element could be written in a **short-cut** manner based on the noble gas that comes just before this element in the periodic table of elements.

Examples of short cut electronic configuration:
$_{26}$Fe: [Ar] $4s^2 3d^6$
$_{38}$Sr: [Kr]$5s^2$

There are **two main *exceptions*** to *electron configurations*, which are **chromium and copper**:

$_{24}$Cr: $1s^2 2s^2 2p^6 3s^2 3p^6 4s^1 3d^5$ (the d sublevel is half filled)
$_{29}$Cu: $1s^2 2s^2 2p^6 3s^2 3p^6 4s^1 3d^{10}$ (the d sublevel is completely filled)

A transition metal is an element where the atom has an **incomplete d subshell** or which can give rise to cations with an incomplete d subshell. For example, the electronic configuration of titanium is:

$_{22}$Ti: $1s^2 2s^2 2p^6 3s^2 3p^6 4s^2 3d^2$. The subshell d is incomplete. Therefore, titanium is a transition metal.

To obtain **cations from transition metals**, electrons in the **ns** orbital are removed **before** any electrons in the **(n − 1)d** orbitals. For example:

$_{22}$Ti$^+$: $1s^2 2s^2 2p^6 3s^2 3p^6 4s^1 3d^2$ (an electron is removed from 4s).
$_{22}$Ti$^+$: $1s^2 2s^2 2p^6 3s^2 3p^6 3d^2$ (two electrons are removed from 4s).
$_{22}$Ti^{3+}: $1s^2 2s^2 2p^6 3s^2 3p^6 3d^1$ (two electrons are removed from 4s; thereafter, an electron is removed from 3d).

The most outer sublevels (for the highest n) are called **valence shells**. Electrons in the most outer sublevels are called the valence electrons. For **transition metals**, the valence shell is **ns (n – 1)d.**
   Examples:
      $_{11}$Na: $1s^2\, 2s^2\, 2p^6\, 3s^1$. The valence shell of Na is 3s and the number of valence electrons is one.

Note that the number of valence electrons in the cation Na^+ is eight.
      $_{22}$Ti: $1s^2 2s^2 2p^6 3s^2 3p^6 4s^2 3d^2$.
      Ti is a transition metal, and its valence shell is 4s3d and the number of valence electrons is four.

PRACTICE ON CHAPTER 6

Q6.1   Complete the following sentences
   In the Bohr model, the electron is considered as _____ moving in circular orbits around the nucleus of the hydrogen atom. When the electron moves from a higher energy orbit to a lower energy orbit, it _____ a photon (light). When an electron moves from a lower energy orbit to a higher energy orbit, it _____ a photon. The energy of the _____ equals the difference in energy between the two orbits. de Broglie states that the electron has _____ behavior as well. However, we cannot observe both the wave nature and the particle nature of the electron at the same time (the so called _____ principle). In Schrödinger's theory, the _____ is defined as a region in space in which there is a _____ of finding an electron. The atomic orbital is specified by three _____ (n, ℓ and m_ℓ) and the electron is specified by the _____ quantum number m_s (or s).

Q6.2   Match the followings

a. The wave-particle duality of light	**1.** The position and speed of an electron cannot be determined simultaneously with the same accuracy.
b. The wave-particle duality of matter	**2.** The atomic orbital is a region in space in which there is a high probability of finding an electron.
c. Heisenberg's uncertainty principle	**3.** Electron has a wave behavior.
d. Photoelectric effect	**4.** The hydrogen atom has only certain allowable energy levels for the electron orbits.
e. Schrödinger's theory	**5.** Light consists of particles called photons.
f. Bohr model	**6.** Light shining on a plate of a metal causes the emission of a flow of electrons.

Q6.3   Choose the correct answer
   1.   **"Energy is quantized" means that**
      a) atoms can gain or lose fixed quantities of energy by absorbing or emitting radiation.
      b) light consists of particles called photons.
      c) electrons have a wave behavior.
      d) light shining on a plate causes the emission of a flow of electrons.
   2.   **The wave-particle duality of light means that**
      a) light consists of particles called photons that have wave-like properties.
      b) electrons behave as a wave.
      c) light shining on certain metal plates causes a flow of electrons.
      d) the wave nature and the particle nature of the electron cannot be observed at the same time.
   3.   **Which of the followings behaves as a wave?**
      a) Tennis ball
      b) Electron
      c) School bus
      d) All of these

4. **In the photoelectric effect, which of the following is ejected from the surface metal as a result of incident light?**
   a) Protons
   b) Neutrons
   c) Electrons
   d) Photons

5. **When an electron moves from a higher energy level to a lower energy level, it**
   a) emits photons
   b) absorbs photons
   c) its energy remains constant
   d) it emits neutrons

6. **The wavelength (λ) of the line corresponding to the transition of an electron from n = 3 to n = 2 in the hydrogen atom is**
   (C = 3.00 × 10^8 m s^{-1} , h = 6.63 × 10^{-34} Js and R$_E$ = 2.18 × 10^{-18} J)
   a) 486 nm
   b) 656 nm
   c) 1250 nm
   d) 32 nm

7. **The photoelectric effect is observed for all lights.**
   a) True
   b) False

8. **The emission spectrum of hydrogen atom is continuous.**
   a) True
   b) False

9. **The frequency of a light is inversely proportional to its wavelength.**
   a) True
   b) False

10. **The energy of an atom is constant.**
    a) True
    b) False

11. **Hydrogenic species have only one electron in the valence shell.**
    a) True
    b) False

Q6.4 Choose the correct answer

1. **Which of the following quantum numbers (n, ℓ, m$_\ell$, m$_s$) describe an electron in an orbital?**
   a) (3, 4, 1, 1/2)
   b) (3, 1, 2, 1/2)
   c) (3, 0, 0, –1/3)
   d) (3, 2, 1, 1/2)

2. **Which of the following quantum numbers (n, ℓ, m$_\ell$, m$_s$) do not describe an electron in an orbital**
   a) (2, 1, 1, 1/2)
   b) (3, 2, 1, –1/2)
   c) (4, 3, –2, –1/4)
   d) (3, 1, 0, –1/2)

3. **The name of the sub-level with n = 3 and ℓ = 2 is**
   a) 2d
   b) 3d
   c) 3p
   d) 2s

4. **When the principal quantum number n = 3, the orbital quantum number ℓ takes the values**
   a) 0 and 1
   b) –3, –1, –2, 0, 1, 2 and 3

c) 0, 1 and 2

d) −1/2 and 1/2

5. **When the principal quantum number n equals 2, the orbital quantum number ℓ takes _____ possible values.**

   a) 2

   b) 5

   c) 3

   d) 16

6. **When the orbital quantum number $\ell = 1$, the magnetic quantum number m_ℓ takes the values**

   a) 0 and 1

   b) −1, 0 and 1

   c) 0

   d) −1/2 and 1/2

7. **What is the number of sublevels in the principal level n = 2?**

   a) 1

   b) 2

   c) 4

   d) 9

8. **What is the number of orbitals in the principal level n = 2?**

   a) 1

   b) 2

   c) 4

   d) 9

9. **How many orbitals are in the p sublevel?**

   a) 1

   b) 3

   c) 5

   d) 7

10. **How many electrons can f sublevel hold?**

    a) 2

    b) 6

    c) 10

    d) 14

11. **Which of the followings about d orbitals is correct?**

    a) They are found in all principal energy levels

    b) There are five types of d orbitals

    c) They are spherical in shape

    d) Each d orbital can hold up to three electrons

12. **The quantum number that takes only the values −1/2 and 1/2 is the magnetic quantum number (m_ℓ).**

    a) True

    b) False

Q6.5 Choose the correct answer

1. **Choose the correct order of the following sublevels from the lowest to the highest energy.**

   a) $3p < 4p < 3d < 4s$

   b) $3p < 3d < 4s < 4p$

   c) $3d < 3p < 4s < 4p$

   d) $3p < 4s < 3d < 4p$

2. **Which of the following electron distributions in the orbitals is correct?**

   a)

   b)

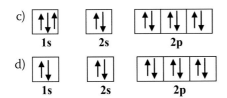

3. **The electronic configuration of $_{17}Cl$ is**
    a) $1s^22s^22p^63s^23p^6$
    b) $1s^22s^22p^63s^23p^5$
    c) $1s^22s^22p^63p^53s^2$
    d) $1s^22p^62s^23p^53s^2$

4. **The electronic configuration of chrome $_{24}Cr$ is**
    a) $1s^22s^22p^63s^23p^64s^23d^4$
    b) $1s^22s^22p^63s^23p^63d^6$
    c) $1s^22s^22p^63s^23p^64s^13d^5$
    d) $s^22s^22p^63s^23p^64s^13d^54p^1$

5. **The electronic configuration of copper $_{29}Cu$ is**
    a) $1s^22s^22p^63s^23p^64s^13d^{10}$
    b) $1s^22s^22p^63s^23p^63d^6$
    c) $1s^22s^22p^63s^23p^64s^23d^9$
    d) $s^22s^22p^63s^23p^64s^13d^94p^1$

6. **Iron (Fe) is a transition metal. The electronic configuration of $_{26}Fe^{2+}$ is**
    a) $1s^22s^22p^63s^23p^64s^23d^6$
    b) $1s^22s^22p^63s^23p^64s^23d^4$
    c) $1s^22s^22p^63s^23p^64s^13d^5$
    d) $1s^22s^22p^63s^23p^63d^6$

7. **The electronic configuration of $_{12}Mg^{2+}$ is**
    a) $1s^22s^22p^63s^1$
    b) $1s^22s^22p^6$
    c) $1s^22s^22p^53s^2$
    d) $2s^22p^63s^23p^2$

8. **The electronic configuration of $_8O^{2-}$ is**
    a) $1s^22s^22p^6$
    b) $1s^22s^22p^4$
    c) $1s^22s^22p^2$
    d) $1s^22p^62s^2$

9. **How many unpaired electrons are there in $_{20}Ca^{2+}$?**
    a) 0
    b) 2
    c) 10
    d) 20

10. **How many unpaired electrons are there in $_{22}Ti^+$?**
    a) 0
    b) 2
    c) 3
    d) 20

11. **How many unpaired electrons are there in $_{11}Na$?**
    a) 0
    b) 1
    c) 2
    d) 11

12. **The ions $_8O^{2-}$, $_{12}Mg^{2+}$ and $_9F^-$**
    a) are isotopes
    b) have isoelectronic configurations
    c) are compounds
    d) are called sublevels

13. **The short-cut electronic configuration of $_{30}$Zn is**
    a) $1s^2 2s^2 2p^6 3s^2 3p^6 4s^2 3d^{10}$
    b) $[_{18}\text{Ar}]\ 3d^{10}\ 4s^2$
    c) $[_2\text{He}]\ 3d^{10}\ 4s^2$
    d) $[_{36}\text{Kr}]\ 3d^{10}\ 4s^2$

14. **The valence shell of the chlorine $_{17}$Cl atom is**
    a) 1s 2s 2p 3s 3p
    b) 1s 2s
    c) 3p
    d) 3s 3p

15. **The valence shell of the bromine $_{35}$Br atom is**
    a) 4s
    b) 3d 4p
    c) 4s 4p
    d) 4s 3d

16. **The valence shell of the nickel $_{28}$Ni atom is**
    a) 1s
    b) 4s 3d
    c) 4s
    d) 3d

17. **How many valence electrons in sodium $_{11}$Na?**
    a) 1 electron
    b) 2 electrons
    c) 8 electrons
    d) 11 electrons

18. **How many valence electrons are in the sodium cation $_{11}$Na$^+$?**
    a) 1 electron
    b) 2 electrons
    c) 8 electrons
    d) 11 electrons

19. **How many valence electrons are present in iron $_{26}$Fe?**
    a) 2 electrons
    b) 6 electrons
    c) 8 electrons
    d) 26 electrons

20. **How many valence electrons are present in bromine $_{35}$Br?**
    a) 5 electrons
    b) 7 electrons
    c) 10 electrons
    d) 35 electrons

21. **A transition metal has an incompletely filled d sublevel.**
    a) True
    b) False

22. **Zinc ($_{30}$Zn) is a transition metal.**
    a) True
    b) False

23. **Titanium ($_{22}$Ti) is a transition metal.**
    a) True
    b) False

24. **The following electron distribution in the atomic orbitals is correct:**

↑↓		☐		↑↓	↑↓	↑↓
$1s^2$		$2s^2$			$2p^6$	

    a) True
    b) False

Calculations

Q6.6 **Red light has a frequency $\nu = 4.57 \times 10^{14}$ s^{-1}. Calculate the wavelength of red light?**
Speed of light C $= 3.00 \times 10^8$ m s^{-1}.

Q6.7 **The wavelength of blue light is 434 nm. What is the energy of one photon of blue light?**
Speed of light C $= 3.00 \times 10^8$ m s^{-1} and Planck's constant h $= 6.63 \times 10^{-34}$ J s.

Q6.8 **Calculate the energy ΔE_{atom} and the wavelength of the line that corresponds to the transition of electrons from n = 1 to n = 4 in the hydrogen atom.**
Speed of light C $= 3.00 \times 10^8$ m s^{-1} and Planck's constant h $= 6.63 \times 10^{-34}$ Js.

Q6.9 **Calculate the wavelength of the line that corresponds to the transition of electrons from n = 1 to n = 4 in the lithium ($_3$Li) atom.**
Speed of light C $= 3.00 \times 10^8$ m s^{-1}, Planck's constant h $= 6.63 \times 10^{-34}$ Js and the Rydberg constant R $= 1.09677 \times 10^7$ m^{-1}

Q6.10 **An electron moves near an atomic nucleus in a circular orbit. The orbit radius is $0.529 \times 10^{-10} \pm 10\%$ m. Calculate the uncertainty in its velocity (ΔV).**
Mass of the electron: m $= 9.11 \times 10^{-31}$ kg.
Planck's constant: h $= 6.63 \times 10^{-34}$ J s.

ANSWERS TO QUESTIONS

Q6.1

In the Bohr model, the electron is considered to be a **particle** moving in circular orbits around the nucleus of the hydrogen atom. When the electron moves from a higher energy orbit to a lower energy orbit, it **emits** a photon (of light). When an electron moves from a lower energy orbit to a higher energy orbit, it **absorbs** a photon. The energy of **the photon** equals the difference in energy between the two orbits. de Broglie states that the electron also shows **a wave** behavior. However, we cannot observe both the wave nature and the particle nature of the electron at the same time (the so-called **Heisenberg's uncertainty principle**). In Schrödinger's theory, the **atomic orbital** is defined as a region in space in which there is a **high probability** of finding an electron. The atomic orbital is specified by three **quantum numbers** (n, ℓ and m$_\ell$) and the electron is specified by **the spin** quantum number m$_s$ (or s).

Q6.2 (a, 5); (b, 3); (c, 1); (d, 6); (e, 2); (f, 4)

Q6.3
1. a
2. a
3. b
4. c
5. a
6. b
7. b
8. b
9. a
10. b
11. a

Q6.4
1. d
2. c
3. b
4. c
5. a
6. b

7. b
8. c
9. b
10. d
11. b
12. b

Q6.5

1. d
2. d
3. b
4. c
5. a
6. d
7. b
8. a
9. a
10. c
11. b
12. b
13. b
14. d
15. c
16. b
17. a
18. c
19. c
20. b
21. a
22. b
23. a
24. b

Calculations

Q6.6

$$\lambda = 656 \times 10^{-9}\,\text{m} \; = \; 656\;\text{nm}.$$

Q6.7

$$E_{\text{photon}} = 4.58 \times 10^{-19}\,\text{J}$$

Q6.8

$\Delta E_{\text{atom}} = -2.04 \times 10^{-19}\,\text{J}$ (Note that there is no negative energy. Here, the negative energy (sign –) means that the electron loses energy)

$$\lambda = 9.75 \times 10^{-7}\,\text{m} = 97.5\;\text{nm}$$

Q6.9

$$\lambda = 1.08061 \times 10^{-8}\,\text{m} = 11\;\text{nm}$$

Q6.10

The uncertainty in the velocity of the electron ΔV:

$$\Delta V \geq 1.09 \times 10^{7}\,\text{m s}^{-1}.$$

KEY EXPLANATIONS

Q6.3

1. (and 2) "Energy is quantized" means that atoms can gain or lose only certain quantities of energy by absorbing or emitting radiation.

    The wave particle duality of light means that light consists of particles called photons that also have wave-like properties.

    The wave-particle duality of matter implies that electrons have a wave behavior.

    The photoelectric effect: light shining on a metal plate causes the emission of a flow of electrons.

3. Every particle has a wave nature as well. However, it is only truly evident when a particle is very light, such as an electron ($m = 9.11 \times 10^{-31}$ kg).

    For the school bus and the tennis ball, the corresponding wavelengths are so incredibly small that such wave could not be detected.

4. In the word "photoelectric", "photo-" means light and "-electric" is derived from electron. Light shining on a metal surface causes the emission of a flow of electrons.

5. Generally, it is easy to move from a higher energy point to a lower one. In that case, we do not need energy. When an electron moves from a higher energy level to a lower energy level, it loses energy by emitting light (photon). It is difficult, however, to move from a lower energy point to a higher point. In that case, energy is needed. When an electron moves from a lower energy level to a higher energy level, it absorbs energy (photon).

6. An electron moves from $n = 3$ to $n = 2$ in the hydrogen atom

    We can use the Rydberg formula for determining the wavelength λ:

$$\frac{1}{\lambda} = R \times \left(\frac{1}{n_1^2} - \frac{1}{n_2^2} \right)$$

where R is the Rydberg constant ($R = 1.09677 \times 10^7$ m^{-1}) and $n_2 > n_1$.

$$\frac{1}{\lambda} = R \times \left(\frac{1}{n_1^2} - \frac{1}{n_2^2} \right) = 1.09677 \times 10^7 \times \left(\frac{1}{2^2} - \frac{1}{3^2} \right) = 1.52329 \times 10^6 \text{ m}^{-1}$$

and $\lambda = 6.56474 \times 10^{-7}$ m $= 656$ nm

or we can calculate ΔE_{atom}, and then deduce the wavelength λ:

The electron moves from a higher energy orbit ($n_{initial} = 3$) to a lower energy orbit ($n_{final} = 2$). It emits a photon (of light).

$$\Delta E_{atom} = h \frac{C}{\lambda} = -2.18 \times 10^{-18} \times \left(\frac{1}{n_{final}^2} - \frac{1}{n_{initial}^2} \right)$$

$$\Delta E_{atom} = h \frac{C}{\lambda} = -2.18 \times 10^{-18} \times \left(\frac{1}{2^2} - \frac{1}{3^2} \right)$$

$$\Delta E_{atom} = -3.03 \times 10^{-19} \text{ J}$$

Note that there is no negative energy and the sign (–) herein means that the atom loses energy.

The negative sign (–) should not be taken into account when calculating the wavelength of the emitted light since wavelength cannot be negative:

$$\lambda = \frac{hC}{\Delta E_{atom}} = \frac{6.63 \times 10^{-34} \times 3.00 \times 10^8}{3.03 \times 10^{-19}} = 6.56 \times 10^{-7} \text{ m} = 656 \text{ nm}$$

7. A photon may penetrate the metal plate and transfer its energy to an electron. If the energy of the photon is at least as high as the energy required for ejecting the

electron from the metal plate, the electron will have enough energy to emerge from the metal. If the energy of the photon is less than the energy required for ejecting the electron from the metal plate, the electron cannot be removed from the metal. This means that the photoelectric effect is not observed for all electromagnetic radiation. For example, red light ($\lambda = 700$ nm) cannot emit electrons since its energy ($h\nu$) is lower than the energy required for ejecting the electron from the metal plate ($h\nu_0$).

8. The emission spectrum of light is continuous. However, the emission spectrum of hydrogen atom is discontinuous (line spectrum).

9. The frequency (ν) of a light is inversely proportional to the wavelength (λ), i.e., when the wavelength decreases, the frequency increases: $\left(\nu = h\dfrac{C}{\lambda} \right)$.

10. The energy of an atom is not constant. Indeed, atoms can gain or lose certain quantities of energy by absorbing or emitting a photon:

$$\Delta E_{atom} = E_{photon} = h = h\frac{C}{\lambda}$$

11. Hydrogenic species, such as He^+ and Li, are atoms or ions that, like a hydrogen atom, have only one electron in the valence shell.
    For the hydrogenic species:

$$\frac{1}{\lambda} = R \times Z^2 \left(\frac{1}{n_1^2} - \frac{1}{n_2^2} \right) \text{ and } \Delta E = 2.18 \times 10^{-18} \times Z^2 \left(\frac{1}{n_1^2} - \frac{1}{n_2^2} \right)$$

where Z is the atomic number of the hydrogenic species, R is the Rydberg constant ($R = 1.09677 \times 10^7$ m^{-1}) and $n_2 > n_1$.

Q6.4

1. $n = 0, 1, 2, \ldots$
   ℓ takes the values from zero to (n – 1)
   m_ℓ takes the values from $-\ell$ to ℓ including zero.
   m_s takes the values 1/2 and –1/2
   a) (3, 4, 1, 1/2): $n = 3$, $\ell = 4$, $m_\ell = 1$ and $m_s = 1/2$
      For $n = 4$, ℓ takes the values 0, 1, 2 and 3. Here, $\ell = 4$, which is not allowed. Therefore, the quantum numbers (3, 4, 1, 1/2) cannot describe an electron in an orbital.
   b) (3, 1, 2, 1/2): $n = 3$, $\ell = 1$, $m_\ell = 2$ and $m_s = 1/2$
      For $\ell = 1$, m_ℓ takes the values –1, 0 and 1. Here, $m_\ell = 2$, which is not allowed. Therefore, the quantum numbers (3, 1, 2, 1/2) cannot describe an electron in an orbital.
   c) (3, 0, 0, –1/3): $n = 3$, $\ell = 0$, $m_\ell = 0$ and $m_s = 1/3$
      The spin quantum number m_s takes only the values 1/2 and –1/2. The value –1/3 is not allowed for m_s. Therefore, the quantum numbers (3, 0, 0, –1/3) cannot describe an electron in an orbital;
   d) (3, 2, 1, 1/2): $n = 3$, $\ell = 2$, $m_\ell = 1$ and $m_s = 1/2$
      For $n = 3$, ℓ takes the values 0, 1 and 2. Here, $\ell = 2$, which is allowed.
      For $\ell = 2$, m_ℓ takes the values –2, –1, 0, 1 and 2. Here, $m_\ell = 1$, which is allowed.
      m_s takes only the values 1/2 and –1/2. Here, $m_s = 1/2$, which is allowed. Therefore, the quantum numbers (3, 2, 1, 1/2) can describe an electron in an orbital.

2. a) (2, 1, 1, 1/2): $n = 2$, $\ell = 1$, $m_\ell = 1$ and $m_s = 1/2$
      For $n = 2$, ℓ takes the values 0 and 1. Here, $\ell = 1$, which is allowed.
      For $\ell = 1$, m_ℓ takes the values –1, 0 and 1. Here, $m_\ell = 1$, which is allowed.
      m_s takes only the values 1/2 and –1/2. Here, $m_s = 1/2$, which is allowed. Therefore, the quantum numbers (2, 1, 1, 1/2) can describe an electron in an orbital.
   b) (3, 2, 1, –1/2): $n = 3$, $\ell = 2$, $m_\ell = 1$ and $m_s = -1/2$
      For $n = 3$, ℓ takes the values 0, 1 and 2. Here, $\ell = 2$, which is allowed.

For $\ell = 2$, m_ℓ takes the values –2, –1, 0, 1 and 2. Here, $m_\ell = 1$, which is allowed. Therefore, the quantum numbers (3, 2, 1, –1/2) can describe an electron in an orbital.

   c) (4, 3, –2, –1/4): n = 4, $\ell = 3$, $m_\ell = -2$ and $m_s = -1/4$

The spin quantum number m_s takes only the values 1/2 and –1/2. The value –1/4 is not allowed for m_s. Therefore, the quantum numbers (4, 3, –2, –1/4) cannot describe an electron in an orbital;

   d) (3, 1, 0, –1/2): n = 3, $\ell = 1$, $m_\ell = 0$ and $m_s = -1/2$

For n = 3, ℓ takes the values 0, 1 and 2. Here, $\ell = 1$, which is allowed.

For $\ell = 1$, m_ℓ takes the values –1, 0 and 1. Herein, $m_\ell = 0$ which is allowed.

The spin quantum number m_s takes only the values 1/2 and –1/2. Here, $m_s = -1/2$, which is allowed. Therefore, the quantum numbers (3, 1, 0, –1/2) can describe an electron in an orbital.

3. When $\ell = 0$, we have the s sublevel,
    When $\ell = 1$, we have the p sublevel,
    When $\ell = 2$, we have the d sublevel,
    When $\ell = 3$, we have the f sublevel.
    The name of the sub-level with n = 3 and $\ell = 2$ is 3d

4. ℓ takes the values from 0 to (n – 1).
    For n = 3, $\ell = 0$, 1 and 2.

5. For n = 2, $\ell = 0$ and 1 (two possible values)

6. The orbital quantum number m_ℓ takes the values from $-\ell$ to ℓ, including zero.
    When $\ell = 1$, m_ℓ takes the values –1, 0 and 1.

7. The number of sublevels in a principal level is given by the angular momentum quantum number ℓ.
    For n = 2, $\ell = 0$ and 1. There are two possible values for ℓ, so there are two sublevels for n = 2 (sublevel 2s and sublevel 2p).

8. The number of orbitals in a sublevel is given by the magnetic quantum number m_ℓ.
    For n = 2, $\ell = 0$ and 1.
    For $\ell = 0$, $m_\ell = 0$ (one orbital)
    For $\ell = 1$, $m_\ell = -1$, 0 and 1 (three orbitals)
    Therefore, there are (1 + 3 =) 4 orbitals in the principal level n = 2.

9. For p sublevel, $\ell = 1$ and $m_\ell = -1$, 0 and 1. Therefore, there are three orbitals in the p sublevel.

10. For f sublevel, $\ell = 3$ and $m_\ell = -3$, –2, –1, 0, 1, 2 and 3. Therefore, there are seven orbitals in the f sublevel. Each orbital can hold two electrons. Therefore, the f sublevel can hold $(7 \times 2 =)$ 14 electrons.

11. For the sublevel d, $\ell = 2$ and $m_\ell = -2$, –1, 0, 1 and 2. Therefore, there are five orbitals in the d sublevel. Each orbital can hold two electrons. The d orbitals are found for principal levels n > 2. The d orbitals look like eggs stacked on one another (a) or like a dumbbell but with a donut around the middle (b).

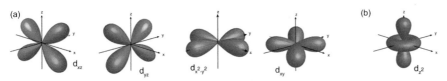

a) Sublevel d like eggs stacked on one another (image from Wikimedia Commons) and b) Sublevel d with dumbbell-like shape, with donut around the middle (image from Wikimedia Commons).

The quantum number that takes only the values –1/2 and 1/2 is the electron spin quantum number (m_s). The magnetic quantum number (m_ℓ) takes the values between $-\ell$ and ℓ, including zero.

Q6.5

1. The general pattern for filling the sublevels from the lowest to the highest energy is:
    1s < 2s < 2p < 3s < 3p < 4s < 3d < 4p < 5s < 4d < 5p < 6s < 4f < 5d < 6p < 7s …

and it could be obtained easily from the following scheme:

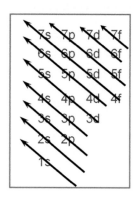

2. a)

1s   2s   2p

    Two electrons in the same orbital 2s have the same spin quantum number. This distribution does not respect Pauli's exclusion principle. Therefore, this electron distribution is incorrect.

    b)

1s   2s   2p

    An orbital in the 2p sublevel is empty; however, electrons are paired in this sublevel. According to Hund's rule, electrons remained impaired as far as it is possible. This distribution does not respect Hund's rule. Therefore, this electron distribution is incorrect.

    c)

1s   2s   2p

    There are three electrons in the same orbital 1s. According to the Pauli's exclusion principle, an orbital can hold a maximum of two electrons with opposite spin quantum numbers. This distribution does not respect Pauli's exclusion principle. Therefore, this electron distribution is incorrect.

    d)

1s   2s   2p

    There are two electrons with opposite spin quantum numbers in each orbital. This distribution respects the Aufbau principle, Pauli's exclusion principle and Hund's rule. Therefore, this electron distribution is correct.

3. Chlorine $_{17}Cl$ contains 17 electrons. The electronic configuration of chlorine is $1s^2 2s^2 2p^6 3s^2 3p^5$

4. (and 5). There are two main *exceptions* to *electron configuration* which are chromium and copper:

$_{24}Cr$: $1s^2 2s^2 2p^6 3s^2 3p^6 4s^1 3d^5$ (the d sublevel is half filled)

$_{29}Cu$: $1s^2 2s^2 2p^6 3s^2 3p^6 4s^1 3d^{10}$ (the d sublevel is completely filled)

An *electron* from the 4s orbital moves to a 3d orbital, resulting in a completely filled d sublevel in the case of Cu and a half-filled d sublevel in the case of Cr, leading to more stable atoms.

6. For transition metals, cations are obtained by removing electrons from the ns orbital before any electrons in the (n − 1)d orbitals:

$_{26}Fe$: $1s^2 2s^2 2p^6 3s^2 3p^6 4s^2 3d^6$

$_{26}Fe^{2+}$: $1s^2 2s^2 2p^6 3s^2 3p^6 3d^6$ (two electrons are removed from 4s).

7. Magnesium contains twelve electrons. The magnesium cation $_{12}Mg^{2+}$ contains ten electrons. The electronic configuration of $_{12}Mg^{2+}$ is $1s^2 2s^2 2p^6$

8. Oxygen atom contains eight electrons. The oxygen anion $_8O^{2-}$ contains ten electrons. The electronic configuration of $_8O^{2-}$ is $1s^2 2s^2 2p^6$

9. $_{20}$**Ca^{2+}:**$1s^22s^22p^63s^23p^6$

$3s$ $\quad$ $3p$

All orbitals are completely filled. There are no unpaired electrons in Ca^{2+}

10. $_{22}$Ti: $1s^22s^22p^63s^23p^64s^23d^2$

The electronic configuration of titanium shows an incomplete sublevel d. Therefore, Ti is a transition metal. For transition metals, cations are obtained by removing electrons from the ns orbital before any electrons in the (n − 1)d orbitals

$_{22}$Ti^{2+}: $1s^22s^22p^63s^23p^63d^2$ (two electrons are removed from 4s).

$3d$

There are two unpaired electrons in Ti^{2+}

11. $_{11}$**Na:**$1s^22s^22p^63s^1$

$3s$

There is one unpaired electron in Na

12. $_8$O^{2-}:$1s^22s^22p^6$

$_{12}$Mg^{2+}:$1s^22s^22p^6$

$_9$F$^-$:$1s^22s^22p^6$

The ions $_8$O^{2-}, $_{12}$Mg^{2+} and $_9$F$^-$ have the same electronic configuration, i.e., they have isoelectronic configurations.

13. To find the short-cut electronic configuration of an element, find the noble gas that has an atomic number just lower than that of the element. For zinc, the atomic number is Z = 30. The noble gas that has an atomic number just lower than that of zinc is argon Ar (Z = 18). Therefore, the shortcut electronic configuration of $_{30}$Zn is [$_{18}$Ar] $3d^{10}4s^2$.

14. $_{17}$Cl: $1s^22s^22p^63s^23p^5$

The valence shell of chlorine is 3s3p which contains seven valence electrons.

15. $_{35}$Br: $1s^22s^22p^63s^23p^64s^23d^{10}4p^5$

The valence shell of bromine is 4s4p which contains seven valence electrons.

16. $_{28}$Ni: $1s^22s^22p^63s^23p^64s^23d^8$

The electronic configuration of nickel shows an incomplete sublevel d. Therefore, Ni is a transition metal. For transition metals, the valence shell is ns(n − 1)d. Therefore, the valence shell of nickel is 4s3d which contains ten valence electrons.

17. $_{11}$Na: $1s^22s^22p^63s^1$

The valence shell of sodium is 3s which contains one valence electron.

18. $_{11}$Na$^+$: $1s^22s^22p^6$

The valence electrons in sodium cation Na$^+$ are situated in the sublevels 2s2p. The number of valence electrons is eight.

19. $_{26}$Fe: $1s^22s^22p^63s^23p^64s^23d^6$

The electronic configuration of iron (Fe) shows an incomplete sublevel d. Therefore, Fe is a transition metal. For transition metals, the valence shell is ns(n − 1)d. Therefore, the valence shell of iron is 4s3d which contains eight valence electrons.

20. $_{35}$Br: $1s^22s^22p^63s^23p^64s^23d^{10}4p^5$

Note that bromine contains a completely filled d sublevel, so bromine is not a transition metal. The valence shell of bromine is 4s4p which contains seven valence electrons.

21. A transition metal is an element whose atom has an incomplete d subshell or which can give rise to cations with an incomplete d subshell.

22. $_{30}$Zn: $1s^22s^22p^63s^23p^64s^23d^{10}$

Zinc contains a completely filled d sublevel. Zinc is not a transition metal.

23. $_{22}$Ti: $1s^22s^22p^63s^23p^64s^23d^2$

Titanium (Ti) contains an incompletely filled d sublevel. Ti is a transition metal.

24.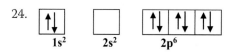

The sublevel 2s is not filled with electrons; however, sublevel 2p, which has a higher energy than 2s, is filled. According to the Aufbau principle, electrons enter orbitals in order of ascending energy. Thus, this distribution does not respect the Aufbau principle. Therefore, this electron distribution is incorrect.

Calculations

Q6.6

The wavelength of this radiation is

$$\lambda = \frac{C}{\nu} = \frac{3.00 \times 10^8}{4.57 \times 10^{14}} = 6.56 \times 10^{-7} \text{ m} = 656 \text{ nm}.$$

Q6.7

$$E_{photon} = h \frac{c}{\lambda}$$

$$E_{photon} = 6.63 \times 10^{-34} \frac{3.00 \times 10^8}{434 \times 10^{-9}} = 4.58 \times 10^{-19} \text{J}$$

Q6.8

The electron moves from a lower energy orbit ($n_{initial} = 1$) to a higher energy orbit ($n_{final} = 4$). It absorbs a photon (light).

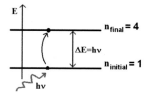

Transition of an electron from n = 1 to n = 4

$$\Delta E_{atom} = h \frac{C}{\lambda} = -2.18 \times 10^{-18} \times \left(\frac{1}{n_{final}^2} - \frac{1}{n_{initial}^2} \right)$$

$$\Delta E_{atom} = h \frac{C}{\lambda} = -2.18 \times 10^{-18} \times \left(\frac{1}{4^2} - \frac{1}{1^2} \right)$$

$$\Delta E_{atom} = -2.04 \times 10^{-18} \text{ J}$$

Note that there is no negative energy. Here, the negative energy (sign –) means that the electron loses energy and the sign (–) should not be taken into account when calculating the wavelength of the emitted light (there is no negative wavelength):

$$\lambda = \frac{hC}{\Delta E_{atom}} = \frac{6.63 \times 10^{-34} \times 3.00 \times 10^8}{2.04 \times 10^{-18}} = 9.75 \times 10^{-8} \text{ m} = 97.5 \text{ nm}$$

Q6.9

For the hydrogenic species:

$$\frac{1}{\lambda} = R \times Z^2 \left(\frac{1}{n_1^2} - \frac{1}{n_2^2} \right)$$

where Z is the atomic number of the hydrogenic species and $n_2 > n_1$.
Lithium $(Z = 3)$ is a hydrogenic species

$$\frac{1}{\lambda} = R \times Z^2 \left(\frac{1}{n_1^2} - \frac{1}{n_2^2} \right) = 1.09677 \times 10^7 \times 3^2 \times \left(\frac{1}{1^2} - \frac{1}{4^2} \right) = 9.25400 \times 10^7 \text{ m}^{-1}$$

The wavelength of the line corresponding to the transition of electron from $n = 1$ to $n = 4$ in the lithium $(_3\text{Li})$ atom is

$$\lambda = \frac{1}{9.25400 \times 10^7} = 1.08061 \times 10^{-8} \text{m} = 11 \text{nm}$$

Q6.10

The uncertainty in position of the electron is $10\% = 0.1$, so

$$\Delta x = 0.529 \times 10^{-10} \times 0.1 = 5.29 \times 10^{-12} \text{m}.$$

From the uncertainty equation $\left(\Delta x \cdot m\Delta V \geq \dfrac{h}{4\pi} \right)$, we deduce the uncertainty in the velocity of the electron ΔV:

$$\Delta V \geq \frac{h}{4\pi m \Delta x}$$

$$\Delta V \geq \frac{6.63 \times 10^{-34}}{4\pi \times 9.11 \times 10^{-31} \times 5.29 \times 10^{-12}}$$

$$\Delta V \geq 1.09 \times 10^7 \text{m s}^{-1}.$$

Note that if the velocity of an electron is 6×10^6 m s^{-1}:

$$\Delta V \geq \frac{1.09 \times 10^7}{6 \times 10^6} \times 100 = 177.7\%$$

This means that we cannot know both the position and the speed of a particle at a given time with the same accuracy. Indeed, when the position is known with a relatively high accuracy
$(\Delta x = 10\%)$, the velocity is not known with the same accuracy $(\Delta V \geq 177.7\%)$.

Periodic Table of Elements and Properties of Atoms

7

7.1 OBJECTIVES

At the end of the present chapter, the student will be able to:

1. Explain how elements are arranged in the periodic table of elements.
2. Determine the position of an element in the periodic table.
3. Recognize the different categories of elements in the periodic table of elements.
4. Describe the properties of atoms and how these properties vary along a period or along a group in the periodic table of elements.

7.2 THE PERIODIC TABLE OF ELEMENTS

In the periodic table of elements, the elements are arranged in order of **increasing atomic number (Z). The vertical columns** of the table (numbered from 1 to 18) are called **groups** or **families**. Elements in the same group have similar but not identical characteristics.

The horizontal rows of the table (numbered from 1 to 7) are called **periods** (Figure 7.1).

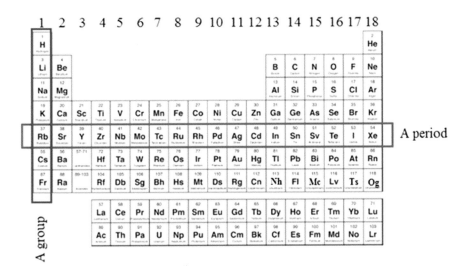

FIGURE 7.1   The periodic table of elements

Elements in the same group have the same number of valence electrons (same number of electrons in the outermost shell), and hence similar chemical properties. The valence electrons are the electrons that are transferred or shared when atoms bond together.

Examples:

$_1$H: 1s^1
$_3$Li: [He] 2s^1
$_{11}$Na: [Ne] 3s^1

DOI: 10.1201/9781003257059-7

$_{19}$K: [Ar] 4s^1
$_{37}$Rb: [Kr] 5s^1
$_{55}$Cs: [Xe] 6s^1
$_{87}$Fr: [Rn] 7s^1

All these elements have one valence electron, and they belong to Group 1.

Each period contains elements with the same valence shell.

Examples:

$_5$B: 1s^2**2s^{2}2p^1**
$_6$C: 1s^2**2s^{2}2p^2**
$_7$N: 1s^2**2s^{2}2p^3**
$_8$O: 1s^2**2s^{2}2p^4**
$_9$F: 1s^2**2s^{2}2p^5**
$_{10}$Ne: 1s^2**2s^{2}2p^6**

All these elements have the same valence shell 2s2p and they belong to the second period.

In the periodic table of elements, the elements can be arranged according to their electronic configuration. For example, the position of an element in the periodic table can be defined easily using its electronic configuration, and more specifically, using its valence shell and valence electrons. Indeed, the number of valence electrons indicates the group in which the element is situated, and the valence shell gives the period in which the element is situated (Figure 7.2):

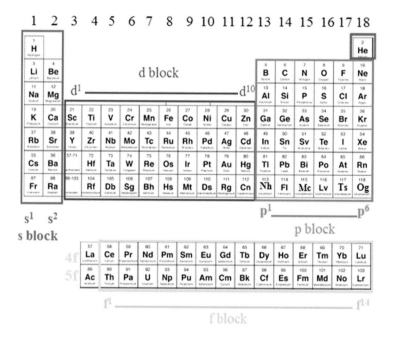

FIGURE 7.2   The s, p, d, and f blocks in the periodic table of elements

Practice 7.1 In which group and period are the elements $_{17}$Cl, $_{20}$Ca, $_{41}$Nb and $_{58}$Ce situated?

Answer:

○ $_{17}$Cl: 1s^{2}2s^{2}2p^6**3s^{2}3p^5**

The valence shell is 3s3p. Cl is situated in period 3.

The number of valence electrons is seven. There are ten groups of block d. Therefore, Cl is situated in the group (7+10=) 17. Note that there are five electrons in sublevel p; therefore, Cl is situated in Group 5 of block p (which is the same as Group 17 of the periodic table).

○ $_{20}$Ca: $1s^2 2s^2 2p^6 3s^2 3p^6 \mathbf{4s^2}$

The valence shell is 4s. Ca is situated in period 4.

The number of valence electrons is two. There are two electrons in sublevel s, therefore, Ca is situated in Group 2 of block s (which is the same as Group 2 of the periodic table).

○ $_{41}$Nb: $1s^2 2s^2 2p^6 3s^2 3p^6 4s^2 3d^{10} 4p^6 \mathbf{5s^2 4d^3}$

Nb has an incomplete d subshell. The valence shell is 5s4d. Nb is situated in period 5.

The number of valence electrons is five. There are ten groups of block d, so that Nb is situated in Group 5. Note that there are three electrons in sublevel d, so that Nb is situated in Group 3 of block d (which is the same as the Group 5 of the periodic table).

○ $_{58}$Ce: $1s^2 2s^2 2p^6 3s^2 3p^6 4s^2 3d^{10} 4p^6 5s^2 4d^{10} 5p^6 6s^2 \mathbf{4f^1 5d^1}$

The valence shell is 6s4f5d. Ce is situated in the first period of block f (the second period of block f is 5f). There is one electron in the sublevel f, therefore, Ce is situated in Group 1 of block f. Note that the electronic configuration of lanthanum is La: $[Xe]5d^1 6s^2$ and the electronic configuration of cerium is Ce: $[Xe]4f^1 5d^1 6s^2$. Therefore, in the periodic table, cerium appears next to lanthanum (Figure 7.3).

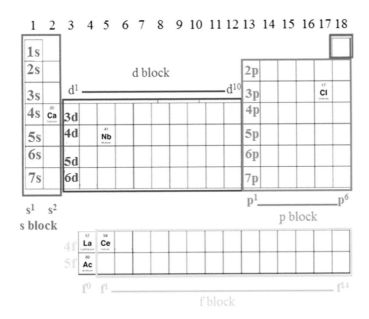

FIGURE 7.3   Positions of the elements (Cl, Ca, Nb and Ce) in the periodic table of elements, determined using their electronic configurations

7.3 METALS, NON-METALS, AND METALLOIDS

Elements in the periodic table of elements can be divided into three main categories: metals, non-metals and metalloids (Figure 7.4). Most elements are metals. Metalloids (B, Si, Ge, As, Sv, Te and Po) have properties of both metals and non-metals. The physical and chemical properties of metals, non-metals and metalloids are summarized in Table 7.1.

Metals can be divided into alkali metals, alkaline earth metals, transition metals, lanthanides and actinides. Lanthanides and actinides form the rare earth elements. Non-metals include halogens and noble gases (Figure 7.5).

TABLE 7.1   Physical and chemical properties of metals, non-metals, and metalloids

Metals	Non-metals	Metalloids
Physical properties	**Physical properties**	**Physical properties**
Generally, solids at room temperature.	Generally, solids or liquids at room temperature.	Generally, solids at room temperature.
Good conductors of heat and electricity	Poor conductors of heat and electricity	Conduct heat and electricity better than non-metals, but not as well as metals.
High density	Low density	
High melting point	Low melting point	
Ductile	Not ductile and break easily	Ductile
Malleable	Not malleable	Malleable
Chemical properties	**Chemical properties**	**Chemical properties**
Easily lose electrons	Gain electrons	Chemically, they behave mostly as non-metals. They can form alloys with metals.
Corrode easily (e.g., iron rusting)	When two or more non-metals bond with each other, they form a covalent compound.	Readily form glasses
Since metals tend to lose electrons and non-metals tend to gain electrons, metals and non-metals form ionic compounds with each other, such as Fe_2O_3 and NaCl.		React with the halogens to form compounds

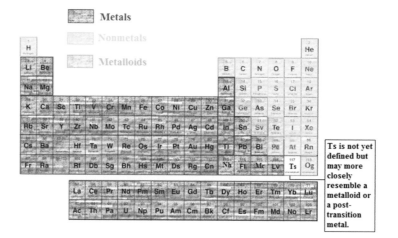

FIGURE 7.4   Metals, non-metals, and metalloids in the periodic table of elements

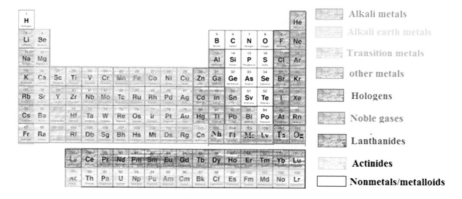

FIGURE 7.5   Metals are divided into alkali metals, alkaline earth metals, transition metals, lanthanides and actinides. Non-metals include halogens and noble gases.

7.3.1 ALKALI METALS

Atoms of the alkali metals present a valence shell (ns) and have one valence electron. The alkali family is found in the first column (Group 1) of the periodic table of elements. They include, among others, lithium (Li), sodium (Na) and potassium (K). Note that alkali metals do not include hydrogen, even though hydrogen belongs to the first group. Hydrogen is a nonmetal. The valence shell of the alkali family is incomplete (ns^1). They are the most reactive metals. Elements that are reactive bond easily with other elements to make compounds. That is why, in nature, the alkali elements are not found free; they are always bonded with another element.

7.3.2 ALKALINE EARTH METALS

Alkaline earth metals present a valence shell (ns) and have two valence electrons. They are found in Group 2 of the periodic table of elements. The valence shell of the alkaline earth family is completely filled (ns^2). Alkaline earth metals include, among others, beryllium (Be), magnesium (Mg) and calcium (Ca).

7.3.3 TRANSITION METALS

Transition metals are elements where the atom has an incomplete d subshell, or which can give rise to cations with an incomplete d subshell. Transition metals are found from Group 3 to Group 12 in the periodic table of elements. They include iron (Fe), copper (Cu), nickel (Ni), gold (Au) and cobalt (Co), among others. They are good conductors of heat and electricity. They can lose many electrons when they form bonds with other atoms. Many transition metals combine chemically with oxygen to form oxides.

7.3.4 HALOGEN FAMILY

Halogens are found in Group 17. Halogens are fluorine (F), chlorine (Cl), bromine (Br), iodine (I), astatine (At) and tennessine (Ts). Halogens present a valence shell (ns np) and have seven valence electrons (ns^2np^5). They require only one additional electron to form a full octet. This characteristic makes them more reactive than other non-metal groups. The name "halogen" means "salt-producing". Indeed, when halogens react with metals, they form many salts such as NaCl, KBr, CaF and KI.

7.3.5 NOBLE GASES

The last group of the periodic table of elements (group 18) is the group of noble gases, which are helium (He), neon (Ne), argon (Ar), krypton (Kr), xenon (Xe), radon (Ra) and oganesson (Og). Noble gases present a valence shell (ns np) and have eight valence electrons, except for He which shows the valence shell electrons $1s^2$. Thus, their valence shell is completely filled (ns^2np^6 or $1s^2$ for He). For this reason, they are inactive and because they do not readily combine with other elements to form compounds, the noble gases are also called inert gases.

7.3.6 LANTHANIDES AND ACTINIDES

Rare earth elements referred to as the lanthanide series are a group of 15 elements, from lanthanide (La; atomic number 57) to lutetium (Lu; atomic number 71), plus scandium and yttrium which exhibit similar properties to the lanthanides. They fill their 4f sublevel progressively.

Actinides, from actinium (Ac, atomic number 89) to lawrencium (Lr, atomic number 103), are typical metals which fill their 5f sublevel progressively and have properties of both the d-block and the f-block elements. Furthermore, they are radioactive.

GET SMART

WHAT MAKES AN ELEMENT REACTIVE?
An element is reactive when its valence shell is incomplete. Indeed, all atoms (except hydrogen) want to have eight electrons in their very outermost energy level. So, atoms bond until this level is complete. Atoms with few valence electrons lose them during bonding. Atoms with six or seven valence electrons gain electrons during bonding.

7.4 PROPERTIES OF ATOMS

Several properties of atoms depend on the strength of attraction between the valence electrons and the nucleus of the atom. This attraction will depend on:

○ The positive charge that attracts the other valence electrons.
○ The distance of electrons from the nucleus.

7.4.1 EFFECTIVE NUCLEAR CHARGE AND SHIELDING EFFECT

The effective nuclear charge is the net charge an electron experiences in a polyelectronic atom (with multiple electrons). In other words, the effective nuclear charge is the amount of attraction exerted on the valence electrons by the positive nucleus. The effective nuclear charge can be approximated by the equation:

$$Z_{eff} = Z - S$$

where Z is the atomic number of the atom and S is the number of shielding electrons, also called inner electrons or core electrons. The inner electrons have a repelling effect on the valence electrons. This effect, called shielding effect, reduces the full attractive force of the positive nucleus exerted on the valence electrons. When the number of inner electrons increases, repulsions increase pushing valence electrons further out. Conversely, when the number of inner electrons decreases, repulsions decrease, and the valence electrons become closer to the nucleus.

Practice 7.2 Calculate the effective nuclear charge for $_{11}Na$ and $_{11}Na^+$.

Answer:
$_{11}Na$: $1s^2 2s^2 2p^6 3s^1$, Z = 11; the number of valence electrons is one and the number of inner electrons is S = 10, so $Z_{eff} = 11 - 10 = 1$.
$_{11}Na^+$: $1s^2 2s^2 2p^6$, Z = 11; the number of valence electrons is eight and the number of inner electrons is S = 2, so $Z_{eff} = 11 - 2 = 9$.
Therefore, the valence electron in Na is less attracted by the nucleus than the valence electrons in Na$^+$.
The effective nuclear charge of some elements which belong to the same period and other elements of the same group are summarized in Table 7.2.

TABLE 7.2   Effective nuclear charge Z_{eff} of some elements of the same period and other elements of the same group

Elements of the same period			Elements of the same group		
Element/electronic configuration	S	Z_{eff}	Element/electronic configuration	S	Z_{eff}
$_{11}Na$: $1s^2 2s^2 2p^6 3s^1$	10	1	$_3Li$: $1s^2 2s^1$	2	1
$_{13}Al$: $1s^2 2s^2 2p^6 3s^2 3p^1$	10	3	$_{11}Na$: $1s^2 2s^2 2p^6 3s^1$	10	1
$_{15}P$: $1s^2 2s^2 2p^6 3s^2 3p^3$	10	5	$_{19}K$: $1s^2 2s^2 2p^6 3s^2 3p^6 4s^1$	18	1
$_{17}Cl$: $1s^2 2s^2 2p^6 3s^2 3p^5$	10	7	$_{37}Rb$: $1s^2 2s^2 2p^6 3s^2 3p^6 4s^2 3d^{10} 4p^6 5s^1$	36	1

7.4.2 ATOMIC SIZE

The atomic size (or radius r) is defined as the half-distance (d/2) of closest approach between two identical atoms (Figure 7.6).

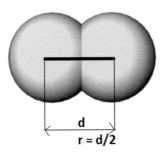

FIGURE 7.6   Atomic size (or radius r) of an atom (molecule drawn with Molview software).

Elements of the same period have the same valence shell and the same number of shielding (core) electrons. Therefore, when we move from the left to the right along a period, the number of protons increases, and the electrons added go to the same energy level as the previous atom. As a result, the effective nuclear charge (Z_{eff}) increases. This means that the valence electrons are attracted closer to the nucleus resulting in a smaller size of the atom.

When we move from the top to the bottom along a group in the periodic table of elements, electrons are added to a new higher energy level, putting them farther from the nucleus. The increase in the core electrons increases the repulsions, pushing the valence electrons further out. The increase of the number of core electrons is counterbalanced by the increase of the number of protons and therefore the effective nuclear charge remains constant. However, the fact that the valence electrons are farther from the nucleus results in the increase of the atomic size.

When a neutral atom loses one or more electrons (a cation), the number of protons exceeds the number of electrons. As a result, the attraction force between the positive nucleus and the valence electrons increases. That is, the valence electrons are attracted closer to the nucleus resulting in a smaller size of the cation. Therefore, a cation is always smaller than the atom from which it is formed. However, an anion is larger than the atom from which it is formed. Indeed, adding electrons, to a neutral atom, results in greater repulsion between the electrons, and, consequently, the ionic radius increases.

To summarize:

1. When we move from the left to the right along a period, in the periodic table of elements, the atomic size decreases.
2. When we move from the top to the bottom along a group in the periodic table, the atomic size increases.
3. A cation is smaller than the atom from which it is formed.
4. An anion is larger than the atom from which it is formed.

Practice 7.3 For each set of elements, indicate the element which has the greatest atomic radius.

a) $_6$C and $_9$F
b) $_3$Li and $_{19}$K
c) Na and Na$^+$
d) Cl and Cl$^-$

Answer:

a) $_6$C: $1s^2 2s^2 2p^2$, C belongs to period 2 and $_9$F: $1s^2 2s^2 2p^5$, F belongs also to period 2, with F having a greater atomic number Z than C. Thus, F is found to the right of C when moving along period 2. Alternatively, when we move from the left to the right along a period, the atomic size decreases; therefore, C has a greater atomic radius than F.

b) $_3$Li: $1s^2 2s^1$, Li belongs to Group 1 and $_{19}$K: $1s^2 2s^2 2p^6 3s^2 3p^6 4s^1$, K belongs also to Group 1, with K having a greater atomic number than Li. Thus, Li is found at the top of Group 1 and K at the bottom of the same group. Alternatively, when we move from the top to the bottom along a group, the atomic size increases, so that K has a greater atomic radius than Li.

c) Na (which is a neutral atom) has a greater atomic size than the cation Na$^+$. Indeed, the atomic radii of Na and Na$^+$ are 1.57 Å and 0.98 Å, respectively.

d) The anion Cl⁻ has a greater atomic size than the neutral atom Cl. Indeed, the atomic radii of Cl⁻ and Cl are 1.81 Å and 0.99 Å, respectively.

7.4.3 IONIZATION ENERGY

The ionization energy (denoted by I) is the amount of energy it takes to detach one electron from a neutral gas atom. An atom has low ionization energy if it is easy to detach an electron from it. The larger the atom, the easier it is to detach an electron. An atom has high ionization energy if it is difficult to detach an electron from it. The smaller the atom is, the closer the valence electrons are from the nucleus and the harder it is to detach an electron from the atom. In other words, atoms with a high atomic radius have a lower ionization energy than atoms with a low atomic radius.

The amount of energy it takes to detach one electron from a neutral gas atom (X) is called the first ionization energy (I_1):

$$X_{(g)} \rightarrow X^+_{(g)} + 1e^- \left(\textbf{First ionization energy } \mathbf{I_1}\right).$$

The amount of energy it takes to detach one electron from a gas cation (X⁺) is called the second ionization energy (I_2):

$$X^+_{(g)} \rightarrow X^{2+}_{(g)} + 1e^- \left(\textbf{Second ionization energy } \mathbf{I_2}\right).$$

Note that the second ionization energy is not the energy it takes to remove two electrons at the same time but it is the energy it takes to remove one electron from an atom that has already lost one electron before.

The amount of energy it takes to detach one electron from a divalent gas cation (X²⁺) is called the third ionization energy (I_3):

$$X^{2+}_{(g)} \rightarrow X^{3+}_{(g)} + 1e^- \left(\textbf{Third ionization energy } \mathbf{I_3}\right).$$

When an atom or an ion loses an electron, the attraction between the nucleus and the electrons increases and the atom/ion becomes smaller. Therefore, for a gaseous atom ($X_{(g)}$), we have, in order of decreasing radius: $X_{(g)} > X^+_{(g)} > X^{2+}_{(g)}$, or atoms with high atomic radius have lower ionization energy, so that, in order of increasing ionization energy, we have $I_1 < I_2 < I_3$.

Note that, when we move from the left to the right along a period, in the periodic table of elements, the atomic size decreases, the attraction between the nucleus and the valence electrons increases and it becomes harder to remove an electron from the atom. Consequently, the ionization energy increases. When we move from the top to the bottom along a group, the number of levels increases so that the atomic radius increases and the attraction between the nucleus and the valence shell electrons decreases. Consequently, it is easier to remove an electron from the atom and, therefore, the ionization energy decreases.

To summarize:

1. Atoms with a high atomic radius have a lower ionization energy.
2. When we move from the left to the right along a period, the ionization energy increases.
3. When we move from the top to the bottom along a group, the ionization energy decreases.
4. $I_1 < I_2 < I_3$.

Practice 7.4 For each pair of elements, indicate the element that has the higher ionization energy.

a) $_6$C and $_9$F
b) $_3$Li and $_{19}$K

Answer:

a) Atoms with a higher atomic radius have a lower ionization energy. Fluorine (F) has a higher atomic radius than carbon (C), therefore F has a lower ionization energy than C.

b) Lithium (Li) has a lower atomic radius than potassium (K), therefore Li has a higher ionization energy than C.

7.4.4 ELECTRON AFFINITY

The electron affinity of an atom, designated by A, is defined as the amount of energy released or spent when an electron is added to a neutral gaseous atom to form an anion:

$$X_{(g)} + 1e^- \rightarrow X^-_{(g)}$$

When it is hard to attach an electron to an atom, the electron affinity of the atom is low. However, when an atom has a high electron affinity, it is easy to add an electron to this atom. In other words, atoms that have high electron affinities are more likely to gain electrons than atoms with low electron affinities.

The electron affinities have generally negative values since the atom releases energy but, in a few cases, they can have positive values. The more negative the electron affinity value, the greater is the atom's affinity for electrons. For example, when an electron is added to a neutral atom ($X_{(g)} + 1e^- \rightarrow X^-_{(g)}$; first electron affinity) energy is released. Thus, the first electron affinity has a negative value. However, more energy is required to add an electron to an anion ($X^-_{(g)} + 1e^- \rightarrow X^{2-}_{(g)}$; second electron affinity) and the second electron affinity has a positive value. Indeed, the energy needed to add an electron to the anion is more than the energy released from the electron attachment process.

> **Example:** $O_{(g)} + 1e^- \rightarrow O^-_{(g)}$: first electron affinity $A_1 = -34$ kcal/mol < 0.
> $O^-_{(g)} + 1e^- \rightarrow O^{2-}_{(g)}$: second electron affinity $A_2 = 171$ kcal/mol > 0.

The electron affinity of an atom depends on many parameters, such as the atomic number, the distance between the nucleus and the subshell (s, p, d, or f) to which the extra electron is added and the effective nuclear charge of the atom. Generally, the absolute value of the electron affinity increases when we move from the left to the right along a period. However, some irregularities are observed such as the decrease of the electron affinity (in absolute value) when we move from carbon C ($A = -28.8$ kcal mol^{-1}) to nitrogen N ($A = -4.6$ kcal mol^{-1}). When we move from the top to the bottom, along a group, the absolute value of the electron affinity decreases.

Note that halogens present a high electron affinity since, when an electron is added to a halogen, its electronic configuration becomes the same as the noble gas that comes just after it in terms of atomic number. Thus, the halogen becomes chemically very stable since it has a complete valence shell. For the same reason, the electron affinity of noble gases is null ($A = 0$ kcal mol^{-1}). Metals can easily lose their valence electrons to form cations and, when an electron is added to a metal element, energy is needed to gain that electron ($A > 0$). Metals are known to have low electron affinities. However, non-metals can easily gain electrons and their electron affinities are higher than those of metals.

7.4.5 ELECTRONEGATIVITY

The electronegativity (denoted by E_N) is the ability of an atom to attract a pair of electrons towards itself in a chemical bond. That atom A is more electronegative than atom B means that, when A and B form a chemical bond, the pair of electrons is attracted towards atom A. Therefore, the electronegativity of an atom can only be assessed relative to that of another element.

> **Example:** Cl is more electronegative than H (H :Cl; the pair of electrons is attracted towards Cl); however, Cl is less electronegative than F (F: Cl; the pair of electrons is attracted towards F).

According to Milliken, the electronegativity of an atom depends on its ionization energy and electron affinity. Indeed, when an atom has a high ionization energy and a high electron affinity, it strongly retains its electrons, and it can readily attract electrons from another atom. As a result, its electronegativity is high. For example, metals lose their electrons easily, so they have low electro-negativities. However, non-metals tend to attract electrons, so they have high electronegativities.

Milliken proposed the following equation to determine the electronegativity (E_N) of an atom:

$$E_N = k\left(I + |A|\right)$$

where $k = 0.18$ if the ionization energy (I) and the electron affinity (A) of the atom are expressed in eV atom^{-1}.

Note that Pauling proposed another equation, based on the bond energy in the diatomic molecule AB, for determining the difference of electronegativity ($\Delta E_N = E_N(A) - E_N(B)$) between two atoms A and B:

$$\Delta E_N = k\left[E_{A-B} - \sqrt{(E_{A-A} \times E_{B-B})}\right]$$

where E_{A-B} is the bond energy between the atoms A and B, E_{A-A} and E_{B-B} are the bond energies A-A and B-B, respectively and $k = 0.0208$ if the bond energies are expressed in kcal mol^{-1}. The bond energy is defined as the amount of energy required to break apart a mole of molecules into its component atoms. Note that when the atoms A and B have the same electronegativity:

$$\Delta E_N = 0 \; and \; E_{A-B} = \sqrt{(E_{A-A} \times E_{B-B})}$$

This means that, when the atoms A and B have the same electronegativity, the bond energy E_{A-B} equals the average of the bond energies A-A and B-B. When A and B have different electronega-tivities, the bond energy $E_{A-B} > \sqrt{(E_{A-A} \times E_{B-B})}$.

When we move from the left to the right along a period in the periodic table of elements, the electronegativity increases. Indeed, since there is an increase in the effective nuclear charge, there is a greater attraction of the outer shell electrons to the nucleus. When we move from the top to the bottom along a group in the periodic table of elements, the electrons are further from the nucleus and there is a weaker attraction. Thus, the electronegativity decreases.

Practice 7.5 Arrange the following elements $_{12}$Mg, $_{16}$S, $_{13}$Al and $_{17}$Cl from the highest to the lowest in terms of:

a) **electron affinity**
b) **ionization energy**
c) **electronegativity**

Answer:
The elements $_{12}$Mg, $_{16}$S, $_{13}$Al and $_{17}$Cl belong to the same period and in order of increasing atomic number (from the left to right along the period), we have:

$$_{12}Mg < _{13}Al < _{16}S < _{17}Cl$$

a)   From the highest to the lowest affinity, we have $_{12}$Mg $< _{13}$Al $< _{16}$S $< _{17}$Cl
b)   From the highest to the lowest ionization energy, we have $_{12}$Mg $< _{13}$Al $< _{16}$S $< _{17}$Cl
c)   From the highest to the lowest electronegativity, we have $_{12}$Mg $< _{13}$Al $< _{16}$S $< _{17}$Cl
     Note that chlorine Cl has the highest affinity, so it can attract electrons of another atom easily. Chlorine has also the highest ionization energy, so it strongly retains its electrons. Consequently, Cl has the highest electronegativity.

7.4.6 ACIDIC AND BASIC TRENDS OF OXIDES AND HYDRIDES

Oxygen and hydrogen are very reactive and form compounds with most of the other elements. Except for noble gases, elements form binary compounds with oxygen called oxides (MxOy).

Metals form basic oxides, e.g., Na_2O which dissolves in water to form basic solutions:

$$Na_2O + H_2O \rightarrow 2Na^+ + 2OH^-$$

Non-metals form acidic oxides, e.g., SO_3 which dissolves in water to form acidic solutions:

$$SO_3 + 2H_2O \rightarrow HSO_4^- + H_3O^+$$

Some oxides, such as Al_2O_3, which is insoluble in water, may show an amphoteric character. An amphoteric compound is a molecule or an ion that can react both as an acid and as a base. For example, Aluminum oxide Al_2O_3 shows a basic character in the presence of H^+ ions and an acidic character in the presence of OH^- ions. Indeed, aluminum oxide reacts with hot dilute hydrochloric acid to give aluminum chloride and water according to the following reaction:

$$Al_2O_3 + 6HCl \rightarrow 2AlCl_3 + 3H_2O \quad \left(\text{Basic character}\right)$$

And it reacts with sodium hydroxide solution to produce sodium aluminate and water according to the following reaction:

$$Al_2O_3 + 2NaOH \rightarrow 2NaAlO_2 + H_2O \quad \left(\text{Acid character}\right)$$

Note that the same element can form different oxides with oxygen that may have different acidities. For example, vanadium (V) can form four oxides (VO, V_2O_3, VO_2 and V_2O_5). The VO and V_2O_3 oxides are basic, VO_2 is an amphoteric oxide and V_2O_5 is acidic.

When we move from the right to the left along a period or when we move from the top to the bottom along a group in the periodic table of elements, the alkalinity of oxides increases (and their acidity decreases). The more metallic an element (the metallic character of an element depends on its ability to lose the valence electrons), the more basic is its oxide.

Elements can also form binary compounds with hydrogen called hydrides (MxHy), e.g., LiH, CaH_2, B_2H_6, H_2Se and HBr. The alkali metals (Group 1) and some of the alkali earth metals (such as Mg, Ca, Sr, and Ba) form ionic hydrides. The ionic hydrides such as LiH and CaH_2 show a basic character. Hydrides formed with elements of Groups 13 to 17, such as B_2H_6, CH_4, NH_3, H_2Te and HI, show a covalent structure. Their acidity increases with the atomic number Z. When we move from the left to the right along a period or when we move from the top to the bottom along a group in the periodic table of elements, the acidity of hydrides increases (and their basicity decreases).

GET SMART

WHAT ARE ACIDS AND BASES?

There are three definitions of acids and bases:

1. Arrhenius definition: An acid is a substance that dissociates in water to give H_3O^+ ions and a base is a substance that gives OH^- when dissociated in water.
2. Bronsted-Lowry definition: An acid is a species that can donate a proton (H^+) and a base is a species that can accept a proton.
3. Lewis definition: An acid is an electron pair accepter and a base is an electron pair donor.

The pH ($10^{-pH} = [H^+]$) of the solution expresses the degree of acidity. When pH = 7, the solution is neutral. When the pH increases to 14, the basicity increases. When the pH decreases to 0, the acidity increases.

Practice 7.6 Using the periodic table of elements, determine

a) **which is the more acidic oxide, Na_2O or Cl_2O_7?**
b) **which is the more acidic hydride, HF or HCl?**

Answer:

    a) Sodium (Na) and chlorine (Cl) belong to the same period (period 3). The atomic number of Na is 11 and the atomic number of Cl is 17. Sodium is at the left and chlorine is at the right of the period. When we move from the left to the right along a period, the acidity of oxides increases. Therefore, Cl_2O_7 is more acidic than Na_2O.

    b) Fluorine (F) and chlorine (Cl) belong to the same group (Group 17). The atomic number of F is 9 and the atomic number of Cl is 17. Fluorine is at the top and chlorine is at the bottom of the group. When we move from the top to the bottom along a group, the acidity of hydrides increases. Therefore, HCl is more acidic than HF.

CHECK YOUR READING

How are elements arranged in the periodic table of elements?

What do elements in the same group have in common?

What do elements in the same period have in common?

How can the position of an element in the periodic table be defined?

What are the different categories of elements in the periodic table of elements? How are these categories distinguished?

What are the different properties of atoms? How do these properties vary along a period or along a group in the periodic table of elements?

What are the periodic trends of the acidity and basicity of oxides and hydrides?

SUMMARY OF CHAPTER 7

In the periodic table of elements, elements are arranged in order of **increasing atomic number (Z).**

The vertical columns of the table (numbered from 1 to 18) are called **groups or families**.

The horizontal rows of the table (numbered from 1 to 7) are called **periods**.

Elements in the same group have the same number of valence electrons (same number of electrons in the outermost shell), and hence similar chemical properties.

Elements in the same period have the same valence shell organization.

The number of valence electrons indicates the group in which the element is situated, and the valence shell gives the period in which the element is situated.

    Example: $_{13}Al$: $1s^22s^22p^6\mathbf{3s^23p^1}$

    The valence shell is 3s3p. Al is situated in period 3. The number of valence electrons is three. There are ten groups of block d. Therefore, Cl is situated in the group (3+10=) 13. Note that there is one electron in sublevel p, therefore Al is situated in Group 1 of block p (which is the same as group 13 of the periodic table).

Elements in the periodic table of elements can be divided into three main categories: **metals, non-metals and metalloids**. Most elements are metals. Metalloids have properties of both metals and nonmetals. Non-metals include halogens and noble gases.

Metals can be divided into alkali metals, alkaline earth metals, transition metals, lanthanides and actinides.

    Alkali metals: (ns^1). The alkali family (i.e., Li, Na and K) is found in Group 1 of the periodic table of elements.

    Alkaline earth metals: (ns^2). They are found in Group 2 of the periodic table (i.e., B, Mg and Ca).

    Transition metals are elements where the atoms have an incomplete d subshell, or which can give rise to cations with an incomplete d subshell. Transition metals (i.e., Fe, Cu and Ni) are found from Group 3 to Group 12 in the periodic table of elements.

    Halogens (ns^2np^5) are found in Group 17. Halogens are F, Cl, Br, I, At and Ts. Halogens are very reactive elements.

Noble gases: (ns^2np^6) are found in the last group of the periodic table of elements (group 18). Their valence shell is completely filled. They include He, Ne, Ar, Kr, Xe and Ra.

Lanthanides are a group of 15 elements, from lanthanide (atomic number 57) to lutetium (atomic number 71) in the periodic table of elements, plus scandium and yttrium. They fill their 4f sublevel progressively.

Actinides, from actinium (atomic number 89) to lawrencium (atomic number 103) in the periodic table, are typical metals which fill their 5f sublevel progressively.

Properties of atoms:

The effective nuclear charge Z_{eff} is the net charge that an electron experiences in a poly-electronic atom (with multiple electrons). The effective nuclear charge is approximated by the equation: $Z_{eff} = Z - S$ where Z is the atomic number of the atom and S is the number of shielding electrons, also called inner electrons or core charges.

The inner electrons have a repelling effect on the valence electrons called **shielding effect**. This effect reduces the full attractive force of the positive nucleus exerted on the valence electrons.

The atomic size (or radius) is defined as the half-distance of the closest approach between two identical atoms.

The ionization energy (denoted by I) is the amount of energy it takes to detach one electron from a neutral gas atom. An atom has a low ionization energy if it is easy to detach an electron from it. An atom has a high ionization energy if it is difficult to detach an electron from it. Atoms with high atomic radius have lower ionization energy.

$$X_{(g)} \rightarrow X^+_{(g)} + 1e^- \left(\text{First ionization energy } I_1 \right).$$

$$X^+_{(g)} \rightarrow X^{2+}_{(g)} + 1e^- \left(\text{Second ionization energy } I_2 \right).$$

$$X^{2+}_{(g)} \rightarrow X^{3+}_{(g)} + 1e^- \left(\text{Third ionization energy } I_3 \right).$$

$I_1 < I_2 < I_3$.

The electron affinity of an atom, designated by A, is defined as the amount of energy released or spent when an electron is added to a neutral gaseous atom to form an anion:

$$X_{(g)} + 1e^- \rightarrow X^-_{(g)}$$

When it is hard to attach an electron to an atom, the electron affinity of the atom is low. Atoms that have high electron affinities are more likely to gain electrons than atoms with low electron affinities.

The first electron affinity has a negative value. However, the second electron affinity has a positive value.

Example: $O_{(g)} + 1e^- \rightarrow O^-_{(g)}$: first electron affinity $A_1 = -34$ kcal mol$^{-1} < 0$.
$O^-_{(g)} + 1e^- \rightarrow O^{2-}_{(g)}$: second electron affinity $A_2 = 171$ kcal mol$^{-1} > 0$.

The electronegativity (denoted by E_N) is the ability of an atom to attract a pair of electrons towards itself in a chemical bond.

Example: Cl is more electronegative than H (H: Cl; the pair of electrons is attracted towards Cl); however, Cl is less electronegative than F (F: Cl; the pair of electrons is attracted towards F).

When we move from the left to the right along a period in the periodic table of elements, the effective nuclear charge, the ionization energy, the electron affinity and the electronegativity all increase. However, the atomic radius decreases. Note that some irregularities are observed for the electron affinity, such as the decrease (in absolute value) when we move from C (A = −28.8 kcal mol^{-1}) to N (A = −4.6 kcal mol^{-1}).

When we move from the top to the bottom of a column, along a group in the periodic table of elements, the ionization energy, the electron affinity and the electronegativity all decrease. However, the atomic radius increases. The effective nuclear charge remains constant.

The evolution of the atomic properties along a group or a period in the periodic table of elements are summarized in Figure 7.7.

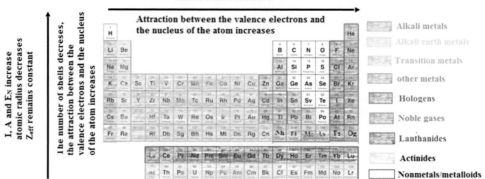

FIGURE 7.7   Evolution of the atomic properties along a group or a period in the periodic table of elements

PRACTICE ON CHAPTER 7

Q7.1   **Complete the following sentences**

In the periodic table of elements, elements are arranged in order of _____ atomic number (Z). The vertical columns of the table (numbered from 1 to 18) are called _____ or _____ Elements in the same _____ have the same number of valence electrons and hence similar chemical properties. The horizontal rows of the table (numbered from 1 to 7) are called .... Each period contains elements with the same _____

Elements in the periodic table of elements can be divided into three main categories that are _____ and _____ Metals are generally _____ at room temperature and can _____ easily electrons. Non-metals are generally _____ or _____ at room temperature and they _____ electrons. Metalloids are generally _____ at room temperature and they can form compounds with _____ .

Q7.2   **Match the followings**

a. Alkali metals	**1.** have an incomplete d subshell
b. Alkaline earth metals	**2.** are situated in the last group.
c. Transition metals	**3.** fill their 5f sublevel progressively
d. Halogens	**4.** fill their 4f sublevel progressively
e. Noble gases	**5.** have the valence shell electrons ns^2np^5.
f. Lanthanide	**6.** have the valence shell electrons ns^2
g. Actinides	**7.** have the valence shell electrons ns^1

Q7.3   **Choose the correct answer**

1.   **What do elements in the same period have in common?**
   a) Atomic mass
   b) Number of valence electrons
   c) Valence shell
   d) Total number of electrons

2.   **What do elements in the same group have in common?**
   a) Atomic mass
   b) Number of valence electrons

    c) Valence shell

    d) Total number of electrons

3. **What family of elements is in the right-most column of the periodic table?**

    a) Halogens

    b) Alkali metals

    c) Transition metals

    d) Noble gases

4. **What family of elements is in the left-most column of the periodic table?**

    a) Halogens

    b) Alkali metals

    c) Alkali earth metals

    d) Noble gases

5. **What is the name of the family of elements that has a full valence shell?**

    a) Halogens

    b) Alkali metals

    c) Alkali earth metals and noble gases

    d) Transition metals

6. **Elements can be divided into metals, non-metals and metalloids.**

    a) True

    b) False

7. **Elements of the same group have comparable chemical properties.**

    a) True

    b) False

8. **Halogens and noble gases are metals.**

    a) True

    b) False

9. **Lanthanides and actinides are metalloids.**

    a) True

    b) False

Q7.4   Choose the correct answer

1. **In the periodic table of elements, sodium $_{11}$Na is situated in**

    a) Group 1 and period 2.

    b) Group 1 and period 3.

    c) Group 3 and period 1.

    d) The last group and period 3

2. **In the periodic table of elements, zinc $_{30}$Zn is situated in**

    a) Group 12 and period 4.

    b) Group 10 and period 3.

    c) Group 4 and period 12.

    d) Group 3 and period 7.

3. **In the periodic table of elements, chlorine $_{17}$Cl is situated in**

    a) Group 5 and period 3.

    b) Group 3 and period 5.

    c) Group 17 and period 5.

    d) Group 17 and the period 3.

4. **In the periodic table of elements, xenon $_{54}$Xe is situated in**

    a) Group 5 and period 7.

    b) Group 18 and period 4.

    c) Group 18 and period 5.

    d) Group 15 and period 3.

5. **Which one of the following sets of elements belongs to the alkali metals?**

    a) $_{22}$Ti, $_{24}$Cr and $_{26}$Fe.

    b) $_{4}$Be, $_{12}$Mg and $_{20}$Ca.

    c) $_{3}$Li, $_{11}$Na and $_{19}$K.

    d) $_{9}$F, $_{17}$Cl and $_{35}$Br.

6. **Which one of the following sets of elements belongs to the transition metals?**
   a) $_{22}$Ti, $_{24}$Cr and $_{26}$Fe.
   b) $_{4}$Be, $_{12}$Mg and $_{20}$Ca.
   c) $_{3}$Li, $_{11}$Na and $_{19}$K.
   d) $_{9}$F, $_{17}$Cl and $_{35}$Br.

7. **Which one of the following sets of elements belongs to the alkali earth metals?**
   a) $_{22}$Ti, $_{24}$Cr and $_{26}$Fe.
   b) $_{4}$Be, $_{12}$Mg and $_{20}$Ca.
   c) $_{3}$Li, $_{11}$Na and $_{19}$K.
   d) $_{9}$F, $_{17}$Cl and $_{35}$Br.

8. **Which one of the following sets of elements belongs to the halogens?**
   a) $_{22}$Ti, $_{24}$Cr and $_{26}$Fe.
   b) $_{4}$Be, $_{12}$Mg and $_{20}$Ca.
   c) $_{3}$Li, $_{11}$Na and $_{19}$K.
   d) $_{9}$F, $_{17}$Cl and $_{35}$Br.

9. **Which one of the following sets of elements belongs to the noble gases?**
   a) $_{22}$Ti, $_{24}$Cr and $_{26}$Fe.
   b) $_{2}$He, $_{10}$Ne and $_{18}$Ar.
   c) $_{3}$Li, $_{11}$Na and $_{19}$K.
   d) $_{9}$F, $_{17}$Cl and $_{35}$Br.

10. **What is the chemical symbol of hydrogen?**
    a) Hy
    b) He
    c) H
    d) Hn

11. **Hydrogen belongs to the alkali metals.**
    a) True
    b) False

12. **Zinc ($_{30}$Zn) is a transition metal.**
    a) True
    b) False

Q7.5   Match each atomic property (on the left side) by its definition (on the right side)

a. Electron affinity	**1.** The net charge an electron experiences in a polyelectronic atom
b. Atomic radius	**2.** The amount of energy it takes to detach one electron from a neutral gas atom.
c. Electronegativity	**3.** The half-distance of the closest approach between two identical atoms
d. Ionization energy	**4.** The ability of an atom to attract a pair of electrons towards itself in a chemical bond.
e. Effective nuclear charge	**5.** The amount of energy released or spent when an electron is added to a neutral gas atom to form an anion

Q7.6   Choose the correct answer

1. **In the periodic table of elements, when we move from the top to the bottom in a group, the effective nuclear charge**
   a) increases
   b) decreases
   c) remains constant
   d) increases or decreases, depending on the atomic number Z.

2. **In the periodic table of elements, when we move from the top to the bottom in a group, the atomic size**
   a) increases
   b) decreases
   c) remains constant
   d) increases or decreases, depending on the atomic number Z.

3. **In the periodic table of elements, when we move from the left to the right in a period, the ionization energy**
    a) increases
    b) decreases
    c) remains constant
    d) increases or decreases, depending on the atomic number Z.

4. **In the periodic table of elements, when we move from the right to the left in a period, the electronegativity**
    a) increases
    b) decreases
    c) remains constant
    d) increases or decreases, depending on the atomic number Z.

5. **In the periodic table of elements, when we move from the bottom to the top in a group, the electron affinity**
    a) increases
    b) decreases
    c) remains constant
    d) increases or decreases, depending on the atomic number Z.

6. **The correct order from the lowest to the highest ionization energy is**
    a) Second ionization energy I_2 < first ionization energy I_1 < third ionization energy I_3.
    b) Second ionization energy I_2 < third ionization energy I_3 < first ionization energy I_1.
    c) First ionization energy I_1 < second ionization energy I_2 < third ionization energy I_3.
    d) Third ionization energy I_3 < second ionization energy I_2 < first ionization energy I_1.

7. **Which of the following observations about the electron affinity are correct?**
    a) The second electron affinity is positive, and the first electron affinity is negative.
    b) The second electron affinity is negative, and the first electron affinity is positive.
    c) The first and the second electron affinities are both positive.
    d) The first and the second electron affinities are both negative.

8. **The effective nuclear charge (Z_{eff}) of oxygen ($_8O$) is**
    a) 2
    b) 4
    c) 6
    d) 8

9. **The effective nuclear charge (Z_{eff}) of oxygen anion ($_8O^{2-}$) is**
    a) 2
    b) 4
    c) 6
    d) 8

10. **The effective nuclear charge (Z_{eff}) of magnesium ($_{12}Mg$) is**
    a) 2
    b) 4
    c) 6
    d) 10

11. **The effective nuclear charge (Z_{eff}) of magnesium cation ($_{12}Mg^{2+}$) is**
    a) 2
    b) 4
    c) 6
    d) 10

12. **Arrange the elements $_3Li$, $_{11}Na$, $_{19}K$ and $_{37}Rb$ from the highest to the lowest electronegativity.**
    a) $_3Li > _{11}Na > _{19}K > _{37}Rb$
    b) $_{37}Rb > _{19}K > _{11}Na > _3Li$
    c) $_3Li > _{19}K > _{11}Na > _{37}Rb$
    d) $_{37}Rb > _{11}Na > _{19}K > _3Li$

13. **Arrange the elements $_3$Li, $_{11}$Na, $_{19}$K and $_{37}$Rb from the highest to the lowest in terms of atomic size (radius).**
    a) $_3$Li $> _{11}$Na $> _{19}$K $> _{37}$Rb
    b) $_{37}$Rb $> _{19}$K $> _{11}$Na $> _3$Li
    c) $_3$Li $> _{19}$K $> _{11}$Na $> _{37}$Rb
    d) $_{37}$Rb $> _{11}$Na $> _{19}$K $> _3$Li

14. **Arrange the elements $_3$Li, $_{11}$Na, $_{19}$K and $_{37}$Rb from the lowest to the highest in terms of ionization energy.**
    a) $_3$Li $< _{11}$Na $< _{19}$K $< _{37}$Rb
    b) $_{37}$Rb $< _{19}$K $< _{11}$Na $< _3$Li
    c) $_3$Li $< _{19}$K $< _{11}$Na $< _{37}$Rb
    d) $_{37}$Rb $< _{11}$Na $< _{19}$K $< _3$Li

15. **Arrange the elements $_{13}$Al, $_{14}$Si, $_{15}$P and $_{16}$S from the lowest to the highest in terms of atomic radius.**
    a) $_{13}$Al $< _{14}$Si $< _{15}$P $< _{16}$S
    b) $_{13}$Al $< _{15}$P $< _{14}$Si $< _{16}$S
    c) $_{16}$S $< _{15}$P $< _{14}$Si $< _{13}$Al
    d) $_{16}$S $< _{14}$Si $< _{15}$P $< _{13}$Al

16. **Arrange the elements $_{13}$Al, $_{14}$Si, $_{15}$P and $_{16}$S from the highest to the lowest in terms of electronegativity.**
    a) $_{13}$Al $> _{14}$Si $> _{15}$P $> _{16}$S
    b) $_{13}$Al $> _{15}$P $> _{14}$Si $> _{16}$S
    c) $_{16}$S $> _{15}$P $> _{14}$Si $> _{13}$Al
    d) $_{16}$S $> _{14}$Si $> _{15}$P $> _{13}$Al

17. **Elements that show high electron affinity and high ionization energy have a low electronegativity.**
    a) True
    b) False

18. **Halogens such as $_{17}$Cl, $_9$F and $_{53}$I present a low electron affinity.**
    a) True
    b) False

19. **An atom has low ionisation energy if it is easy to detach an electron from it.**
    a) True
    b) False

20. **Atoms with high atomic radius have low ionization energy.**
    a) True
    b) False

21. **Metals lose electrons, non-metals gain electrons and metalloids chemically behave mostly as non-metals.**
    a) True
    b) False

ANSWERS TO QUESTIONS

Q7.1

In the periodic table of elements, elements are arranged in order of **increasing** atomic number (Z). The vertical columns of the table (numbered from 1 to 18) are called **groups** or **families**. Elements in the same **group** have the same number of valence electrons and hence similar chemical properties. The horizontal rows of the table (numbered from 1 to 7) are called **periods**. Each period contains elements with the same **valence shell**.

Elements in the periodic table of elements can be divided into three main categories that are **metals, non-metals,** and **metalloids**. Metals are generally **solids** at room temperature and can easily **lose** electrons. Non-metals are generally **solids** or **liquids** at room temperature, and they **gain** electrons. Metalloids are generally **solids** at room temperature, and they can form compounds with **halogens**.

Q7.2 (a, 7); (b, 6); (c, 1); (d, 5); (e, 2); (f, 4); (g,3)

Q7.3

1. c
2. b
3. d
4. b
5. c
6. a
7. a
8. b
9. b

Q7.4

1. b
2. a
3. d
4. c
5. c
6. a
7. b
8. d
9. b
10. c
11. b
12. b

Q7.5 (a, 5); (b, 3); (c, 4); (d, 2); (e, 1)

Q7.6

1. c
2. a
3. a
4. b
5. b
6. c
7. a
8. c
9. c
10. a
11. d
12. a
13. b
14. b
15. c
16. c
17. b
18. b
19. a
20. a
21. a

KEY EXPLANATIONS

Q7.3

3. The right-most column of the periodic table is column 18. This column contains the Group 18 elements, the noble gases family.
4. The left-most column of the periodic table is the first column. This column contains the Group 1 elements, the alkali metals.
5. Halogens are F, Cl, Br, I, At and Ts. Halogens present a valence shell (ns np) and have seven valence electrons (ns^2np^5).

6. Noble gases present a valence shell (ns np) and have eight valence electrons. Thus, their valence shell is completely filled (ns^2np^6). For this reason, they are inactive and, because they do not readily combine with other elements to form compounds, the noble gases are called inert gases. Noble gases include He, Ne, Ar, Kr, Xe, Ra and Og.

7. Halogens (Group 17) and noble gases (Group 18) are both non-metals.

8. Lanthanides and actinides are both metals.

Q7.4

1. $_{11}$Na: [Ne] $\mathbf{3s^1}$
   The valence shell is 3s. Na is situated in period 3.
   The number of valence electrons is one. There is one electron in sublevel s, therefore Na is situated in the first group of block s (which is the same as Group 1 of the periodic table of elements).

2. $_{30}$Zn: $1s^22s^22p^63s^23p^6\mathbf{4s^23d^{10}}$
   The valence shell is 4s3d. Zn is situated in period 4.
   The number of valence electrons is twelve. Therefore, Zn is situated in Group 12. Note that there are ten electrons in sublevel d, therefore Zn is situated in Group 10 of block d (which is the same as Group 12 of the periodic table of elements).

3. $_{17}$Cl: [Ne] $\mathbf{3s^23p^5}$
   The valence shell is 3s3p. Cl is situated in period 3.
   The number of valence electrons is seven. There are ten groups of block d. Therefore, Cl is situated in the group (7+10 =) 17. Note that there are five electrons in sublevel p, therefore Cl is situated in Group 5 of block p (which is the same as Group 17 of the periodic table of elements).

4. $_{54}$Xe: $1s^22s^22p^63s^23p^64s^23d^{10}4p^65s^24d^{10}\mathbf{5p^6}$
   The valence shell is 5p. Xe is situated in period 5. Note that there are six electrons in sublevel p, therefore Xe is situated in Group 6 of block p, which is the same as Group 18 of the periodic table of elements. Indeed, there are ten groups of block d and two groups of block s before the six groups of block p. Therefore, Xe is situated in Group (2+10+6 =) 18.

5. Atoms of the alkali metals present a valence shell (ns) and have one valence electron (ns^1):
   $_3$Li: $1s^22s^1$
   $_{11}$Na: $1s^22s^22p^63s^1$
   $_{19}$K: $1s^22s^22p^63s^23p^64s^1$
   Li, Na and K belong to the alkali family. The alkali family is found in Group 1 of the periodic table of elements.

6. Transition metals are elements in which the atoms have an incomplete d subshell, or they can give rise to cations with an incomplete d subshell:
   $_{22}$Ti: $1s^22s^22p^63s^23p^64s^23d^2$
   $_{24}$Cr: $1s^22s^22p^63s^23p^64s^13d^5$
   $_{26}$Fe: $1s^22s^22p^63s^23p^64s^23d^6$
   Ti, Cr and Fe are transition metals. Transition metals are found from Group 3 to Group 9 in the periodic table of elements.

7. Alkaline earth metals present a valence shell (ns) and have two valence electrons (ns^2). Their valence shell (ns) is completely filled:
   $_4$Be: $1s^22s^2$
   $_{12}$Mg: $1s^22s^22p^63s^2$
   $_{20}$Ca: $1s^22s^22p^63s^23p^64s^2$
   Alkaline earth metals are found in Group 2 of the periodic table of elements. The valence shell of the alkali family is completely filled. They include beryllium (Be), Mg and calcium (Ca), among others.

8. Halogens present a valence shell (ns np) and have seven valence electrons (ns^2np^5):
   $_9$F: $1s^22s^22p^5$
   $_{17}$Cl: $1s^22s^22p^63s^23p^5$
   $_{35}$Br: $1s^22s^22p^63s^23p^64s^23d^{10}4p^5$

Halogens are found in Group 17 in the periodic table of elements. Halogens are F, Cl, Br, I, At and Ts.

9. Noble gases present a valence shell (ns np) and have eight valence electrons. Thus, their valence shell is completely filled (ns^2np^6), except for He ($1s^2$):

   $_2$He: $1s^2$

   $_{10}$Ne: $1s^22s^22p^63s^23p^5$

   $_{18}$Ar: $1s^22s^22p^63s^23p^5$

   The last group of the periodic table of elements (Group 18) contains the noble gases. They include He, Ne, Ar, Kr, Xe and Ra.

10. There are no elements with the chemical symbols Hy and Hn. He is the chemical symbol of helium and H is the chemical symbol of hydrogen.

11. Alkali metals do not include hydrogen, even though hydrogen belongs to Group 1. Hydrogen is a non-metal.

12. Transition metals are elements whose atoms have an incomplete d subshell, or they can give rise to cations with an incomplete d subshell:

    $_{30}$Zn: $1s^22s^22p^63s^23p^64s^23d^{10}$

    The d subshell is fully complete. Note that, for Zn^{2+}, the electrons are removed from the subshell 4s since, for the d block elements, sublevel ns has greater energy than sublevel (n-1)d:

    $_{30}Zn^{2+}$: $1s^22s^22p^63s^23p^63d^{10}$

    According to the above-mentioned definition of transition metal, Zn does not count as a transition metal.

Q7.6

8. The effective nuclear charge $Z_{eff} = Z - S$ where Z is the atomic number of the element and S is the number of inner electrons:

   $_8$O: $1s^22s^22p^4$; the number of valence electrons is six and $Z = 8$, so $S = 2$ and $Z_{eff} = 8 - 2 = 6$

9. $_8O^{2-}$: $1s^22s^22p^6$; the number of valence electrons is eight, the number of inner electrons is $S = 2$ and $Z = 8$, so $Z_{eff} = 8 - 2 = 6$.

10. $_{12}$Mg: $1s^22s^22p^63s^2$; the number of valence electrons is two, the number of inner electrons is $S = 10$ and $Z = 12$, so $Z_{eff} = 12 - 10 = 2$.

11. $_{12}Mg^{2+}$: $1s^22s^22p^6$; the number of valence electrons is eight, the number of inner electrons is $S = 2$ and $Z = 12$, so $Z_{eff} = 12 - 2 = 10$.

12. $_3$Li, $_{11}$Na, $_{19}$K and $_{37}$Rb belong to the same group, Group 1 ($_3$Li is at the top and $_{37}$Rb is at the bottom of this group). Electronegativity decreases from the top to the bottom along a group; therefore, in order of decreasing electronegativity: $_3$Li $> _{11}$Na $> _{19}$K $> _{37}$Rb .

13. The atomic radius decreases from the bottom to the top along a group, therefore in order of decreasing atomic radius: $_{37}$Rb $> _{19}$K $> _{11}$Na $> _3$Li.

14. Ionization energy decreases from the top to the bottom along a group, therefore in order of increasing ionization energy: $_{37}$Rb $< _{19}$K $< _{11}$Na $< _3$Li.

15. $_{13}$Al, $_{14}$Si, $_{15}$P and $_{16}$S belong to the same period ($_{13}$Al is at the left and $_{16}$S is at the right of this period). The atomic radius decreases from the left to the right along a period. Therefore, in order of increasing atomic radius we have: $_{16}$S $< _{15}$P $< _{14}$Si $< _{13}$Al.

16. The electronegativity increases from the left to the right along a period. Therefore, in order of decreasing electronegativity we have: $_{16}$S $> _{15}$P $> _{14}$Si $> _{13}$Al.

17. According to Milliken, the electronegativity of an atom depends on its ionization energy and electron affinity. Indeed, when an atom has a high ionization energy and a high electron affinity, it strongly retains its electrons, and it can readily attract electrons of another atom. As a result, its electronegativity is high.

18. Halogens such as $_{17}$Cl, $_9$F and $_{53}$I present a high electron affinity since, when an electron is added to a halogen, its electronic configuration becomes the same as the noble gas that comes just after it in the periodic table. Thus, this halogen becomes chemically very stable since it has a complete valence shell.

19. The ionization energy is the amount of energy it takes to detach one electron from a neutral gas atom. An atom has a low ionisation energy if it is easy to detach an electron from it.

20. The smaller the atom, the closer the valence electrons are from the nucleus and the harder it is to detach an electron from the atom. The larger the atom, the easier it is to detach an electron. In other words, atoms with a high atomic radius have lower ionization energy than atoms with a low atomic radius.

Chemical Bonding and Molecular Geometry

8

8.1 OBJECTIVES

At the end of the present chapter, the student will be able to:

1. Recognize the different types of chemical bonding.
2. Draw the Lewis structure of different atoms and ions.
3. Recognize the different molecular geometries.
4. Determine the geometry of molecules.

8.2 CHEMICAL BONDING

Chemical bonding (intramolecular forces) is defined as the attractive forces that hold two or more chemical constituents (atoms and ions) together in different chemical species. The electrons involved in bonding are usually those in the outermost (valence) shell.

There are three types of chemical bonds:

1. **Covalent bonds** result from sharing electrons between the atoms. The covalent bond is usually found between non-metals. Depending on the electronegativities of the atoms involved in the bond, the covalent bond can be polar or non-polar (Figure 8.1). Polar covalent bonds occur when the difference in electronegativity between the two constituent atoms is between 0.4 and 2.

   Examples of polar covalent bonds: The bond between hydrogen and chlorine in HCl is a polar covalent bond. Chlorine (Cl) is more electronegative than hydrogen (H). The shared pair of electrons between Cl and H is attracted toward the chlorine atom. Because of this, the hydrogen atom has a partially positive charge (δ^+) and the chlorine atom has a partially negative charge (δ^-). In addition, the bond between hydrogen and bromine in HBr, the bond between hydrogen and fluorine in HF and the bond between hydrogen and oxygen in water H_2O are polar covalent bonds.

   Examples of non-polar covalent bonds: The bond between two iodine atoms in I_2 is a non-polar covalent bond. The bonds between hydrogen (H) and carbon (C) in CH_4 are non-polar covalent bonds.

   Note that the non-polar covalent bond is found in homonuclear molecules such as Br_2, Cl_2, O_2, I_2 and in hydrocarbons (C_nH_{2n+2}).

$$ I \text{ ———} \overset{\textbf{.}}{\underset{\textbf{.}}{} } \text{——— } I \qquad\qquad H \text{ ———} \overset{\textbf{.}}{\underset{\textbf{.}}{} } \text{——— } Cl $$

FIGURE 8.1 Non-polar covalent bond in I_2 and polar covalent bond in HCl

    Note that the greater the difference in electronegativities between the atoms of the molecule, the more ionic the bond is.

2. **An ionic bond** results from the transfer of electrons from a metal to a non-metal. The ionic bond involves the attraction between two oppositely charged ions (an anion and a cation). Electrons are transferred so that the anion and cation each have a noble gas electron configuration.

DOI: 10.1201/9781003257059-8

Examples: The bond between sodium (Na) and chlorine (Cl) in NaCl is an ionic bond (Figure 8.2). Electrons are transferred between Na and Cl to form ions and the bond is established between Na$^+$ and Cl$^-$. The NaCl compound obtained is an ionic compound.

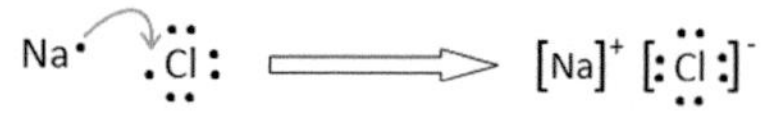

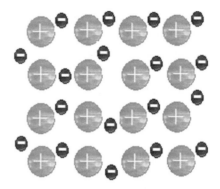

FIGURE 8.2   Ionic bond in NaCl: electrons are transferred between Na and Cl to form ions and the bond is established between Na$^+$ and Cl$^-$.

3. **A metallic bond** is formed between positively charged atoms in which the free electrons are shared among a structure of positively charged ions (Figure 8.3). Metallic bonding is the main type of chemical bond that forms between metal atoms. Metallic bonds are weaker than ionic and covalent bonds.

FIGURE 8.3   Metallic bond: metal atoms share their electrons.

8.3 LEWIS DOT REPRESENTATIONS OF ATOMS

Lewis dot representation or an electron dot diagram is a simplistic way of showing the valence electrons of an atom, that uses dots around the chemical symbol of the element with no more than two dots on a side. The valence electrons are electrons that are transferred or involved in chemical bonding.

To represent the Lewis dot diagram of an atom, follow the following steps:

1. Write out the electronic configuration of the atom or the ion
   Example: $_{17}$Cl: $1s^22s^22p^63s^23p^5$
2. Find the number of valence electrons
   Example: The number of valence electrons in Cl is seven.
3. Draw the valence electrons as dots around the chemical symbol of the atom or the ion.
   Example: The seven valence electrons of Cl are drawn in the following configuration (Figure 8.4):

FIGURE 8.4   Lewis dot diagram of chlorine Cl.

Note that:

1. Elements in the same group of the periodic table of elements have similar Lewis electron dot diagrams because they have the same valence shell electron configuration.
   Example: $_7$N: $1s^22s^22p^3$
   $_{15}$P: $1s^22s^22p^63s^23p^3$

$_7$N and $_{15}$P belong to the same group (Group 15) and have the same number of valence electrons, which is five. They have similar Lewis dot diagrams. The Lewis dot diagrams of these atoms are represented in Figure 8.5.

$$\cdot\overset{\cdot\cdot}{\underset{}{N}}\cdot \qquad \cdot\overset{\cdot\cdot}{\underset{}{P}}\cdot$$

FIGURE 8.5   Lewis dot diagrams of atoms N and P.

2. Electron dot diagrams for ions are like those of atoms, except that some electrons have been removed from a neutral atom to obtain a cation, while some electrons have been added to a neutral atom to obtain an anion. Conventionally, when we draw electron dot diagrams for ions, we show the original valence shell of the atom and not the valence shell of the ion.

Example 1: $_{11}$Na: $1s^22s^22p^63s^1$
$_{11}$Na$^+$: $1s^22s^22p^6$

The Lewis dot diagrams of Na atom and Na$^+$ cation are, respectively (Figure 8.6):

$$\overset{\cdot}{Na} \qquad Na^+$$

FIGURE 8.6   Lewis dot diagrams of Na atom and Na$^+$ cation.

Note that for Na$^+$, there are no dots around the Na symbol since we show the original valence shell of the atom 3s, which does not contain electrons, and not the valence shell 2s 2p of the ion which contains eight electrons.

In comparing the electron configurations and electron dot diagrams for the Na atom and the Na$^+$ cation, we note that the Na atom has a single valence electron in its Lewis dot diagram, while the Na$^+$ cation has lost that one valence electron.

Example 2: $_8$O: $1s^22s^22p^4$
$_8$O^{2-}: $1s^22s^22p^6$

The Lewis dot diagrams of O atom and O^{2-} anion are as follows (Figure 8.7):

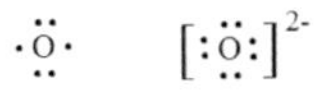

$$\cdot\overset{\cdot\cdot}{\underset{\cdot\cdot}{O}}\cdot \qquad \left[:\overset{\cdot\cdot}{\underset{\cdot\cdot}{O}}:\right]^{2-}$$

FIGURE 8.7   Lewis dot diagrams of O atom and O^{2-} anion.

In comparing the electron configurations and electron dot diagrams for the O atom and the O^{2-} anion, we note that the O atom has six valence electrons in its Lewis dot diagram, while the O^{2-} anion has eight valence electrons since it gained two electrons.

3. For atoms with partially filled d (transition metals) or f subshells (actinides and lanthanides), the electrons in the d or f sublevels are typically omitted from Lewis electron dot diagrams.

Example: $_{26}$Fe: [Ar] $4s^23d^6$
$_{26}$Fe^{2+}: [Ar] $3d^6$

To draw the Lewis dot diagrams for the Fe atom and the Fe^{2+} cation, the electrons in the d sublevel are omitted. The Lewis dot diagrams of Fe atom and Fe^{2+} cation are, respectively (Figure 8.8):

$$Fe: \qquad Fe^{2+}$$

FIGURE 8.8   Lewis dot diagrams of Fe atom and Fe^{2+} cation.

Note that there are no dots around the Fe symbol in the Lewis dot diagram of Fe^{2+}. In comparing the electron configurations and electron dot diagrams for the Fe atom and the Fe^{2+} cation, we note that the Fe atom has two valence electrons in the s sublevel, while the Fe^{2+} cation has lost these two valence electrons. By putting the two electrons together on the same side in the Lewis dot diagram of Fe atom, we emphasize the fact that these two electrons are both in the 2s subshell.

8.4 LEWIS STRUCTURES

8.4.1 OCTET RULE

The Octet rule states that, in chemical bonding, atoms tend to gain, lose or share electrons until they are surrounded by eight valence electrons (or four electron pairs). Note that there are many exceptions to the octet rule:

○ Atoms from period 3 (such as Na, Mg, Al, Si, P and S) can accommodate more than an octet.
○ Be, B and N can accommodate less than an octet to form covalent compounds.
○ Beyond period 3, the d-orbitals are low enough in energy to participate in bonding and accept the extra electron density.

8.4.2 RULES FOR DRAWING LEWIS STRUCTURES

To draw the Lewis structures, follow the following steps:

1. Identify the metal and non-metal in the molecule and add up the number of valence electrons for each of the atoms in the molecule. If the molecule is an ion, then add one electron for each negative charge and subtract one electron for each positive charge.
   Example: In NH_3, both H and N are non-metals; there is a covalent bond between N and H.
   $_7$N: [He] $3s^23p^3$, the number of valence electrons in N is five.
   $_1$H: $1s^1$, the number of valence electrons in H is one. Because we have three atoms of H in NH_3, so the number of valence electrons for all H atoms is three.
   The total number of valence electrons is $(5 + 3 \times 1 =)$ eight electrons.
2. Draw a trial structure. The trial structure has single bonds connecting the atoms to the central atom in the molecule (Figure 8.9). Double or triple bonds are not part of this step. The central atom is the atom which has the lowest electronegativity. Hydrogen (H) always goes on the outside.

FIGURE 8.9   Trial structure of NH_3.

Each bond (presented by a continuous line) involves two electrons, resulting in six electrons involved in this trial structure.
3. Subtract the number of electrons in your trial structure from the number of valence electrons (determined in step 1) to obtain the remaining valence electrons. These are the electrons that need to be in your trial structure but are not. It is possible to have the number of electrons in this step to be equal to zero.
   Example: The number of electrons in the trial structure (used in bonds) of NH_3 is $T_t = 6$ and the total number of valence electrons is $T_v = 8$.
   Therefore, the remaining valence electrons is $T_v - T_t = 8 - 6 =$ two electrons.
4. Take the electrons from step 3 and put them on the trial structure. The goal is to give every atom (except hydrogen) an octet. Whatever you do, do not put down more electrons than you are allowed.
   Example: The two electrons (from step 3) are put on the trial structure (Figure 8.10):

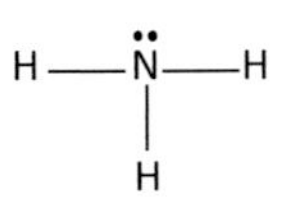

$$H —— \overset{\cdot\cdot}{\underset{|}{N}} —— H$$
$$H$$

FIGURE 8.10   Lewis structure of NH_3.

5. If you cannot give every atom an octet in step 4, then rearrange the electrons necessary to form double or triple bonds. This step is a last resort.

   Example: When drawing the Lewis structure of O_2, at step 4, oxygen atoms did not have an octet. In that case, the electrons are rearranged to form a double bond (Figure 8.11):

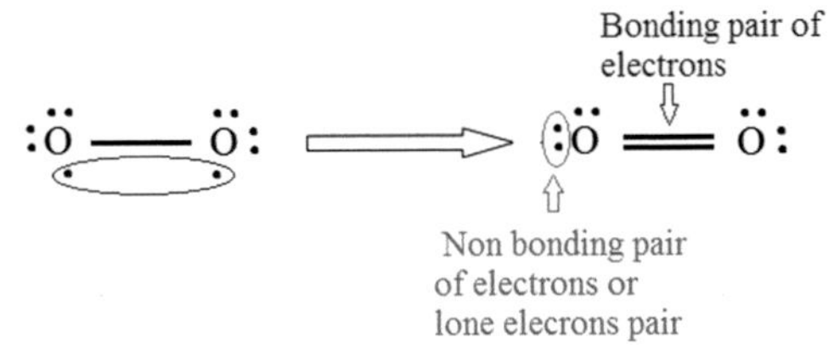

Bonding pair of electrons

$$:\overset{\cdot\cdot}{O} —— \overset{\cdot\cdot}{O}: \implies (:)\overset{\cdot\cdot}{O} == \overset{\cdot\cdot}{O}:$$

Non bonding pair of electrons or lone elecrons pair

FIGURE 8.11   When drawing the Lewis structure of O_2, oxygen atoms did not have an octet and the electrons are rearranged to form a double bond.

Practice 8.1 Draw the Lewis structure of

a) F_2
b) CO_2
c) SO_2

Answer:

a) 1. F is a non-metal; there is a covalent bond in F_2.
       $_9F$: [He] $3s^23p^5$, the number of valence electrons in F is seven.
       The total number of valence electrons is $7 \times 2 = 14$ electrons.
    2. Trial structure of F_2: F—F, the number of electrons shared is two.
    3. The remaining valence electrons are $14 - 2 = 12$ electrons.
    4. The 12 remaining valence electrons are put on F atoms as lone pairs (Figure 8.12):

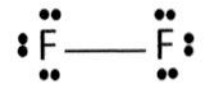

$$:\overset{\cdot\cdot}{\underset{\cdot\cdot}{F}} —— \overset{\cdot\cdot}{\underset{\cdot\cdot}{F}}:$$

FIGURE 8.12   Lewis structure of F_2.

b) 1. Both O and C are non-metals; CO_2 is a covalent compound.
       $_8O$: [He] $2s^22p^4$, the number of valence electrons in O is six.
       $_6C$: [He] $2s^22p^2$, the number of valence electrons in C is four.
       The total number of valence electrons is $6 \times 2 + 4 = 16$ electrons.
    2. Carbon is less electronegative than oxygen, so carbon is the central atom. The trial structure of CO_2 is O—C—O and the number of electrons shared (used in the two bonds) is $2 \times 2 = 4$.
    3. The remaining valence electrons are $16 - 4 = 12$ electrons.
    4. The remaining valence electrons are put first on the most electronegative atom, then on the central atom, if electrons remain. The 12 remaining valence electrons are put on O atoms as lone pairs (Figure 8.13):

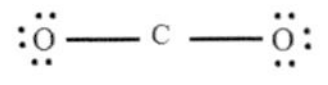

$$:\overset{\cdot\cdot}{\underset{\cdot\cdot}{O}} —— c —— \overset{\cdot\cdot}{\underset{\cdot\cdot}{O}}:$$

FIGURE 8.13   The remaining valence electrons are put on O atoms as lone pairs.

5. The carbon does not have an octet. We rearrange the electrons to form double bonds and to provide an octet for all atoms (Figure 8.14):

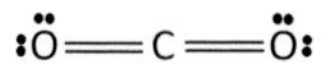

$$\ddot{\text{:O}}=\text{C}=\ddot{\text{O:}}$$

FIGURE 8.14 Lewis structure of CO_2.

c) 1. Both O and S are non-metals; SO_2 is a covalent compound.
   $_8$O: [He] $2s^22p^4$, the number of valence electrons in O is six.
   $_{16}$S: [Ne] $3s^23p^4$, the number of valence electrons in S is six.
   The total number of valence electrons is $6 \times 2 + 6 = 18$ electrons.
2. Sulfur (S) is less electronegative than O, so S is the central atom. The trial structure of SO_2 is O—S—O and the number of electrons shared (used in bonds) is $2 \times 2 = 4$.
3. The remaining valence electrons are $18 - 4 = 14$.
4. The remaining valence electrons are put first on the most electronegative atom, then on the central atom if electrons remain. Twelve remaining valence electrons are put on the O atoms and two valence electrons are put on the S atom as lone-pair electrons (Figure 8.15):

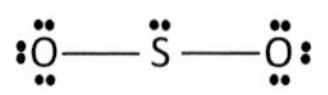

$$\ddot{\text{:O}}—\ddot{\text{S}}—\ddot{\text{O:}}$$

FIGURE 8.15 The remaining valence electrons are put first on the most electronegative atom (O), then on the central atom (S).

5. The sulfur does not have an octet. We rearrange the electrons to form double bonds and to provide an octet for all atoms (Figure 8.16):

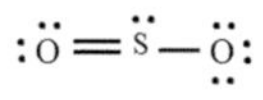

$$:\ddot{\text{o}}=\ddot{\text{s}}—\ddot{\text{o}}:$$

FIGURE 8.16 Lewis structure of SO_2.

8.4.3 LEWIS STRUCTURES OF IONIC COMPOUNDS

Electrons are transferred from a metal to a non-metal to form an ionic compound. In the Lewis structure of the ionic compounds, the non-metal (anion) is put in brackets.

Examples:
The Lewis structure of NaCl is (Figure 8.17):

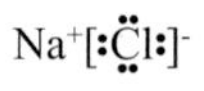

$$\text{Na}^+[\ddot{\text{:Cl:}}]^-$$

FIGURE 8.17 Lewis structure of NaCl.

The Lewis structure of CaF_2 is (Figure 8.18):

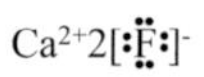

$$\text{Ca}^{2+}2[\ddot{\text{:F:}}]^-$$

FIGURE 8.18 Lewis structure of CaF_2.

The Lewis structure of K_2O is (Figure 8.19):

$$2\text{K}^+[\ddot{\text{:O:}}]^{2-}$$

FIGURE 8.19 Lewis structure of K_2O.

8.4.4 FORMAL CHARGE

The formal charge (FC) is the charge that an atom would have if the electronegativity differences were ignored. The sum of all formal charges must equal the overall charge of the molecule or the ion. The best Lewis structure has the fewest number of non-zero formal charges. The FC is given by the following equation:

$$FC = n\left(Ve^-\right) - \left[n\left(NBe^-\right) + \frac{n\left(Be^-\right)}{2}\right]$$

where n(Ve⁻) is the number of valence electrons,
   n(NBe⁻) is the number of non-bonding electrons
   and n(Be⁻) is the number of bonding electrons

Note that the FC can also be given simply by the following equation:

$$FC = n\left(Ve^-\right) - \left[b + d\right]$$

where n(Ve⁻) is the number of valence electrons of the atom, b is the number of bonds (the double bond is counted as two bonds and the triple bonds is counted as three bonds) and d is the number of dots (non-bonded electrons).

Practice 8.2 Draw the best Lewis structure of SO_4^{2-}

Answer:

1. Both O and S are non-metals; the bond between S and O is a covalent bond.
   $_8$O: [He] $2s^2 2p^4$, the number of valence electrons in O is six.
   $_{16}$S: [Ne] $3s^2 3p^4$, the number of valence electrons in S is six.
   The total number of valence electrons is $4 \times 6 + 6 + 2 = 32$ electrons (24 from 4 O, 6 from S and 2 since SO_4^{2-} is an anion with a charge of –2).
2. S is less electronegative than O, so S is the central atom. The trial structure of SO_4^{2-} is (Figure 8.20):

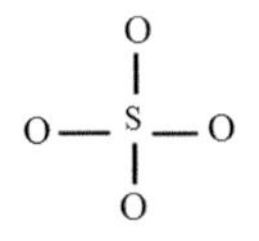

FIGURE 8.20   Trial structure of SO_4^{2-}.

   and the number of electrons shared (used in bonds) is $2 \times 4 = 8$.
3. The remaining valence electrons are $32 - 8 = 24$.
4. The 24 remaining valence electrons are put on O atoms as lone pairs (Figure 8.21). We obtain (structure 1):

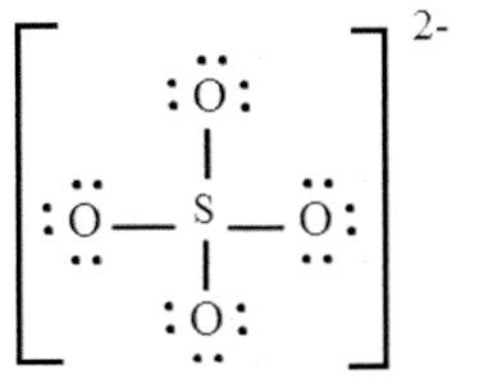

FIGURE 8.21   The remaining valence electrons are put on O atoms as lone pairs (Structure 1).

   Sulfur can accommodate more than one octet by using its empty valence d orbital. An alternative Lewis structure of SO_4^{2-} can be given by (structure 2) (Figure 8.22):

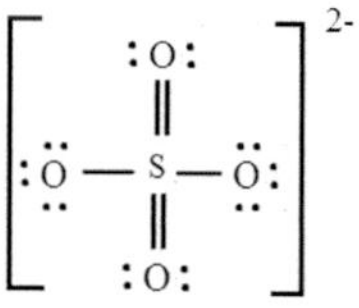

FIGURE 8.22 Alternative Lewis structure of SO_4^{2-} (Structure 2).

5. The best Lewis structure has the fewest number of non-zero FCs. The FC is given by the following equation:
   $$FC = n(Ve^-) - [n(NBe^-) + n(Be^-)/2]$$
   Let us calculate the FCs for each atom of the two structures of SO_4^{2-} (Figures 8.23 and 8.24):

	Structure 1		**Structure 2**
atom	FC	atom	FC
S	$6 - 0 - (8/2) = 2$	S	$6 - 0 - (12/2) = 0$
O (4 atoms)	$6 - 6 - (2/2) = -1$	O (2 atoms)	$6 - 4 - (4/2) = 0$
		O (2 atoms)	$6 - 6 - (2/2) = -1$

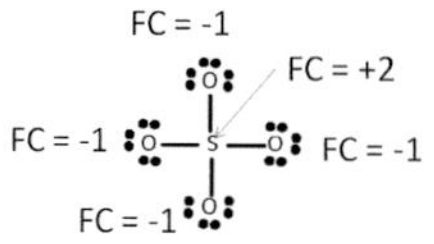

Formal charges of atoms of structure 1.

The best Lewis structure of SO_4^{2-}

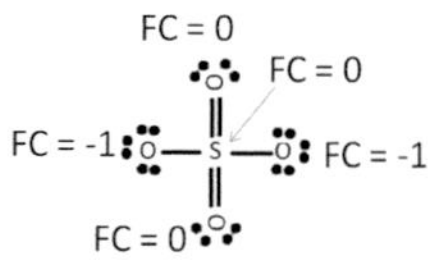

Formal charges of atoms of structure 2.

Note that:

For the two Lewis structures of SO_4^{2-}, the sum of all FCs equals the overall charge of the SO_4^{2-} ion (= –2). For structure 1, the sum of FCs is $(4 \times (-1) + 2) = -2$, and for structure 2, the sum of FCs is $((2 \times (-1) + 2 \times 0 + 0) = -2$.

The better Lewis structure is structure 2, because it has the fewest number of non-zero FCs.

8.4.5 RESONANCE FORMULAS

Resonance is a manner of describing delocalized electrons within certain molecules or polyatomic ions where the bonding cannot be expressed by a single Lewis formula. A molecule or an ion with such delocalized electrons is represented by two or more contributing structures to show bonding, called resonance structures. Double-headed arrows are used to indicate the resonance formulas.

Example: The resonance structure of SO_3 is as follows (Figure 8.25):

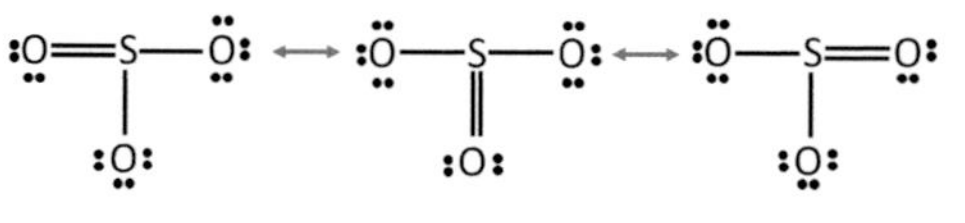

FIGURE 8.25 Resonance structures of SO_3.

8.5 MOLECULAR GEOMETRY

8.5.1 ELECTRON GROUPS AND MOLECULE NOTATION

On a central atom

1. Each lone pair of electrons constitutes one electron group.

2. Each bond constitutes one electron group, regardless of whether it is a single, a double or a triple bond.

   Example: On the central atom N of NO_2, there are three electron groups that are one lone pair of electrons, one single bond and one double bond (Figure 8.26):

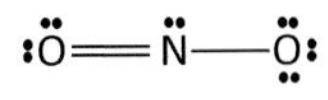

$$:\ddot{O}=\ddot{N}-\ddot{O}:$$

FIGURE 8.26   Lewis structure of NO_2.

Generally, for the notation of a molecule, the letter A is used to represent the central atom, X_m is used to represent the number of atoms (m) bonded to the central atom and E_n is used to represent the number of lone electron pairs (n) on the central atom (lone pairs on bonded atoms are ignored).

   Example: The central atom N in NO_2 is bonded to two atoms of oxygen and there is one lone pair of electrons on it. The NO_2 molecule is denoted by AX_2E.

8.5.2 BASIC GEOMETRIES AND DERIVATIVES

There are five basic molecular geometries corresponding to several electron groups, from two to six, that are linear, trigonal planar, tetrahedral, trigonal bipyramidal and octahedral, respectively. Additional molecular shapes (derivatives) can be obtained by removing corner atoms from the basic shape and replacing them by lone electron pairs (Figures 8.27 to 8.39). The basic molecular geometries and derivatives are shown in Table 8.1.

8.5.3 VALENCE SHELL ELECTRON PAIR REPULSION THEORY

The valence shell electron pair repulsion theory (**VSEPR theory**) is a model used to predict the geometry of a molecule from the number of electron groups of its central atom. Regions of high electron density around the central atom go as far apart as possible to minimize repulsions. Indeed, lone pairs (LP) of electrons (unshared pairs) require a greater volume than shared pairs. This affects the angles between the bonded atoms. There is an ordering of repulsions of electrons around the central atom:

- o Lone pair–lone pair repulsion is strongest.
- o Lone pair–bonding pair (BP) repulsion is intermediate.
- o Bonding pair–bonding pair repulsion is weakest.
   LP/LP > LP/BP > BP/BP

To determine the molecular geometry, follow the following steps:

1. Draw the Lewis structure of the molecule.
2. Determine how many electron groups there are around the central atom (count the total number of electron pairs around the central atom and count multiple bonds as one bonding pair).
3. Find what structure this molecule is based on (arrange the electron pairs in one of the five basic geometries to minimize electron–electron repulsion).
4. Find which atoms are replaced by lone electron pairs.
5. Determine the final structure and the bond angles.

Practice 8.3 Determine the molecular geometries of CO_2 and SO_2.

Answer: The molecular geometries of CO_2 and SO_2 are as follows (Figures 8.14 and 8.40):

TABLE 8.1   Basic molecular geometries and derivatives

Electron groups	Basic geometry	Derivatives		
2	AX_2			
	Linear			
3	AX_3	AX_2E		
	Trigonal planar	Bent (V-shape)		
4	AX_4	AX_3E	AX_2E_2	
	Tetrahedral	Trigonal pyramid	Bent (V-shape)	
5	AX_5	AX_4E	AX_3E_2	AX_2E_3
	Trigonal bipyramidal	See-saw	T-shape	Linear
6	AX_6	AX_5E	AX_4E_2	
	Octahedral	Square pyramidal	Square planar	

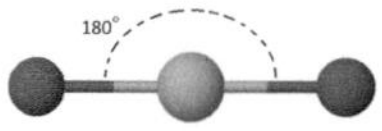

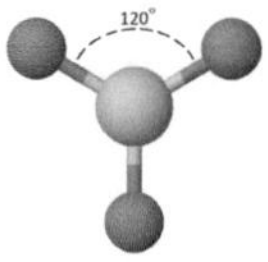

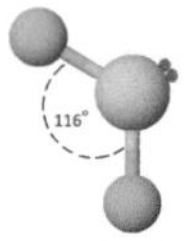

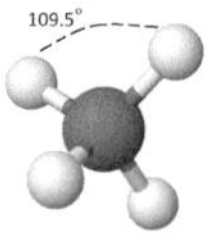

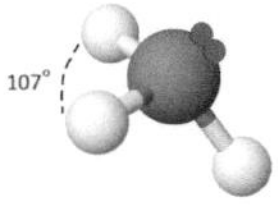

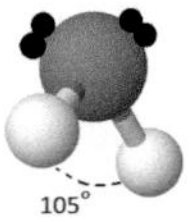

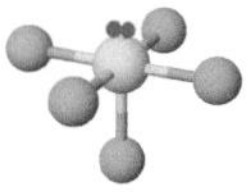

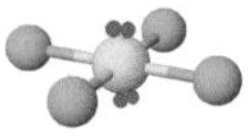

Answer:	CO_2	SO_2
(Best) Lewis structure		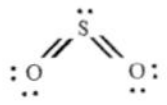
Number of electron groups	Two (two double bonds)	Three (two double bonds and one lone electron pair)
The molecule is based on a	Linear structure	Trigonal planar
Number of lone electron pairs	None	One lone electron pair
Final structure	Linear (AX_2) 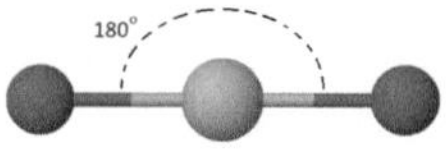	Bent or V-shape (AX_2E) 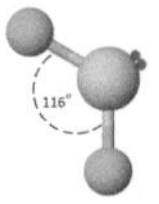

Note that S accommodates more than an octet and the angle between the two bonded atoms (O) in SO_2 is less than 120° because of the repulsion between the lone electron pair and the bonding pairs. If S accommodates an octet, we obtain the Lewis structure shown in Figure 8.16 which is not the best Lewis structure. Indeed, it has the highest number of nonzero formal charges (See the key explanation of question Q.8.4 for detail).

Practice 8.4 Determine the molecular geometries of CH$_4$, PCl$_5$ and SF$_6$.

Answer: The molecular geometries of CH$_4$, PCl$_5$ and SF$_6$ are as follows (Figures 8.41 to 8.43):

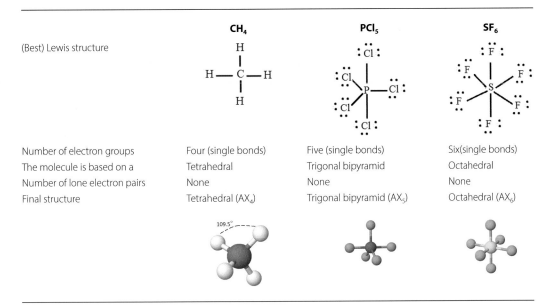

	CH$_4$	PCl$_5$	SF$_6$
Number of electron groups	Four (single bonds)	Five (single bonds)	Six(single bonds)
The molecule is based on a	Tetrahedral	Trigonal bipyramid	Octahedral
Number of lone electron pairs	None	None	None
Final structure	Tetrahedral (AX$_4$)	Trigonal bipyramid (AX$_5$)	Octahedral (AX$_6$)

GET SMART

IS THE VESPR THEORY SUFFICIENT TO DESCRIBE THE O$_2$ MOLECULE?

The oxygen atom has six valence electrons. The molecular notation of O$_2$ is AXE$_2$ (linear). The Lewis structure of O$_2$ is as follows (Figure 8.44):

$$:\ddot{O}=\ddot{O}:$$

FIGURE 8.44   Lewis structure of O$_2$.

According to this Lewis structure, there are no unpaired electrons and the O$_2$ molecule would be diamagnetic (repelled by magnetic fields). However, this is not the case since experience shows that O$_2$ is paramagnetic (attracted by magnetic fields). Therefore, a new theory (linear combination of atomic orbitals–molecular orbitals (LCAO MO) bonding theory), describing the molecular structure of covalent molecules, has been developed and is worth studying (Section 8.6.).

8.6 MOLECULAR ORBITAL THEORY

The molecular orbital bonding theory known as LCAO MO bonding theory describes how atomic orbitals overlap to form bonds. When the bond forms, the probability of finding electrons within the region of space between the two atoms nuclei becomes greater. LCAO MO bonding theory states that each molecular orbital is constructed from a superposition of atomic orbitals that belong to the atoms in the molecule. In other words, atomic orbitals combine to form molecular orbitals. The number of orbitals is always conserved, e.g., two atomic orbitals combine to form two molecular orbitals.

The LCAO MO bonding theory assumes that the valence electrons of the atoms within the molecule become the valence electrons of the entire molecule and the fitting of the molecular orbitals takes place according to the Aufbau "building-up" principle, Pauli's exclusion principle and Hund's rule.

For example, two hydrogen atoms form diatomic molecule H_2. When the hydrogen atoms approach each other, atomic orbitals combine to become molecular orbitals (Figure 8.45). The two atomic orbitals (1s) of the two H atoms become one bonding molecular orbital of lower energy, denoted by σ_{1s}, and one anti-bonding molecular orbital of higher energy, denoted by σ^*_{1s}. It is worth noting that the bonding orbital is stronger than the simple atomic orbital; that is, H_2 is more stable than the two H atoms. The anti-bonding orbital is less stable.

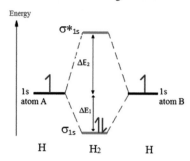

FIGURE 8.45   The two atomic orbitals (1s) of the two atoms of H become one bonding molecular orbital σ_{1s} and one anti-bonding molecular orbital σ^*_{1s}.

Atomic orbitals combine to form molecular orbitals with respect to the following conditions:

1. The atomic orbitals that combine to form the molecular orbitals should have the same energy. For example, an orbital 2p of an atom can combine with another orbital 2p of another atom but a 2p orbital cannot combine with an orbital 1s since they have appreciable energy differences.
2. The combining atoms should have the same symmetry around the molecular axis to achieve a proper combination. For example, the three orbitals (2px, 2py and 2pz) of the sublevel 2p have the same energy but the orbital 2pz of an atom can still combine only with an orbital 2pz of another atom; it cannot combine with the orbitals 2px and 2py because they have different axes of symmetry. Generally, the z-axis is considered to be the molecular axis of symmetry.
   When the atomic orbitals (2p) of two atoms combine, they become:
   - One bonding molecular orbital of lower energy, denoted by σ_{2pz}, and one anti-bonding molecular orbital of higher energy, denoted by σ^*_{2pz}. This corresponds to the combination of 2pz of the atom with another 2pz orbital of another atom (axial overlap).
   - One bonding molecular orbital of lower energy, denoted by π_{2px}, and one anti-bonding molecular orbital of higher energy, denoted by π^*_{2px}. This corresponds to the combination of 2px of the atom with another 2px orbital of another atom (lateral overlap). The same process is obtained when a 2py orbital is combined with another 2py orbital where the molecular orbitals π_{2py} and π_{2py} are formed.
   $$\sigma_{1s} < \sigma^*_{1s} < \sigma_{2s} < \sigma^*_{2s} < \sigma_{2pz} < \pi_{2px} < \pi_{2py} < \pi^*_{2px} < \pi^*_{2py} < \sigma^*_{2pz}$$
3. The atomic orbitals should have a spatial overlap.

For the O_2 molecule, we have, in order of increasing energy of the molecular orbitals (Figure 8.46): Note that, according to the LCAO MO bonding theory, the O_2 molecule has two unpaired electrons which explains why O_2 is a paramagnetic molecule.

The molecular orbital diagram of N_2 (Figure 8.47) is slightly different from that of O_2. Indeed, the molecular orbitals π_{2px} and π_{2py} have lower energies than σ_{2pz}. Thus, we have, in order of increasing energy of the molecular orbitals:

$$\sigma_{1s} < \sigma^*_{1s} < \sigma_{2s} < \sigma^*_{2s} < \pi_{2px} < \pi_{2py} < \sigma_{2pz} < \pi^*_{2px} < \pi^*_{2py} < \sigma^*_{2pz}$$

Homonuclear molecules, such as F_2, Cl_2 and Br_2, have molecular orbital diagrams of the same type as N_2. Heteronuclear molecules and ions such as NO, CN^- and CO also have molecular orbital diagrams like that of N_2. When the electronegativity of one atom is higher than the other atom, the orbitals of the more electronegative atom are lower in energy (Figure 8.48).

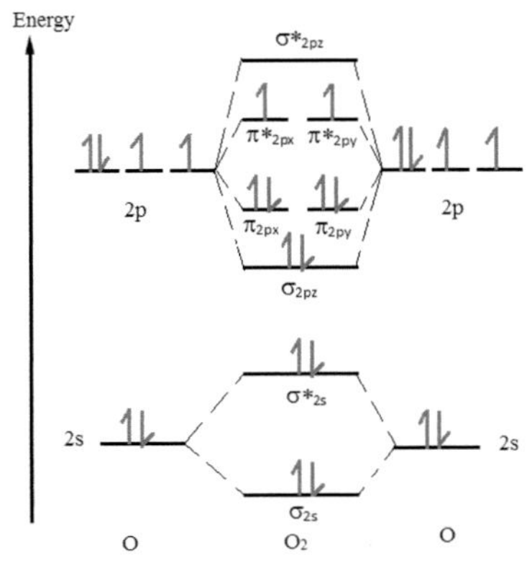

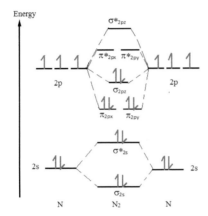

FIGURE 8.46 Molecular orbital diagram for O_2 molecule.

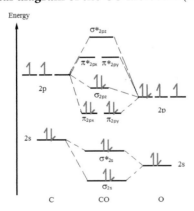

FIGURE 8.47 Molecular orbital diagram for N_2 molecule.

Example: Molecular orbital diagram of the CO molecule (Figure 8.48)

FIGURE 8.48 Molecular orbital diagram for CO molecule.

The bond order (BO) is defined by the following equation:

$$BO = \frac{\left[n(BE) - n(ABE) \right]}{2}$$

where n(BE) is the number of electrons in bonding orbitals and n(ABE) is the number of electrons in anti-bonding orbitals. When BO = 1, the molecule shows a single bond, when BO = 2, the molecule shows a double bond and, when BO = 3, the molecule shows a triple bond. As BO increases, the molecule becomes more stable.

For the H_2 molecule, BO = (2 – 0)/2 = 1 which correlates with a single bond (H—H).

For O_2, BO = (8 – 4)/2 = 2 which correlates with the double bond (O=O).

For N_2, BO = (8 − 2)/2 = 3 which correlates with the triple bond (N≡N). N_2 is the most stable gas and O_2 is more stable than H_2.

8.7 ORBITAL HYBRIDIZATION THEORY

The formation of many molecules cannot be explained by the LCAO MO bonding theory. For this reason, a new theory called the orbital hybridization theory was proposed. Orbital hybridization is the concept of mixing atomic orbitals to form new hybrid orbitals. The hybrid orbitals have different energies and shapes from those of the component atomic orbitals.

8.7.1 HYBRIDIZED sp³ ORBITALS

The electronic structure of carbon (C: $1s^2 2s^2 2p^2$) cannot explain the formation of the CH_4 molecule. Indeed, the presence of two unpaired electrons cannot explain the formation of four single bonds in CH_4. Therefore, and since the four bonds in CH_4 are identical, a new hypothesis is formulated in which the three higher-energy p orbitals and the lower-energy s orbital are hybridized into four equal-energy sp³ orbitals (Figure 8.49). When these sp³ hybrid orbitals overlap with the s orbitals of the hydrogen atoms in CH_4, four (C—H) bonds are obtained.

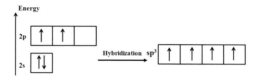

FIGURE 8.49   Hybrid orbitals (sp³) obtained from the atomic orbitals 2s and 2p of carbon.

In CH_4 molecule, the axes of the hybrid orbitals are dispersed according to a tetrahedron. The angle formed (H—Ĉ—H) equals 109.5° (Figure 8.50).

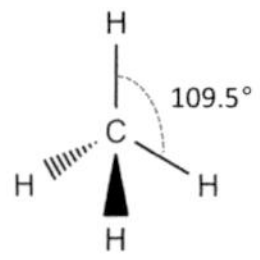

FIGURE 8.50   Tetrahedral molecular geometry of CH_4 molecule.

8.7.2 HYBRIDIZED sp² ORBITALS

Two higher-energy p orbitals and one lower-energy s orbital are hybridized into three equal-energy sp² orbitals. There are two ways to form the hybrid sp² orbitals, achieving two types of bonding:

1. Hybridization of an element with three valence electrons, such as boron ($_5$B: $1s^2 2s^2 2p^1$), results in three sp² hybrid orbitals, with no electrons left over (Figure 8.51):

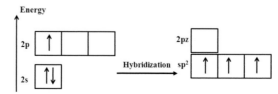

FIGURE 8.51   Hybridization sp² of an element with three valence electrons.

For example, in the molecule BF_3, boron B forms bonds from the three hybrid sp² orbitals. Each B—F bond is the result of the axial overlap between a hybrid sp² orbital of

B with an 2pz orbital of fluorine F. The molecule BF_3 is planar and the angle ($F—\hat{B}—F$) equals 120° (Figure 8.52).

FIGURE 8.52   Planar molecular geometry of BF_3 molecule.

2. Hybridization of an element with more than three valence electrons results in three hybridized sp^2 orbitals and one of the p orbitals remaining unhybridized. This unhybridized p orbital forms a π bond for double bonding (Figure 8.53):

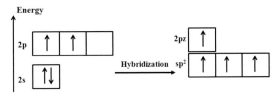

FIGURE 8.53   Hybridization sp^2 of an element with more than three valence electrons.

For example, in the C_2H_4 molecule, each carbon forms two bonds with hydrogen atoms and one bond with the second carbon atom from the three hybrid sp^2 orbitals. The unhybridized 2p orbitals from each carbon atom form a π bond (Figure 8.54). Thus, the C_2H_4 molecule is obtained from:

- Axial overlaps of the 1s orbitals of the hydrogen atoms with the hybridized sp^2 orbitals of the carbon atoms.
- An axial overlap of the sp^2 orbitals of the two carbon atoms which form the σ bond.
- A lateral overlap of the unhybridized 2pz orbitals of the carbon atoms which form the π bond.

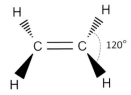

FIGURE 8.54   The molecule C_2H_4 is planar and the angle ($H—\hat{C}—H$) equals 120°.

8.7.3 HYBRIDIZED sp ORBITALS

One higher-energy p orbital and one lower-energy s orbital are hybridized into two equal-energy sp orbitals. There are two ways to form the hybrid sp orbitals, which give two types of bonding:

1. Hybridization of an element with two valence electrons, like beryllium ($_4$Be: $1s^22s^2$), results in two sp hybrid orbitals, with no electrons left over (Figure 8.55):

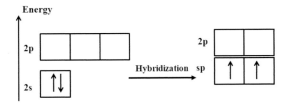

FIGURE 8.55   Hybridization sp of an element with two valence electrons

For example, in the $BeCl_2$ molecule, the beryllium atom (Be) forms two bonds with the two chlorine atoms from the two sp hybrid orbitals. Thus, the $BeCl_2$ molecule is obtained from the axial overlap of the orbital 2pz of the chlorine atoms with the hybridized sp orbitals of the beryllium atom. The $BeCl_2$ molecule is linear.

2. Hybridization of an element with more than two valence electrons, but some of those electrons are left unhybridized. These unhybridized electrons can form π bonds (Figure 8.56):

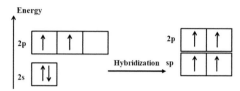

FIGURE 8.56   Hybridization sp of an element with more than two valence electrons.

For example, in the C_2H_2 molecule, each carbon forms one bond with a hydrogen atom and one bond with the second carbon from the two sp hybrid orbitals. The unhybridized 2p orbitals from each carbon form two π bonds which results in a triple bond structure (Figure 8.57). Thus, the C_2H_4 molecule is obtained from:

- Axial overlaps of the 1s orbitals of the hydrogen atoms with the hybridized sp orbitals of the carbon atoms.
- An axial overlap of the sp orbitals of the two carbon atoms, which form the σ bond.
- Lateral overlaps of the two unhybridized 2pz orbitals of the carbon atoms, which form two π bonds.

$$H \longrightarrow C \equiv C \longrightarrow H$$

FIGURE 8.57   The molecule C_2H_2 is linear.

8.7.4 HYBRIDIZED sp³d ORBITALS

One higher-energy d orbital, three lower-energy p orbitals and one lower-energy s orbital are hybridized into five equal-energy sp³d orbitals (Figure 8.58):

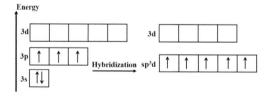

FIGURE 8.58   Hybridization sp³d

For example, in the PCl_5 molecule, the phosphorus atom (P) forms five bonds with the chlorine atoms from the five sp³d hybrid orbitals. Thus, the PCl_5 molecule is obtained from the axial overlap of the 3pz orbital of chlorine atoms with the hybridized sp³d orbitals of the phosphorus atom. The PCl_5 molecule is a triangular bipyramid (Figure 8.59).

FIGURE 8.59   The molecule PCl_5 is triangular bipyramid.

8.7.5 HYBRIDIZED sp³d² ORBITALS

Two higher-energy d orbitals, three lower-energy p orbitals and one lower-energy s orbital are hybridized into six equal-energy sp³d² orbitals (Figure 8.60):

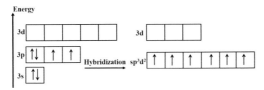

FIGURE 8.60   Hybridization sp³d².

For example, in the SF_6 molecule, the sulfur (S) atom forms six bonds with the fluorine (F) atoms from the six hybridized sp³d² orbitals. Thus, the SF_6 molecule is obtained from the axial overlap of the 2pz orbital of the fluorine atoms with the hybridized sp³d² orbitals of the sulfur atom. The SF_6 molecule is octahedral. The angles F—Ŝ—F equal 90° (Figure 8.61).

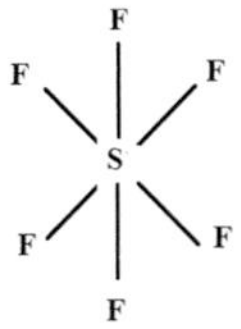

FIGURE 8.61   The molecule SF_6 is octahedral.

8.7.6 HYBRIDIZED sp³d³ ORBITALS

Three higher-energy d orbitals, three lower-energy p orbitals and one lower-energy s orbital are hybridized into seven equal-energy sp³d³ orbitals (Figure 8.62):

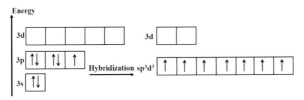

FIGURE 8.62   Hybridization sp³d³.

For example, in the IF_7 molecule, the iodine atom (I) forms seven bonds with the fluorine atoms from the seven hybridized sp³d³ orbitals. Thus, the IF_7 molecule is obtained from the axial overlap of the 2pz orbital of fluorine atoms with the hybridized orbitals sp³d³ of iodine atom. The IF_7 molecule is a pentagonal bipyramid (Figure 8.63).

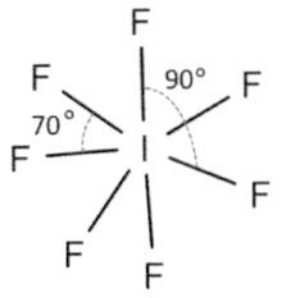

FIGURE 8.63   The molecule IF_7 is a pentagonal bipyramid.

8.7.7 HYBRIDIZED dsp² ORBITALS

The hybridization dsp² is seen in the case of transition metal ions, such as $PtCl_4^{2-}$. For example, one atomic 6s orbital, one atomic 5d orbital and two atomic 6p orbitals of the platinum (Pt^{2+}) ion, are hybridized into four equal-energy dsp² orbitals (Figure 8.64):

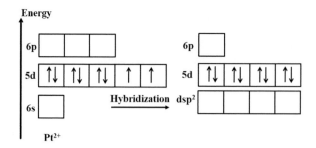

FIGURE 8.64   Hybridization dsp².

For example, the $PtCl_4^{2-}$ ion is obtained from the overlap of the 3pz orbital of chlorine ions (Cl^-), containing a pair of electrons, with the non-filled hybridized dsp^2 orbitals of the platinum ion, Pt^{2+}. The $PtCl_4^{2-}$ ion is square planar (Figure 8.65).

FIGURE 8.65   The $PtCl_4^{2-}$ ion is square planar.

GET SMART

WHAT IS THE DIFFERENCE BETWEEN HYBRID ORBITALS AND MOLECULAR ORBITALS?

When the atomic orbitals of the same atom interact, they form hybrid orbitals, whereas, when the atomic orbitals of two different atoms interact, they form molecular orbitals.

CHECK YOUR READING

What are the different types of chemical bonding?

How is the Lewis dot diagram of an atom represented?

What is the octet rule?

How can a Lewis structure be drawn?

What are resonance structures?

What are the different molecular geometries and their derivatives?

How can the molecular geometry of a molecule be determined?

How does the repulsion between lone electron pairs and bonded pairs affect the angles between the bonded atoms?

What are the different types of orbital hybridization? Give an example of a molecule for each type.

SUMMARY OF CHAPTER 8

The electrons involved in bonding are usually those in the outermost (valence) shell. There are three types of chemical bonds:

1. **A covalent bond** results from sharing electrons between the atoms. The covalent bond is usually found between non-metals. The covalent bond can be polar or non-polar.
   Examples of molecules showing polar covalent bonds: HCl, HBr and H_2O
   Examples of molecules showing non-polar covalent bonds: Homonuclear diatomic molecules such as I_2, Br_2, Cl_2, O_2 and F_2, and hydrocarbons (C_nH_{2n+2}).

$$I \overset{\bullet\bullet}{-\!\!\!-} I \qquad\qquad H \overset{\bullet\bullet}{-\!\!\!-} Cl$$

A non-polar covalent bond in I_2 and a polar covalent bond in HCl.
2. **An ionic bond** results from the transfer of electrons from a metal to a non-metal.
   Example: The bond between sodium (Na) and chlorine (Cl) in NaCl is an ionic bond. The NaCl compound obtained is an ionic compound.

$$Na^{\cdot} \; \overset{\frown}{\cdot}\ddot{\underset{\cdot\cdot}{Cl}}: \quad \Longrightarrow \quad [Na]^+ \; [:\ddot{\underset{\cdot\cdot}{Cl}}:]^-$$

3. **A metallic bond** is formed between positively charged atoms in which the free electrons are shared among a structure of positively charged ions. Metallic bonding is the main type of chemical bond that forms between metal atoms.

Lewis dot representation or electron dot diagram is a simplistic way of showing the valence electrons of an atom that uses dots around the chemical symbol of the element with no more than two dots on a side. To represent the Lewis dot diagram of an atom, carry out the following steps:

1. Write out the electronic configuration of the atom or the ion.
2. Find the number of valence electrons.
3. Draw the valence electrons as dots around the chemical symbol of the atom or the ion.
   Example: The seven valence electrons of Cl are drawn in the following configuration:

$$\cdot\ddot{\underset{\cdot\cdot}{Cl}}:$$

Elements in the same group of the periodic table of elements have similar Lewis electron dot diagrams because they have the same valence shell electron configuration.
   Example: $_7N$ and $_{15}P$ belong to the same group (Group 15) and have similar Lewis dot diagrams:

$$\cdot\ddot{\underset{\cdot}{N}}\cdot \qquad \cdot\ddot{\underset{\cdot}{P}}\cdot$$

Electron dot diagrams for ions are like those of atoms, except that some electrons have been removed from or added to a neutral atom to obtain an ion. Conventionally, when we draw electron dot diagrams for ions, we show the original valence shell of the atom and not the valence shell of the ion.

   Example 1: The Lewis dot diagrams of the Na atom and the Na^+ cation are:
      $\overset{\cdot}{Na}$ and Na^+ (no dots around the Na symbol)
   Example 2: The Lewis dot diagrams of the O atom and the O^{2-} anion are:

$$\cdot\ddot{\underset{\cdot\cdot}{O}}\cdot \qquad [:\ddot{\underset{\cdot\cdot}{O}}:]^{2-}$$

For atoms with partially filled d (transition metals) or f subshells (actinides and lanthanides), the electrons in the d or f sublevels are typically omitted from Lewis electron dot diagrams.
   Example: The Lewis dot diagrams of Fe atom and Fe^{2+} cation are:
      $_{Fe}:$ and Fe^{2+} (no dots around the Fe symbol)
The octet rule states that, in chemical bonding, atoms tend to gain, lose or share electrons until they are surrounded by **eight valence electrons** (or four electron pairs). Note that there are many exceptions to the octet rule:

○ Atoms from period 3 (such as Na, Mg, Al, Si and S) can accommodate more than an octet.
○ Be, B and N can accommodate less than an octet to form covalent compounds.
○ Beyond period 3, the d orbitals are low enough in energy to participate in bonding and accept the extra electron density.

To draw the Lewis structures, carry out the following steps:

1. Identify the metal and non-metal in the molecule and add up the number of valence electrons for each of the atoms in the molecule. If the molecule is an ion, then add one electron for each negative charge and subtract one electron for each positive charge.
2. Draw a trial structure. The trial structure has single bonds connecting the atoms to the central atom in the molecule. The central atom is the atom which has the least electro-negativity. Hydrogen (H) always goes on the outside.
3. Subtract the number of electrons in the trial structure from the total number of valence electrons (calculated in the first step) to obtain the remaining valence electrons.
4. Take the electrons from the third step and put this on the trial structure. The goal is to give every atom (except hydrogen) an octet. Whatever you do, do not put down more electrons than you are allowed. If you cannot give every atom an octet in step 4, then rearrange the electrons that you have to form double or triple bonds.

Examples: Lewis structures of CO_2 and O_2 (Figure 8.66) are as follows:

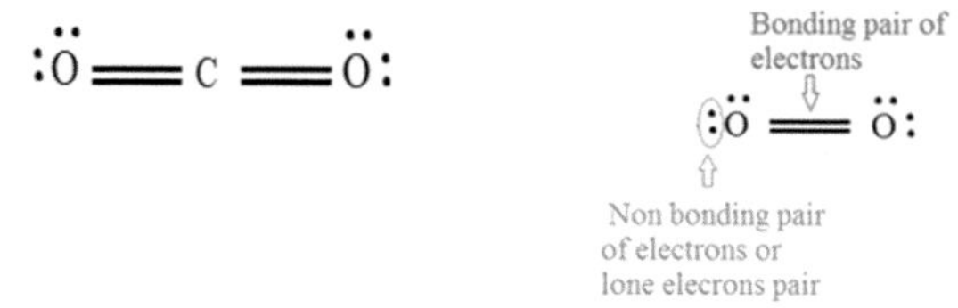

FIGURE 8.66 Lewis structures of CO_2 and O_2.

The formal charge (FC) is the charge that an atom would have if electronegativity differences are ignored. The sum of all FCs must equal the overall charge of the molecule or the ion. The best Lewis structure has the fewest number of non-zero formal charges. The FC is given by the following equation:

$$FC = n(Ve^-) - \left[n(NBe^-) + \frac{n(Be^-)}{2} \right]$$

where $n(Ve^-)$ is the number of valence electrons,
$n(NBe^-)$ is the number of non-bonding electrons
and $n(Be^-)$ is the number of bonding electrons

The FC can also be given simply by the following equation:

$$FC = n(Ve^-) - [b + d]$$

where $n(Ve^-)$ is the number of valence electrons of the atom, b is the number of bonds (the double bond is counted as two bonds and the triple bonds is counted as three bonds) and d is the number of dots (non-bonded electrons).

Example: The best structure of SO_4^{2-}:

Structure 1		Structure 2	
atom	FC	atom	FC
S	$6 - 0 - (8/2) = 2$	S	$6 - 0 - (12/2) = 0$
O (4 atoms)	$6 - 6 - (2/2) = -1$	O (2 atoms)	$6 - 4 - (4/2) = 0$
		O (2 atoms)	$6 - 6 - (2/2) = -1$

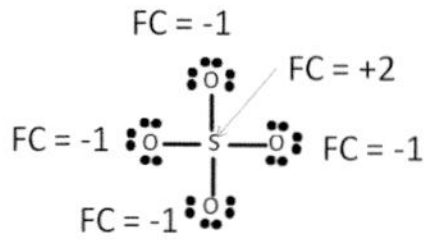

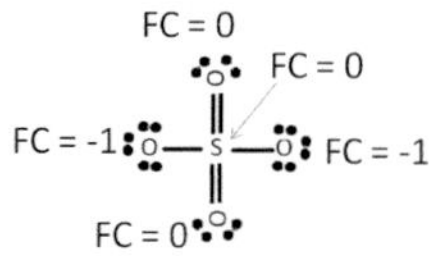

The best Lewis structure of SO_4^{2-}

Resonance is a manner of describing delocalized electrons within certain molecules or poly-atomic ions where the bonding cannot be expressed by a single Lewis formula.

Example:

$$:O\!\!=\!\!S\!-\!\ddot{O}: \longleftrightarrow :\ddot{O}\!-\!S\!-\!\ddot{O}: \longleftrightarrow :\ddot{O}\!-\!S\!\!=\!\!O:$$

On a central atom, each lone pair of electrons constitutes one **electron group,** and each bond constitutes one electron group, regardless of whether it is a single, double or triple bond. Generally, a molecule is denoted by $\mathbf{AX_mE_n}$ where the letter A is used to represent the central atom, X_m is used to represent the number of atoms (m) bonded to the central atom and E_n is used to represent the number of lone electron pairs (n) on the central atom.

> **Example:** The central atom N in NO_2 is bonded to two atoms of oxygen and there is one lone electrons pair on it. The NO_2 molecule is denoted by AX_2E.

The valence shell electron pair repulsion (VSEPR) theory is a model used to predict the geometry of a molecule from the number of electron groups of its central atom. Regions of high electron density around the central atom go as far apart as possible to minimize repulsions.

There are five basic molecular geometries corresponding to several electron groups from two to six: linear, trigonal planar, tetrahedral, trigonal bipyramidal and octahedral, respectively. Additional molecular shapes (derivatives) can be obtained by removing corner atoms from the basic shape:

Electron groups	Notation	Molecular geometry	Examples
2	AX_2	**Linear**	CO_2, $HgCl_2$, ZnI_2, CS_2
3	AX_3	**Trigonal planar**	BF_3, GaI_3
	AX_2E	Bent or V-shape	SO_2, O_3, $SnCl_2$
4	AX_4	**Tetrahedral**	CH_4, CCl_4, BF_4^-
	AX_3E	Trigonal pyramidal	NH_3, OH_3^-
	AX_2E_2	Bent	H_2O, $SeCl_2$
5	AX_5	**Trigonal bipyramidal**	PCl_5, PF_5
	AX_4E	See-saw	$TeCl_4$, SF_4
	AX_3E_2	T-shaped	ClF_3, BrF_3
	AX_2E_3	Linear	I_3^-, ICl_2^-
6	AX_6	**Octahedral**	SF_6, PF_6^-
	AX_5E	Square pyramidal	IF_5, BrF_5
	AX_4E_2	Square planar	ICl_4^-, BrF_4^-

Orbital hybridization is the concept of mixing atomic orbitals to form new hybrid orbitals. The hybrid orbitals have different energies and shapes from those of the component atomic orbitals.

Type of hybridization	Molecular geometry	Examples
sp	Linear	$BeCl_2$, C_2H_2
sp^2	Trigonal planar	$AlCl_3$
sp^3	Tetrahedral	CH_4, H_2O
sp^3d	Trigonal bipyramidal	PCl_5
sp^3d^2	Octahedral	SF_6
sp^3d^3	Pentagonal bipyramidal	IF_7
dsp^2	Square planar	$PtCl_4^{2-}$

PRACTICE ON CHAPTER 8

Q8.1   **Choose the correct answer**
1.   **The ionic bond is formed between**
    a) two metals
    b) a metal and a non-metal
    c) two non-metals
    d) two cations
2.   **The covalent bond is usually found between**
    a) two metals
    b) a metal and a non-metal
    c) two non-metals
    d) two cations
3.   **The metallic bond is found between**
    a) two metals
    b) a metal and a non-metal
    c) two non-metals
    d) two anions
4.   **Which of the following contains an ionic bond?**
    a) HCl
    b) H_2O
    c) Fe_2O_3
    d) HF
5.   **Which of the following contains a covalent bond?**
    a) Fe_2O_3
    b) Cl_2
    c) NaCl
    d) KCl
6.   **Which of the following contains a polar covalent bond?**
    a) HCl
    b) H_2
    c) NaCl
    d) C_2H_6
7.   **Which of the following contains a non-polar covalent bond?**
    a) HCl
    b) HF
    c) NaCl
    d) CH_4
8.   **The intramolecular forces in H_2O are**
    a) ionic bonds
    b) metallic bonds
    c) non-polar covalent bonds
    d) polar covalent bonds
9.   **Hydrocarbons (C_nH_{2n+2}) show polar covalent bonds.**
    a) True
    b) False
10. **Homonuclear molecules, such as H_2, O_2, I_2 and Cl_2, show non-polar covalent bonds.**
    a) True
    b) False

Q8.2   **Draw the Lewis electron dot diagrams of**
    a) Ca and Ca^{2+}
    b) S and S^{2-}
    c) Ti and Ti^{3+}

Q8.3   **Draw the Lewis structure and determine the number of electron groups on the central atom of**
    a) O_3
    b) H_2O
    c) KCl

d) $TeCl_4$
e) BrF_4^-

Q8.4 **Draw the Lewis structure and calculate the formal charge of the atoms of**
a) CH_3O^-
b) CH_2Br
c) SO_2
d) OCN^-

Q8.5 **Determine the resonance structures of NO_2^-.**

Q8.6 **Draw the Lewis structure and determine the molecular geometry of**
a) $BeCl_2$, CO_2 and HCN
b) CH_2O and O_3
c) CH_4, PF_3 and OF_2
d) PCl_5, SF_4, ClF_2 and XeF_2
e) SF_6, IF_5 and XeF_4

Q8.7 **Draw the molecular orbital diagram of**
a) F_2
b) NO
c) HCl

Q.8.8 **Determine the type of hybridization and the geometry of**
a) NH_3
b) H_2O
c) diamond

Q.8.9 **Determine the type of hybridization and the geometry of benzene (C_6H_6) and graphite.**

ANSWERS TO QUESTIONS

Q8.1
1. b
2. c
3. a
4. c
5. b
6. a
7. d
8. d
9. b
10. a

Q8.2
a) The Lewis dot diagrams of $_{20}$Ca and $_{20}$Ca^{2+} are (Figure 8.67):

$$\ddot{C}a \text{ and } Ca^{2+}$$

FIGURE 8.67 Lewis structures of Ca and Ca^{2+}.

b) The Lewis dot diagrams of $_{16}$S and $_{16}$S^{2-} are (Figure 8.68):

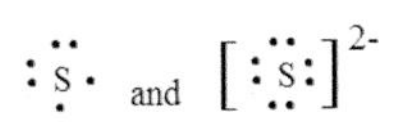

$$:\overset{..}{\underset{.}{S}}\cdot \text{ and } \left[:\overset{..}{\underset{..}{S}}:\right]^{2-}$$

FIGURE 8.68 Lewis structures of S and S^{2-}.

c) The Lewis dot diagrams of $_{22}$Ti and $_{22}$Ti^{3+} are (Figure 8.69):

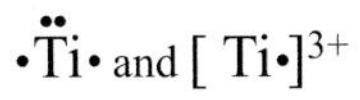

$$\cdot\overset{..}{T}i\cdot \text{ and } [\ Ti\cdot]^{3+}$$

FIGURE 8.69 Lewis structures of Ti and Ti^{3+}.

Q8.3
    a) **Resonance structures of O₃** (Figure 8.70)

$$:\ddot{O}=\ddot{O}-\ddot{O}: \longleftrightarrow :\ddot{O}-\ddot{O}=\ddot{O}:$$

FIGURE 8.70   Resonance structure of O₃.

        There are three electron groups (a lone pair electrons, a single bond and a double bond). The molecular notation is AX_2E.
    b) **Lewis structure of H₂O** (Figure 8.71)

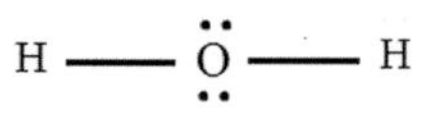

$$H-\ddot{O}-H$$

FIGURE 8.71   Lewis structure of H₂O.

        There are four electron groups (two lone pairs of electrons and two single bonds). The molecular notation is AX_2E_2
    c) **Lewis structure of KCl** (Figure 8.72)

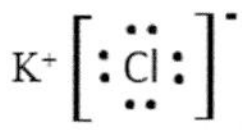

$$K^+\left[:\ddot{Cl}:\right]^-$$

FIGURE 8.72   Lewis structure of KCl.

        KCl is an ionic compound.
    d) **Lewis structure of TeCl₄** (Figure 8.73)

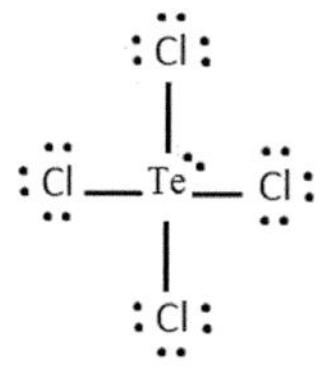

FIGURE 8.73   Lewis structure of TeCl₄.

        There are 5 electron groups (one lone electrons pair and 4 single bonds). The molecular notation is AX_4E
    e) **Lewis structure of BrF₄⁻** (Figure 8.74)

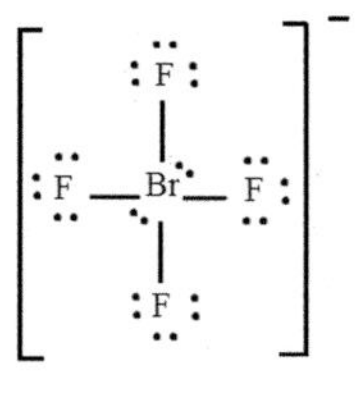

FIGURE 8.74   Lewis structure of BrF₄⁻.

        There are six electron groups (two lone-pair electrons and four single bonds). The molecular notation is AX_4E_2

Q8.4
    a) **Lewis structure and formal charges (FCs) of the atoms of CH₃O⁻** (Figure 8.75)

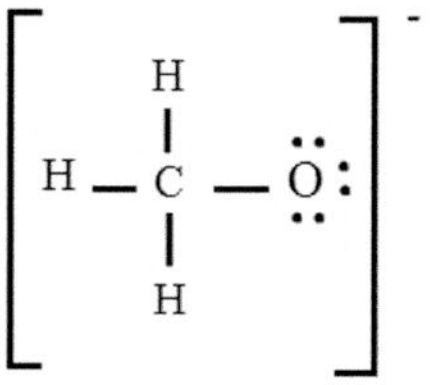

FIGURE 8.75   Lewis structure of CH₃O⁻.

FC of carbon C = 4 − (4 + 0) = 0
FC of hydrogen H = 1 − (1 + 0) = 0 (same FC for the three atoms of H)
FC of oxygen O = 6 − (1 + 6) = −1
Note that the sum of all formal charges equals the overall charge of the CH_3O^- ion
(= −1).

b) **Lewis structure and FCs of the atoms of CH_2Br_2** (Figure 8.76)

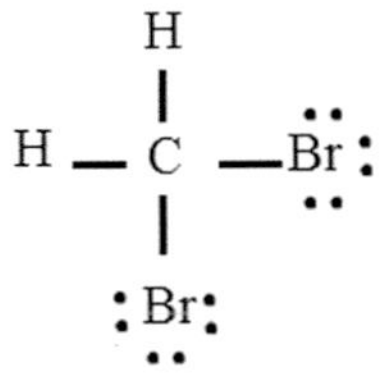

FIGURE 8.76   Lewis structure of CH_2Br_2.

FC of carbon C = 4 − (4 + 0) = 0
FC of hydrogen H = 1 − (1 + 0) = 0 (same FC for the two atoms of H)
FC of bromine Br = 7 − (1 + 6) = 0 (same FC for the two atoms of Br)

c) **Lewis structure and FCs of the atoms of SO_2**

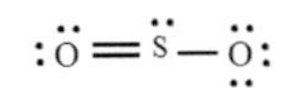

FC of O = 6 − (2 + 4) = 0 (same FC for the two atoms of O)
FC of S = 6 − (4 + 2) = 0

d) **Lewis structure and FCs of the atoms of OCN^-** (Figure 8.77)

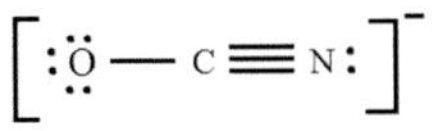

FIGURE 8.77   Lewis structures of OCN^-.

FC of O = 6 − (1 + 6) = −1
FC of C = 4 − (4 + 0) = 0
FC of N = 5 − (3 + 2) = 0

Q8.5 The resonance structures of NO_2^- (Figure 8.78)

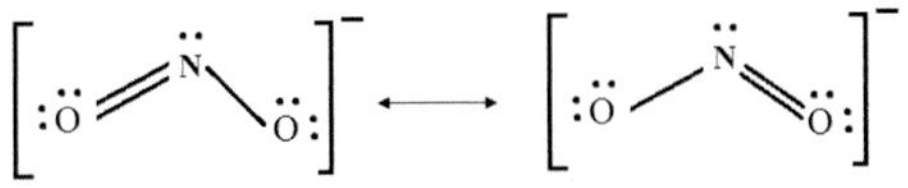

FIGURE 8.78   Resonance structure of NO_2^-.

Q8.6

a) **Lewis structure and molecular geometry of $BeCl_2$** (Figure 8.79), **CO_2** (Figure 8.14)
and HCN (Figure 8.80)

	BeCl₂	**CO₂**	**HCN**
Lewis structure	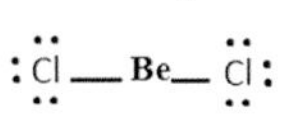	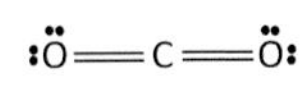	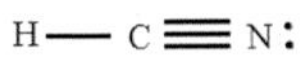
Electron groups and notation	Two groups (two single bonds) AX_2	Two groups (two double bonds) AX_2	Two groups (one single bond and one triple bond) AX_2
Molecular geometry	Linear	Linear	Linear

b)   **Lewis structure and molecular geometry of CH₂O** (Figure 8.81) **and O₃** (Figure 8.82)

	CH₂O	O₃
Lewis structure	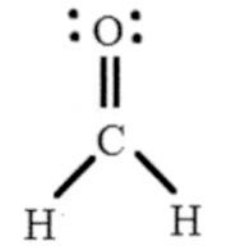	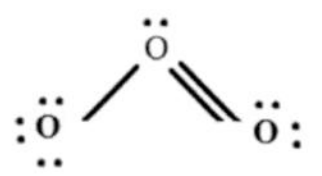
Electron groups and notation	Three groups (two single bonds and one double bonds) AX_3	Three groups (one single bond, one double bond and one lone pair of electrons) AX_2E
Molecular geometry	Trigonal planar	Bent (V-shaped)

c)   **Lewis structure and molecular geometry of CH₄** (Figure 8.41), **PF₃** (Figure 8.83) **and OF₂** (Figure 8.84).

	CH₄	PF₃	OF₂
Lewis structure	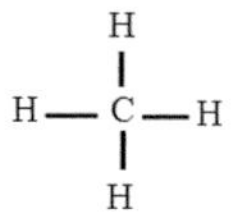	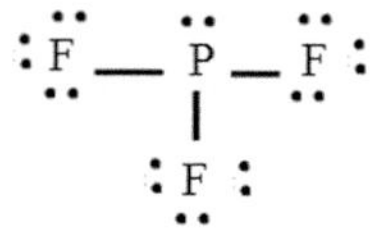	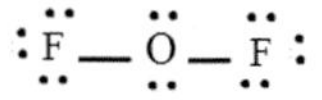
Electron groups and notation	Four groups (four single bonds) AX_4	Four groups (three single bonds and one lone pair of electrons) AX_3E	Four groups (two single bonds and two lone pairs of electrons) AX_2E_2
Molecular geometry	Tetrahedral	Trigonal pyramid	Bent (V-shaped)

d)   **Lewis structure and molecular geometry of PCl₅** (Figure 8.85), **SF₄** (Figure 8.86), **ClF₂** (Figure 8.87) **and XeF₂** (Figure 8.88).

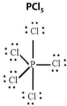

	PCl₅	SF₄	ClF₃	XeF₂
Lewis structure		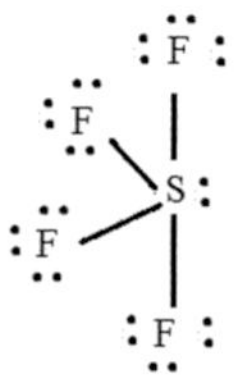	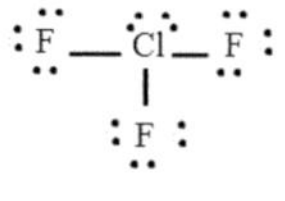	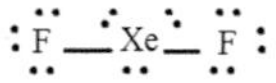
Electron groups and notation	Five groups (five single bonds) AX_5	Five groups (four single bonds and one lone pair of electrons) AX_4E	Five groups (three single bonds and two lone pairs of electrons) AX_3E_2	Five groups (two single bonds and three lone pairs of electrons) AX_2E_3
Molecular geometry	Trigonal bipyramid	See-saw	T-shaped	Linear

e)   **Lewis structure and molecular geometry of SF₆, IF₅ and XeF₄** (Figures 8.89 to 8.91).

	SF₆	IF₅	XeF₄
Lewis structure	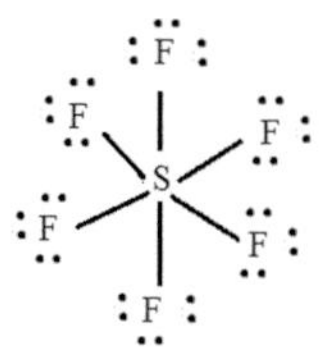	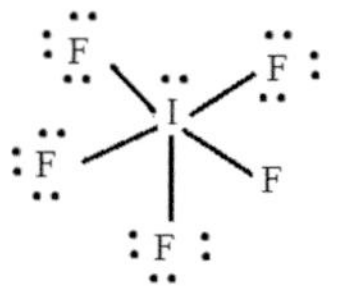	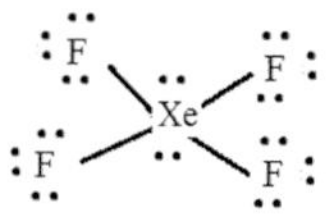
Electron groups and notation	Six groups (six single bonds) AX_6	Five groups (three single bonds and two lone pairs of electrons) AX_5E	Five groups (two single bonds and three lone pairs of electrons) AX_4E_2
Molecular geometry	Octahedral	Square pyramid	Square planar

Q.8.8

a) Hybrid orbitals (sp^3) are obtained from the atomic orbitals 2s and 2p of the nitrogen atom. The three sp^3 hybrid orbitals containing unpaired electrons overlap with the s orbitals of hydrogen atoms in NH_3 to form three (N—H) bonds. The NH_3 molecule has a trigonal pyramid shape.

b) Hybrid orbitals (sp^3) are obtained from the atomic orbitals 2s and 2p of the oxygen atom. The sp^3 hybrid orbitals containing unpaired electrons overlap with the s orbitals of hydrogen atoms in H_2O to form two (O—H) bonds. The molecule of H_2O has a V–shape.

c) Hybrid orbitals (sp^3) are obtained from the atomic orbitals 2s and 2p of the carbon atom. The sp^3 hybrid orbitals containing unpaired electrons overlap with the sp^3 hybrid orbitals of other carbon atoms to form (C—C) bonds. This process can continue indefinitely to lead to the diamond structure consisting of only carbon atoms, all hybridized sp^3.

Q.8.9

In the benzene molecule, each carbon atom has hybridized sp^2 and the chemical bonding is like that of ethylene (C_2H_4). All carbon atoms in benzene (C_6H_6) are sp^2 hybridized and they are all trigonal planar, with bond angles of 120°. The benzene is a flat molecule with a hexagon-like shape.

When the atoms of hydrogen in the benzene molecule are replaced by sp^2 hybridized carbon atoms, an infinite planar structure is obtained. Graphite is obtained by the superposition of several planar structures.

KEY EXPLANATIONS

Q8.1

4. HCl, H_2O and HF show a polar covalent bond. Fe_2O_3 is formed between a metal (Fe) and a non-metal (O). Fe_2O_3 is an ionic compound.

5. Fe_2O_3, NaCl and KCl are ionic compounds. Each one of them is formed between a metal and a non-metal. Cl_2 is a homonuclear molecule. Evidently, the two atoms of Cl have the same electronegativity and are bonded together by a covalent bond.

6. HCl shows a polar covalent bond. Homonuclear molecules, such as H_2, and hydrocarbons, such as C_2H_6, show a non-polar covalent bond. NaCl is formed between a metal (which loses electrons) and a non-metal (which gains electrons) and shows an ionic bond.

7. HCl and HF show a polar covalent bond. Hydrocarbons, such as CH_4, show non-polar covalent bonds. NaCl shows an ionic bond.

8. H_2O shows a polar covalent bond. Both H and O are non-metals and O is more electronegative than H, and the pair of electrons shared between the two atoms is attracted by O.

9. and 10. Hydrocarbons (C_nH_{2n+2}) and homonuclear molecules, such as H_2, O_2, I_2 and Cl_2, show a non-polar covalent bond.

Q8.2

a) $_{20}Ca$: $1s^22s^22p^63s^23p^64s^2$
Calcium has two valence electrons
$_{20}Ca^{2+}$: $1s^22s^22p^63s^23p^6$
$_{20}Ca^{2+}$ has eight valence electrons but, in the Lewis dot diagram for ions, we show the original valence shell of the atom and not the valence shell of the ion. There are no electrons in the original valence shell of calcium. The Lewis dot diagrams of $_{20}Ca$ and $_{20}Ca^{2+}$ are:

$$\ddot{Ca} \text{ and } Ca^{2+}$$

b) $_{16}S$: $1s^22s^22p^63s^23p^4$
Sulfur S has six valence electrons
$_{16}S^{2-}$: $1s^22s^22p^63s^23p^6$
$_{16}S^{2-}$ has eight valence electrons.
The Lewis dot diagrams of $_{16}S$ and $_{16}S^{2-}$ are:

$$:\overset{..}{\underset{.}{S}}\cdot \text{ and } \left[:\overset{..}{\underset{..}{S}}:\right]^{2-}$$

c) $_{22}$Ti: $1s^2 2s^2 2p^6 3s^2 3p^6 4s^2 3d^2$

Titanium has four valence electrons

$_{22}$Ti^{3+}: $1s^2 2s^2 2p^6 3s^2 3p^6 3d^1$ (two electrons are removed from 4s and an electron is removed from 3d).

$_{22}$Ti^{3+} has one valence electron

The Lewis dot diagrams of $_{22}$Ti and $_{22}$Ti^{3+} are:

$\ddot{\cdot}\!\overset{\cdots}{Ti}\!\cdot$ and $[\ Ti\cdot]^{3+}$

Q8.3.

a)   Lewis structure of O_3 (Figure 8.92)

1	**Metal/non-metal and number of valence electrons**	Oxygen is a non-metal
		$_8$O: $1s^2 2s^2 2p^4$; the number of valence electrons is six.
		Since we have three atoms of O, the total number of valence electrons is $3 \times 6 = 18$
2	**Trial structure**	O — O — O (four electrons are shared in the trial structure)
3	**Remaining valence electrons**	$18 - 4 = 14$
4	**Put the remaining electrons on the trial structure**	$\ddot{\underset{\cdot\cdot}{O}} — \ddot{O} — \ddot{\underset{\cdot\cdot}{O}}$
		The central oxygen atom does not have an octet. We rearrange the electrons to form double or triple bonds.
5	**Lewis structure of ozone O_3**	$\ddot{O}\!=\!\ddot{O}\!—\!\ddot{\underset{\cdot\cdot}{O}}\!: \longleftrightarrow :\!\ddot{\underset{\cdot\cdot}{O}}\!—\!\ddot{O}\!=\!\ddot{O}$
		Ozone O_3 has two resonance structures.
6	**Groups on the central atom and molecule notation**	Three (a lone pair of electrons, a single bond and a double bond).
		AX_2E

b)   Lewis structure of H_2O

1	**Metal/non-metal and number of valence electrons**	Both hydrogen and oxygen are non-metals
		$_1$H: $1s^1$; the number of valence electrons is one.
		$_8$O: $1s^2 2s^2 2p^4$; the number of valence electrons is six.
		Since we have two atoms of H and one atom of oxygen, the total number of valence electrons is $2 \times 1 + 6 = 8$
2	**Trial structure**	H — O — H (four electrons are shared in the trial structure)
3	**Remaining valence electrons**	$8 - 4 = 4$
4	**Lewis structure of H_2O**	$H — \ddot{\underset{\cdot\cdot}{O}} — H$
		H has two valence electrons and O has an octet
6	**Groups on the central atom and molecule notation**	Four (two lone pairs of electrons and two single bonds).
		AX_2E_2

c)   Electrons are transferred from a metal to a non-metal to form an ionic compound. In the Lewis structure of the ionic compound, the non-metal (anion) is put in brackets: The Lewis structure of KCl is

$K^+ \left[:\ddot{\underset{\cdot\cdot}{Cl}}: \right]^{\cdot}$

d)   Lewis structure of $TeCl_4$ (Figure 8.93)

1	**Metal/non-metal and number of valence electrons**	Te and Cl are non-metals
		$_{52}$Te: [Kr]$5s^2 5p^4$; the number of valence electrons is six.
		$_{17}$Cl: [Ne]$3s^2 3p^5$; the number of valence electrons is seven.
		Since we have four atoms of Cl and one atom of Te, the total number of valence electrons is $4 \times 7 + 6 = 34$
2	**Trial structure**	

(eight electrons are shared in the trial structure)

| 3 | Remaining valence electrons | $34 - 8 = 26$ |

4 **Put the remaining electrons on the trial structure to obtain the Lewis structure of TeCl₄**

Note that Te has more than an octet

5 **Groups on the central atom and molecule notation**
Five (one lone pair of electrons and four single bonds).
AX_4E

e) Lewis structure of BrF_4^- (Figure 8.94)

1 **Metal/non-metal and number of valence electrons**
Br and F are both non-metals
$_{35}$Br: [Ar]$4s^24p^5$; the number of valence electrons is seven.
$_9$F: [He]$2s^22p^5$; the number of valence electrons is seven.
Since we have four atoms of F, one atom of B, and an electron gained by the molecule (charge −1), the total number of valence electrons is $4 \times 7 + 7 + 1 = 36$

2 **Trial structure**

(8 electrons are shared in the trial structure)

3 **Remaining valence electrons** $36 - 8 = 28$

4 **Put the remaining electrons on the trial structure to obtain the Lewis structure of BrF₄⁻**

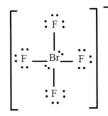

Note that Br has more than an octet

5 **Groups on the central atom and molecule notation**
Six (two lone pairs of electrons and four single bonds).
AX_4E_2

Q8.4
a) Lewis structure and formal charges of the atoms of CH_3O^- (Figure 8.95)

1 **Metal/non-metal and number of valence electrons**
H, C and O are non-metals
$_6$C: [He]$2s^22p^2$; the number of valence electrons is four.
$_8$O: [He]$2s^22p^4$; the number of valence electrons is six.
Since we have an anion (charge −1) containing three atoms of H, one atom of C and one atom of O, the total number of valence electrons is $3 \times 1 + 4 + 6 + 1 = 14$

2 **Trial structure**

```
        H
        |
  H —  C — O
        |
        H
```

(Note that C is less electronegative than O, so C is the central atom, and eight electrons are shared in the trial structure)

3 **Remaining valence electrons** $14 - 8 = 6$

4 **Put the remaining electrons on the trial structure to obtain the Lewis structure of CH₃O⁻**

```
          H            ⁻
          |
    H —  C — O̤ :
          |
          H
```

5 **Formal charge** **FC = n(Ve-) – [b+d]** b: number of bonds d: number of dots (non-bonded electrons).	FC of carbon C = 4 – (4+0) = 0 FC of hydrogen H = 1 – (1+0) = 0 (same FC for the three atoms of H) FC of oxygen O = 6 – (1+6) = –1 Note that the sum of all formal charges equals the overall charge of CH_3O^- ion (= –1).

b)   Lewis structure and formal charges of the atoms of CH_2Br_2 (Figure 8.96)

1	**Metal/non-metal and number of valence electrons**	H, C and Br are non-metals $_6$C: [He]$2s^2 2p^2$; the number of valence electrons is four. $_{35}$Br: [Ar]$4s^2 4p^5$; the number of valence electrons is seven. Hydrogen has one valence electron. Since we have two atoms of H, one atom of C and two atoms of Br, the total number of valence electrons is $2 \times 1 + 4 + 7 \times 2 = 20$
2	**Trial structure**	$$H - \underset{\underset{Br}{|}}{\overset{\overset{H}{|}}{C}} - Br$$ (Note that C is less electronegative than Br, so C is the central atom, and eight electrons are shared in the trial structure)
3	**Remaining valence electrons**	$20 - 8 = 12$
4	**Put the remaining electrons on the trial structure to obtain the Lewis structure of CH_2Br_2**	$$H - \underset{\underset{\ddot{B}\ddot{r}}{|}}{\overset{\overset{H}{|}}{C}} - \ddot{B}\ddot{r}:$$
5	**Formal charge** **FC = n(Ve-) – [b+d]** b: number of bonds d: number of dots (non-bonded electrons).	FC of carbon C = 4 – (4+0) = 0 FC of hydrogen H = 1 – (1+0) = 0 (same FC for the two atoms of H) FC of bromine Br = 7 – (1+6) = 0 (same FC for the two atoms of Br)

c)   Lewis structure and formal charges of the atoms of SO_2 (Figures 8.97 and 8.98)

1	**Metal/non-metal and number of valence electrons**	S and O are non-metals $_8$O: [He]$2s^2 2p^4$; the number of valence electrons is six. $_{16}$S: [Ne]$3s^2 3p^4$; the number of valence electrons is six. Since we have two atoms of O and one atom of S, the total number of valence electrons is $2 \times 6 + 6 = 18$
2	**Trial structure**	$$O - S - O$$ (Note that S is less electronegative than O, so S is the central atom, and four electrons are shared in the trial structure)
3	**Remaining valence electrons**	$18 - 4 = 14$
4	**Put the remaining electrons on the trial structure to obtain the Lewis structure of SO_2**	 The central sulfur atom does not have an octet. We rearrange the electrons to form double or triple bonds. Note that S can accommodate more than an octet.

Formal charge **FC = n(Ve-) – [b+d]** b: number of bonds d: number of dots	$$:\ddot{O} = \overset{..}{S} - \ddot{O}:$$ **Structure 1.** FC of O (showing double bond with S) = 6 – (2+4) = 0 FC of O (showing single bond with S) = 6 – (1+6) = -1 FC of S = 6 – (3+2) = 1	**Structure 2.** FC of O = 6 – (2+4) = 0 (same FC for the two atoms of O) FC of S = 6 – (4+2) = 0 The best Lewis structure of SO_2 (with the fewest number of nonzero FCs).

d)   Lewis structure and formal charges of the atoms of NH_4^+ (Figures 8.99 and 8.100)

1	**Metal/non-metal and number of valence electrons**	H and N are non-metals $_7$N: [He]$2s^2 2p^3$; the number of valence electrons is five. Hydrogen has one valence electron. Since we have a cation (charge +1) containing four atoms of H and one atom of N, the total number of valence electrons is $4 \times 1 + 5 - 1 = 8$

2   **Trial structure**

$$\begin{array}{c} \text{H} \\ | \\ \text{H} - \text{N} - \text{H} \\ | \\ \text{H} \end{array}$$

Eight electrons are shared in the trial structure

3   **Remaining valence electrons** $8 - 8 = 0$

Note that there are no remaining electrons to put on the trial structure

4   **Lewis structure of NH_4^+**

$$\left[\begin{array}{c} \text{H} \\ | \\ \text{H} - \text{N} - \text{H} \\ | \\ \text{H} \end{array}\right]^+$$

5   **Formal charge**
       FC = n(Ve-) – [b + d]
       b: number of bonds
       d: number of dots

FC of H $= 1 - (1 + 0) = 0$ (same FC for the four atoms of H)

FC of N $= 5 - (4 + 0) = 1$

Note that the sum of all formal charges equals the overall charge of NH_4^+ ion $(= +1)$.

e)   Lewis structure and formal charges of the atoms of OCN^- (Figures 8.101 to 8.105)

1   **Metal/non-metal and number of valence electrons**

C, N and O are non-metals

$_8$O: [He]$2s^22p^4$; the number of valence electrons is six.

$_6$C: [He]$2s^22p^2$; the number of valence electrons is four.

$_7$N: [He]$2s^22p^3$; the number of valence electrons is five.

Since we have an anion (charge −1) containing one atom of O, one atom of C and one atom of N, the total number of valence electrons is

$6 + 4 + 5 + 1 = 16$

2   **Trial structure**

$$\text{O} - \text{C} - \text{N}$$

(Note that C is the least electronegative, so C is the central atom, and four electrons are shared in the trial structure)

3   **Remaining valence electrons** $16 - 4 = 12$

4   **Put the remaining electrons on the trial structure to obtain the Lewis structure of OCN^-**

$$:\!\ddot{\text{O}} - \text{C} - \ddot{\text{N}}\!:$$

The central C atom does not have an octet. We rearrange the electrons to form double or triple bonds:

Structure 1

$$\left[:\!\ddot{\text{O}} = \text{C} = \ddot{\text{N}}\!:\right]^-$$

Structure 2

$$\left[:\!\text{O} \equiv \text{C} - \ddot{\text{N}}\!:\right]^-$$

Structure 3

$$\left[:\!\ddot{\text{O}} - \text{C} \equiv \text{N}\!:\right]^-$$

5   **Formal charge**
       FC = n(Ve-) – [b + d]
       b: number of bonds
       d: number of dots

Structure 1:
FC of O $= 6 - (2 + 4) = 0$
FC of C $= 4 - (4 + 0) = 0$
FC of N $= 5 - (2 + 4) = -1$

Structure 2:
FC of O $= 6 - (3 + 2) = 1$
FC of C $= 4 - (4 + 0) = 0$
FC of N $= 5 - (1 + 6) = -2$

Structure 3:

FC of $O = 6 - (1 + 6) = -1$
FC of $C = 4 - (4 + 0) = 0$
FC of $N = 5 - (3 + 2) = 0$

Structures 1 and 3 have the fewest number of non-zero formal charges, but structure 3 has the negative formal charge on one of the more electronegative atom (O). Therefore, structure 3 is the best Lewis structure of OCN^-.

Q8.5 The resonance structures of NO_2^- (Figures 8.106 and 8.107).

1 Metal/non-metal and number of valence electrons	N and O are nonmetals $_8O$: $[He]2s^22p^4$; the number of valence electrons is six. $_7N$: $[He]2s^22p^3$; the number of valence electrons is five. Since we have an anion (charge -1) containing two atoms of O and one atom of N, the total number of valence electrons is $2 \times 6 + 5 + 1 = 18$
2 Trial structure	$$O \underline{\quad} N \underline{\quad} O$$ (Note that N is the least electronegative, so N is the central atom, and four electrons are shared in the trial structure)
3 Remaining valence electrons	$18 - 4 = 14$
4 Put the remaining electrons on the trial structure to obtain the Lewis structure of OCN⁻	 The central N atom does not have an octet. We rearrange the electrons to form double or triple bonds:

Resonance structures:

Q8.6

a) Lewis structure and molecular geometry of $BeCl_2$, CO_2 and HCN (Figures 8.108 to 8.110)

	BeCl₂	**CO₂**	**HCN**
Valence electrons	Be has two valence electrons and Cl has seven valence electrons: The total number of valence electrons is $2 + 7 \times 2 = 16$	C has four valence electrons and O has six valence electrons: The total number of valence electrons is $4 + 6 \times 2 = 16$	H has one valence electron, C has four valence electrons and N has five valence electrons: The total number of valence electrons is $1 + 4 + 5 = 10$
Trial structure/remaining valence electrons	$Cl \underline{\quad} Be \underline{\quad} Cl$ Remaining electrons: $16 - 4 = 12$	$O \underline{\quad} C \underline{\quad} O$ Remaining electrons: $16 - 4 = 12$	$H \underline{\quad} C \underline{\quad} N$ Remaining electrons: $10 - 4 = 6$
Lewis structure			
Electron groups and notation	Two groups (two single bonds) AX_2	Two groups (two double bonds) AX_2	Two groups (one single bond and one triple bond) AX_2
Molecular geometry	Linear	Linear	Linear

b) Lewis structure and molecular geometry of CH_2O and O_3 (Figures 8.111 and 8.112)

	CH₂O	**O₃**
Valence electrons	C has 4 valence electrons, H has one valence electron and O has 6 valence electrons: The total number of valence electrons is $4 + 1 \times 2 + 6 = 12$	O has 6 valence electrons: The total number of valence electrons is $6 \times 3 = 18$

Trial structure/remaining electrons	Remaining electrons: 12 − 6 = 6	O — O — O Remaining electrons: 18 − 4 = 14
Lewis structure		
Electron groups and notation	3 groups (2 single bonds and one double bonds) AX_3	3 groups (One single bond, one double bond and one lone pair electrons) AX_2E
Molecular geometry	Trigonal planar	Bent (V-shaped)

c) **Lewis structure and molecular geometry of CH_4, PF_3 and OF_2** (Figures 8.41, 8.113 to 8.118).

	CH_4	PF_3	OF_2
Valence electrons	C has 4 valence electrons and H has one valence electron: The total number of valence electrons is $4 + 4 \times 1 = 8$	P has 5 valence electrons and F has 7 valence electrons: The total number of valence electrons is $5 + 3 \times 7 = 26$	O has 6 valence electrons, F has 7 valence electrons: The total number of valence electrons is $6 + 2 \times 7 = 20$
Trial structure/ remaining valence electrons	H—C—H (H top and bottom) Remaining electrons: $8 - 8 = 0$	F—P—F (F below) Remaining electrons: $26 - 6 = 20$	F—O—F Remaining electrons: $20 - 4 = 16$
Lewis structure	H—C—H (H top and bottom)	$:\!F\!-\!P\!-\!F\!:$ with $:F:$ below	$:\!F\!-\!O\!-\!F\!:$
Electron groups and notation	4 groups (4 single bonds) AX_4	4 groups (3 single bonds and one lone pair electrons) AX_3E	4 groups (2 single bonds and 2 lone pair electrons) AX_2E_2
Molecular geometry	Tetrahedral	Trigonal pyramid	Bent (V-shaped)

d) **Lewis structure and molecular geometry of PCl_5, SF_4, ClF_2 and XeF_2.**

	PCl_5	SF_4	ClF_3	XeF_2
Valence electrons	P has five valence electrons and Cl has seven valence electrons. The total number of valence electrons is $5 + 5 \times 7 = 40$	S has six valence electrons and F has seven valence electrons. The total number of valence electrons is $6 + 4 \times 7 = 34$	Cl has seven valence electrons and F has seven valence electrons. The total number of valence electrons is $7 + 3 \times 7 = 28$	Xe has eight valence electrons and F has seven valence electrons. The total number of valence electrons is $8 + 2 \times 7 = 22$
Trial structure/ remaining valence electrons	(PCl5 trial structure) Remaining electrons: $40 - 10 = 30$	(SF4 trial structure) Remaining electrons: $34 - 8 = 26$	F—Cl—F (F below) Remaining electrons: $28 - 6 = 22$	F—Xe—F Remaining electrons: $22 - 4 = 18$
Lewis structure	(PCl5 Lewis structure)	(SF4 Lewis structure)	$:\!F\!-\!Cl\!-\!F\!:$ with $:F:$ below	$:\!F\!-\!Xe\!-\!F\!:$

Electron groups and notation	Five groups (five single bonds) AX_5	Five groups (four single bonds and one lone pair electrons) AX_4E	Five groups (three single bonds and two lone pair electrons) AX_3E_2	Five groups (two single bonds and three lone pair electrons) AX_2E_3
Molecular geometry	Trigonal bipyramid	See-saw	T-shaped	Linear

e) **Lewis structure and molecular geometry of SF_6, IF_5 and XeF_4 (Figures 8.119 to 8.121).**

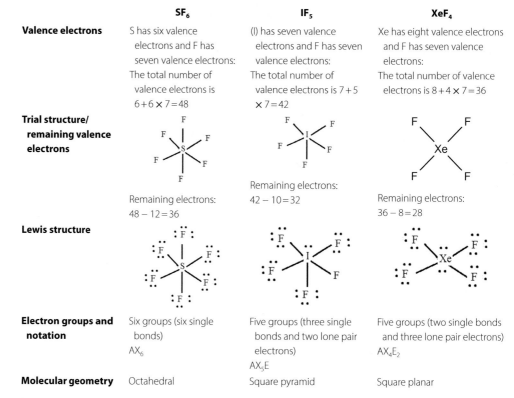

	SF_6	**IF_5**	**XeF_4**
Valence electrons	S has six valence electrons and F has seven valence electrons: The total number of valence electrons is $6 + 6 \times 7 = 48$	(I) has seven valence electrons and F has seven valence electrons: The total number of valence electrons is $7 + 5 \times 7 = 42$	Xe has eight valence electrons and F has seven valence electrons: The total number of valence electrons is $8 + 4 \times 7 = 36$
Trial structure/ remaining valence electrons	Remaining electrons: $48 - 12 = 36$	Remaining electrons: $42 - 10 = 32$	Remaining electrons: $36 - 8 = 28$
Lewis structure			
Electron groups and notation	Six groups (six single bonds) AX_6	Five groups (three single bonds and two lone pair electrons) AX_5E	Five groups (two single bonds and three lone pair electrons) AX_4E_2
Molecular geometry	Octahedral	Square pyramid	Square planar

Q.8.8

a) The electronic configuration of nitrogen N is
$_7N$: $1s^2 2s^2 2p^3$

The three higher-energy p orbitals and the lower-energy s orbital are hybridized into four equal-energy sp^3 orbitals. Three sp^3 hybrid orbitals contain unpaired electrons and one sp^3 hybrid orbitals contain lone pair electrons. The three sp^3 hybrid orbitals containing unpaired electrons overlap with the s orbitals of hydrogen atoms in NH_3 to form three (N—H) bonds (Figure 8.122). The basic geometry of NH_3 is tetrahedral. In the basic geometry, an atom is replaced by a lone electron pair and the bond angles are reduced from 109.5° to 107° due to the lone pair repulsion, and hence the molecule has a trigonal pyramid shape (Figure 8.123).

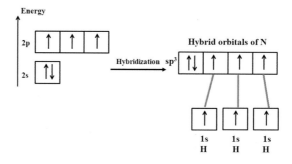

FIGURE 8.122 Hybrid orbitals (sp^3) obtained from the atomic orbitals 2s and 2p of a nitrogen atom. The three sp^3 hybrid orbitals containing unpaired electrons overlap with the s orbitals of hydrogen atoms in NH_3 to form three (N—H) bonds.

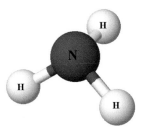

FIGURE 8.123   The NH_3 molecule has a trigonal pyramid shape (drawn with Molview software).

b)   The electronic configuration of oxygen is

$_8O: 1s^2 2s^2 2p^4$

The three higher-energy p orbitals and the lower-energy s orbital are hybridized into four equal-energy sp^3 orbitals. Two sp^3 hybrid orbitals contain unpaired electrons and two sp^3 hybrid orbitals contain lone pair electrons. The sp^3 hybrid orbitals containing unpaired electrons overlap with the s orbitals of hydrogen atoms in H_2O to form two (O—H) bonds (Figure 8.124). The basic geometry of H_2O is tetrahedral. In the basic geometry, two atoms are replaced by two lone electron pairs and the bond angles are reduced from 109.5° to 104.5° due to greater lone pair repulsion and hence the molecule is bent and adopts a V–shape (Figure 8.125).

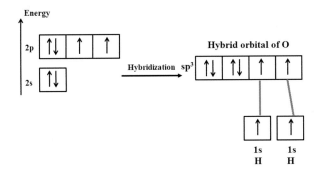

FIGURE 8.124   Hybrid orbitals (sp^3) obtained from the atomic orbitals 2s and 2p of oxygen. The sp^3 hybrid orbitals containing unpaired electrons overlap with the s orbitals of hydrogen atoms in H_2O to form two (O—H) bonds.

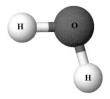

FIGURE 8.125   The molecule of H_2O has a V–shape (drawn with Molview software).

c)   The electronic configuration of carbon is

$_6C: 1s^2 2s^2 2p^2$

The three higher-energy p orbitals and the lower-energy s orbital are hybridized into four equal-energy sp^3 orbitals. Each orbital of the four sp^3 hybrid orbitals contains an unpaired electron. These orbitals overlap with four sp^3 hybrid orbitals of four carbon atoms to form four (C—C) bonds. We obtain a tetrahedral molecule where each of the carbon atoms at the summit of the tetrahedral has three remaining sp^3 hybrid orbitals containing unpaired electron, that can overlap with sp^3 hybrid orbitals of other carbon atoms (Figure 8.126). This process can continue indefinitely to lead to the diamond structure, which is a tridimensional structure (Figure 8.127) where carbon atoms are sp³ hybridized.

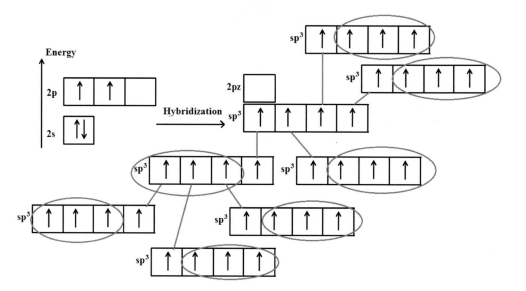

FIGURE 8.126   Hybrid orbitals (sp³) obtained from the atomic orbitals 2s and 2p of carbon atom. The sp³ hybrid orbitals containing unpaired electrons overlap with the sp³ hybrid orbitals of other carbon atoms to form (C—C) bonds. This process can continue indefinitely to lead to the diamond structure consisting of only carbon atoms, all hybridized sp³.

FIGURE 8.127   Tridimensional structure of diamond.

Q.8.9 In a benzene molecule, each carbon is hybridized sp² and the chemical bonding is like that of ethylene (C_2H_4). Indeed, each carbon atom forms two bonds with hydrogen atoms and one bond with the second carbon from the three hybrid orbitals sp². The unhybridized 2p orbitals from each carbon form the π bond. Thus, for two neighboring carbon atoms in the benzene (C_6H_6) molecule, we have (Figure 8.128):

• Axial overlaps of the 1s orbitals of hydrogen atoms with the hybridized orbitals sp² of each carbon atom.
• An axial overlap of the sp² orbitals of two carbon atoms which forms the σ bond.
• A lateral overlap of the unhybridized 2pz orbitals of the carbon atom which forms the π bond.

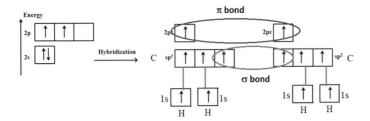

FIGURE 8.128   Axial and lateral overlaps of the hybridized unhybridized orbitals of two neighboring carbon atoms in a benzene molecule

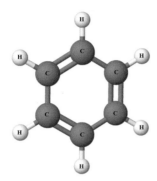

FIGURE 8.129   All carbon atoms in benzene C_6H_6 are sp² hybridized and they are all trigonal planar, with bond angles of 120°. The benzene is a flat molecule with a hexagon-like shape (drawn with Molview software)

When the atoms of hydrogen in the benzene molecule (Figure 8.129) are replaced by sp² hybridized carbon atoms, an infinite planar structure is obtained. Graphite is obtained by the superposition of several planar structures separated by 3.35 Å (1 Å = 10⁻¹⁰ m). The distance between two carbon atoms in the planar structure is 1.42 Å (Figure 8.130).

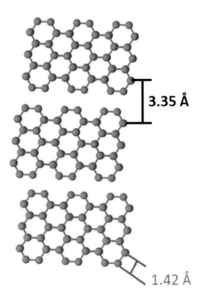

FIGURE 8.130   The structure of graphite. Each sphere represents a carbon atom (drawn with Molview software).

Intermolecular Forces and Properties of Matter

9

9.1 OBJECTIVES

At the end of the present chapter, the student will be able to:

1. Differentiate between intramolecular and intermolecular forces.
2. Identify types of bonds within a molecule and interactions between molecules.
3. Describe the properties of matter based on intermolecular forces.
4. Define the dynamic equilibrium between two phases.
5. Interpret Raoult's law for an ideal solution.

9.2 MOLECULAR FORCES

The atoms of a molecule are held together by intramolecular interactions which include covalent bonds, ionic bonds and metallic bonds. These bonds are described, in detail, in Chapter 8. Intermolecular forces hold molecules to other molecules by intermolecular interactions (Figure 9.1) which are electrostatic in nature and include van der Waals forces and hydrogen bonds. The intramolecular interactions are stronger than the intermolecular forces.

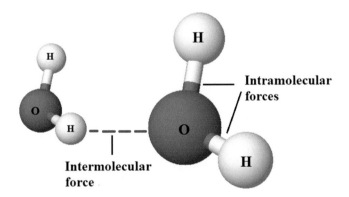

FIGURE 9.1 Intramolecular and intermolecular forces for H_2O molecules (drawn by MolView software).

9.2.1 DIPOLE MOMENT AND POLARIZABILITY

A non-polar molecule can be formed between two identical non-metals that share electrons equally between them, e.g., H_2, O_2, N_2 and Cl_2 (Figure 9.2). Non-polar molecules have spherically symmetrical arrangements of their electronic clouds. When in the presence of an electric field, their electron clouds can be distorted. Because of this, one atom has a partially positive charge $(+\delta)$ and the other atom has a partially negative charge $(-\delta)$. The ease of this distortion is defined as the polarizability (denoted by α) of the atom or molecule. The distortion of the electron cloud causes the originally non-polar molecule or atom to acquire a dipole moment, and an induced dipole is obtained.

DOI: 10.1201/9781003257059-9

Cl : Cl

FIGURE 9.2   There is an equal sharing of electrons in Cl_2. The dipole moment is $\mu = 0$, because the atoms forming the molecule have identical electronegativities. Cl_2 is a non-polar molecule. The two atoms of chlorine are held together by a non-polar covalent bond.

In polar molecules, such as H_2O, HCl, HBr, HF and SO_2, the electrons are unequally shared between the atoms forming the molecule and the atoms are held together by a polar covalent bond (Figure 9.3). For example, in a diatomic molecule A-B, the symmetry of the electronic cloud disappears, and a permanent dipole is obtained. The distribution of the electronic density around the two nuclei becomes more asymmetrical (distorted) and the dipole moment μ ($\mu = \delta \times d$, where δ is the partial charge and d is the length of the bond A-B) is greater as the difference in electronegativity between A and B increases.

FIGURE 9.3   Cl is more electronegative than H. HCl is a polar molecule. H and Cl atoms are held together by a polar covalent bond. The electron pair involved in the covalent bond is close to chlorine and creates a permanent partial negative charge ($-\delta$) on the chlorine atom and a permanent partial positive charge ($+\delta$) on the hydrogen atom. A permanent dipole is obtained ($0 < \delta < q$).

Note that when the dipole moment $\mu = 0$, the atoms forming the molecule have identical electronegativities and they are held together by a non-polar covalent bond. When $0 < \delta < q$, where q is the electric charge, atoms are held together by a polar covalent bond, such as for the HCl molecule. When the dipole moment $\mu = q \times d$, such as in the NaCl molecule (Figure 9.4), a transfer of electrons between the atoms occurs and an ionic bond is formed between them.

$$Na_{\cup}\ddot{\underset{..}{:}}\ddot{Cl}:$$

FIGURE 9.4   In NaCl, the dipole moment $\mu = q \times d$. The sodium atom loses an electron, forming a cation (Na^+) and the chlorine atom loses an electron, forming an anion (Cl^-). An ionic bond is formed between the two ions.

It is interesting to note that the dipole moment of a polyatomic molecule is the vectorial sum of the dipole moment of the different bonds in the molecule. If the individual bond dipole moments cancel one another, the dipole moment of the molecule equals zero.This can explain why certain molecules such as carbon dioxide (CO_2) and hydrocarbons (C_nH_{2n}) are non-polar molecules. For example, in the CO_2 molecule, the two bonds C=O have a non-zero dipole moment because the C and O atoms have different electronegativities. However, the dipole moment of the CO_2 molecule equals zero ($\mu = 0$). This can only be explained when the molecule presents a linear structure. However, the individual bond dipole moments of the different bonds in SO_2 molecule, for example, do not cancel one another since it has a V-shape. The dipole moment of SO_2 molecule is not equal to zero ($\mu \neq 0$) and it is a polar molecule (Figure 9.5). In the same manner, we can explain why O_3 molecule is a polar molecule.

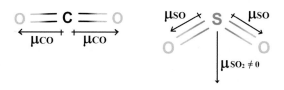

FIGURE 9.5 CO_2 is a non-polar molecule. The two C=O bonds have a non-zero dipole moment ($\mu_{CO} \neq 0$), however $\mu_{CO2} = 0$ because the molecule is linear. SO_2 is a polar molecule that has a V-shape and $\mu_{SO2} \neq 0$.

9.2.2 INTERMOLECULAR INTERACTIONS

Intermolecular forces hold molecules to other molecules and include van der Waals forces, hydrogen bonds, ion–dipole interactions, and ion-induced dipole interactions. The van der Waals forces are classified into the Keesom forces between permanent dipoles, the Debye forces between permanent dipoles and induced dipoles and the London dispersion forces between instantaneous dipoles induced by the movement of electrons.

9.2.2.1 VAN DER WAALS FORCES

o **Keesom forces, or dipole–dipole forces,** have a moderate strength and act between polar molecules in which the permanent dipoles interact with one another. Oppositely charged ends attract and like-charged ends repel. Note that a permanent dipole is obtained when there is a difference in electronegativity between the atoms forming the molecule. For example, in the HCl molecule, the chlorine atom is more electronegative than the hydrogen atom and the electron pair involved in the covalent bond is close to the chlorine atom, which creates a permanent partial negative charge on the chlorine atom and a permanent partial positive charge on the hydrogen atom (Figure 9.6). Thus, a permanent dipole is obtained. Keesom forces are found, for example, in HCl, HBr, and HF.

Keesom forces are affected by the dipole moment, μ, of the molecule, the absolute temperature, T, and the distance, r, between the molecules. Indeed, the Keesom (dipole–dipole) forces increase as the dipole moment of the molecule increases. However, when the temperature increases or when the distance between the molecules increases, the Keesom forces become weaker.

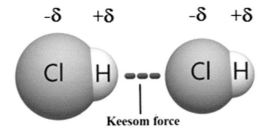

FIGURE 9.6   Keesom forces are found between HCl molecules. Cl is more electronegative than H and the electron pair involved in the covalent bond is close to Cl, which creates a permanent partial negative charge on the Cl atom and a permanent partial positive charge on the H atom (drawn with MolView software).

o **Debye forces** occur between a polar molecule and a non-polar molecule. Debye forces can be described as an induction effect. Indeed, the permanent dipole in the polar molecule induces an electric dipole in the non-polar molecule when the polar and non-polar molecules come extremely close to one another. For example, the intermolecular forces between water (H_2O), which is a polar molecule, and oxygen (O_2), which is a non-polar molecule, are Debye forces. When O_2 comes close to H_2O, the partial negative charge on the oxygen atom of the H_2O molecule repels the electrons in the non-polar O_2 molecule, which becomes temporarily a dipole. Indeed, this electron repulsion induces a temporary partial positive charge in the oxygen side close to H_2O molecule, and a partial negative charge in the oxygen side farther away (Figure 9.7). Another example of an induction interaction between a permanent dipole and an induced dipole is the interaction between hydrochloric acid HCl and argon (Ar). Indeed, when Ar comes extremely close to HCl, it experiences a dipole as its electrons are attracted to the H side of HCl or repelled from the Cl side by HCl.

Debye forces are weaker than Keesom forces because the dipole in the nonpolar molecule involved in the Debye forces is temporary. Debye forces depend on the dipole moment, μ, of the molecules that interact together, the polarizability, α, the absolute temperature, T, and the distance, r, between the molecules. When the dipole moment or the polarizability increases, the Debye forces become stronger. However, when the temperature or the distance increases, Debye forces become weaker.

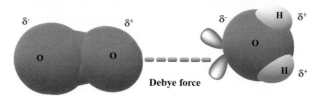

FIGURE 9.7 O_2 is a non-polar molecule. However, when it comes close to a H_2O molecule, the partial negative charge on the oxygen atom of the H_2O molecule repels the electrons in O_2 and the non-polar O_2 molecule becomes temporarily a dipole. The interaction between O_2 and H_2O is an interaction between a permanent dipole (the H_2O molecule) and an induced dipole (O_2) molecule. That is the Debye force.

○ **London dispersion forces** occur between non-polar molecules. Non-polar molecules may have an instantaneous dipole moment even in the absence of a polarizing dipole. With the electrons being in movement, it is not impossible that, at a given instant in time, the electronic cloud deforms, thus creating an instantaneous dipole. The latter can polarize a nearby molecule that will have a temporarily induced dipole moment. Thus, two non-polar molecules can induce polarity on each other because of the movement of electrons. The opposite partial charges of neighboring molecules attract (Figure 9.8). For example, the intermolecular forces between iodine molecules I_2, which are non-polar molecules, are London dispersion forces. The interaction between helium (He) atoms are also London dispersion forces. Indeed, the He atom is non-polar and the distribution of the electrons around the nucleus is spherically symmetrical. However, at a given instant and due to the movement of electrons, the distribution of the electrons may no longer be symmetrical, and the helium atom becomes an instantaneous dipole. This atom can polarize a nearby atom that will have a temporarily induced dipolar moment. The attraction between the neighboring induced dipoles causes atoms to hold together for a short time. London dispersion forces occur in non-polar molecules, such as O_2, I_2, Cl_2, Br_2, F_2, H_2, CO_2 and hydrocarbons, C_nH_{2n+2}.

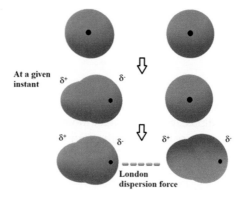

FIGURE 9.8   Two non-polar molecules (or non-polar atoms) can induce polarity in each other because of the movement of electrons. With the electrons being in movement, the electronic cloud may deform at a given instant in time, thus creating an instantaneous dipole. The opposite partial charges of neighboring molecules are attracted, causing neighboring induced dipoles to attract together for a very short time. The dots indicate the location of the nucleus.

London dispersion forces are the weakest of all the intermolecular forces. The strength of London dispersion forces depends essentially on the size of the nonpolar atoms or molecules. Indeed, when the size is greater, the number of electrons is greater, and the electrons are farther from the positive nucleus. The electrons become held less strongly and they move more freely. Consequently, the polarizability of the atom increases, and London dispersion forces become stronger. For example, the London dispersion forces in iodine (I_2) are stronger than in bromine (Br_2) because the atomic size of the iodine (I) atom is greater than that of the bromine (Br) atom. The London dispersion forces in Br_2 are stronger than those in chlorine (Cl_2) or fluorine (F_2) due to the greater size of the Br atom. This explains also why, at room temperature, F_2 and Cl_2 are gases, Br_2 is a liquid and I_2 is a solid.

For hydrocarbons, a molecule with a longer chain has stronger London dispersion forces because there are more points of attachment *via* London dispersion forces (Figure 9.9). For example, butane molecules (C_4H_{10}) show stronger London dispersion forces than do ethane molecules (C_2H_6):

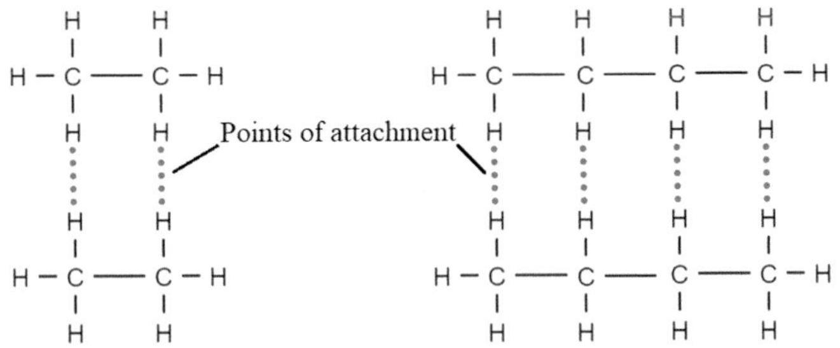

FIGURE 9.9   For hydrocarbons, the molecule with the longer chain has stronger London dispersion forces because there are more points of attachment *via* London dispersion forces.

GET SMART

TABLE 9.1   Energies of interactions between the different dipoles involved in van der Waals forces. Keesom (K), Debye (D) and London (L) are constant

Forces	Energy	The Energy Increases When:
Keesom forces	$E_K = -K\dfrac{\mu^4}{T \times r^6}$	• Dipole moment, μ, increases. • Temperature, T, or distance, r, decreases.
Debye forces	$E_D = -D\dfrac{\alpha \times \mu^2}{r^6}$	• Polarizability, α, or dipole moment, μ, increases. • Distance, r, between the molecules decreases.
London forces	$E_L = -L\dfrac{\alpha^2}{r^6}$	• Polarizability, α, increases. • Distance, r, between the molecules decreases.

9.2.2.2 HYDROGEN BOND

A hydrogen bond is a strong intermolecular force that occurs between a hydrogen atom in one polar molecule and a strongly electronegative atom, such as nitrogen N, oxygen O or fluorine F, on another polar molecule. Therefore, hydrogen bonding occurs between permanent dipoles. For example, a hydrogen bond occurs between water (H_2O) molecules (Figure 9.10), hydrofluoric acid (HF) molecules and ammonia (NH_3) molecules. Hydrogen bonds are responsible for many unusual properties of water, such as abnormally low vapor pressure and high boiling point compared with H_2S, H_2Se and H_2Te, for example. Hydrogen bonds can also occur in molecules of carboxylic acids, aldehydes and alcohols. A hydrogen bond is the strongest molecular force of all the van der Walls forces.

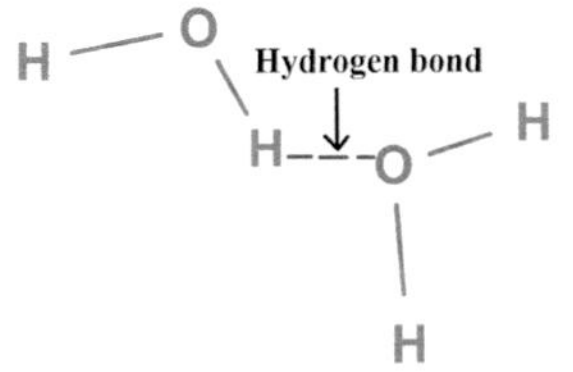

FIGURE 9.10   Hydrogen bond in H_2O.

Note that the hydrogen bond can be observed within a molecule such as the *o*-hydroxybenzal-dehyde molecule, so it can be considered to be an intramolecular force (Figure 9.11).

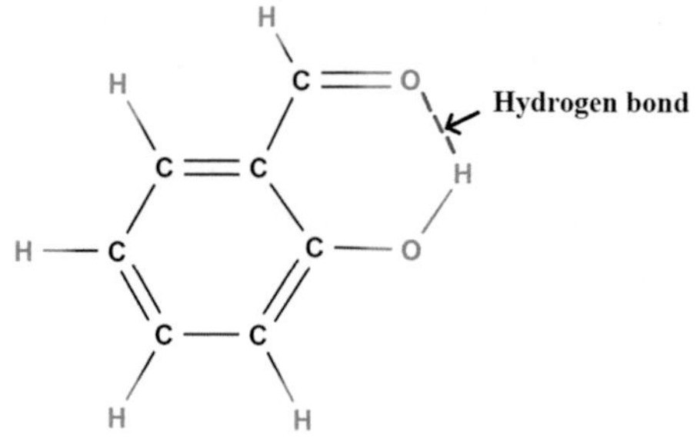

FIGURE 9.11   The hydrogen bond is observed within the *o*-hydroxybenzaldehyde molecule, meaning that it can be considered to be an intramolecular force.

Practice 9.1 Which of the following molecules, C_4H_{10}, NH_3, HF and CH_3COOH, show hydrogen bonds? For those that do, draw two molecules of the substance with the hydrogen bonds between them.

Answer:
C_2H_6 is a hydrocarbon. C_2H_6 is a non-polar molecule so this substance does not exhibit hydrogen bonds which occur between polar molecules.

NH_3, HF, CH_3COOH exhibit hydrogen bonds (Figure 9.12).

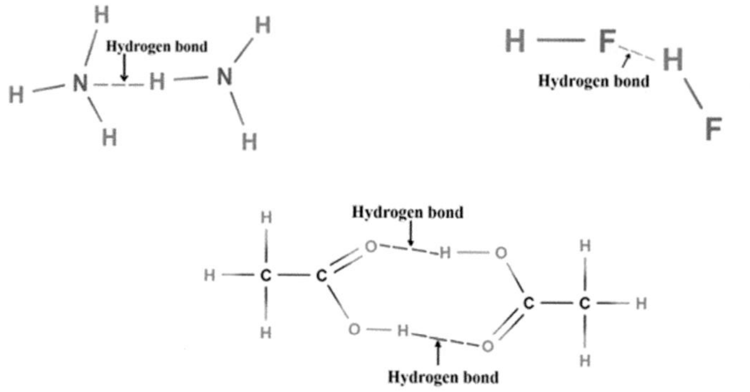

FIGURE 9.12   Hydrogen bonds in NH_3, HF and CH_3COOH.

9.2.2.3 ION–DIPOLE INTERACTION

Ion–dipole forces occur between an ion (cation or anion) and a polar molecule (a dipole) (Figure 9.13). This type of interaction is generally found in solutions where ionic compounds are dissolved in polar liquids, e.g., the dissolution of the salt sodium chloride (NaCl) in water (H_2O). The cation (Na^+) attracts the partial negative charge of the oxygen atom of water, and the anion (Cl^-) attracts the partial positive charge of the hydrogen atoms of water.

FIGURE 9.13   Anion–dipole interaction (on the left) and cation–dipole interaction (on the right).

9.2.2.4 ION-INDUCED DIPOLE INTERACTION

Ion-induced dipole forces occur between an ion (cation or anion) and a non-polar molecule. When the ion becomes close to a nonpolar molecule (or an atom), it causes the distortion of the electron cloud, and the non-polar molecule (or atom) acquires a dipole moment and an induced dipole is obtained. Then, an interaction between the ion and the induced dipole is established (Figure 9.14). For example, in the presence of nitrate (NO_3^-) ion, the non-polar iodine (I_2) molecule becomes polarized. Another example of the ion–dipole induced interaction is between a ferrous ion (Fe^{2+}) and an oxygen (O_2) molecule.

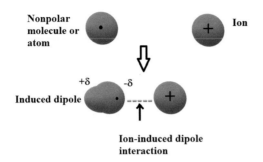

FIGURE 9.14   Ion-induced dipole interaction

GET SMART

The strength of molecular forces from the strongest to the weakest is as follows:

1. For intramolecular bonds:
   Ionic bond > Covalent bond > Metallic bond.
2. For intermolecular forces:
   Ion–dipole interaction > Ion-induced dipole interaction > Hydrogen bond > van der Waals forces.
3. For the van der Waals Forces:
   Keesom forces > Debye forces > London dispersion forces.

Note that the intramolecular forces are stronger than the intermolecular forces.

9.3 PROPERTIES OF MATTER

9.3.1 VAPOR PRESSURE

The condition under which two opposing processes are occurring simultaneously at equal rates in a closed system is called a dynamic equilibrium. For example, in a closed system, when the rate at which the liquid is entering the gas phase is equal to the rate at which the vapor is returning to the liquid phase, the system is at equilibrium. After this time, the liquid level will remain constant. The pressure exerted by the vapor, at a given temperature, on the surface of the liquid at equilibrium is called the vapor pressure.

Vapor pressure is a property of the liquid which depends on neither the volume nor the surface area of the liquid. The vapor pressure is expressed in the SI unit Pascal (Pa) where one Pascal equals one Newton per square meter (1 Pa = N m^{-2}). The vapor pressure of a liquid depends on the intermolecular forces that are established in the liquid, the temperature and the presence of a solute in the liquid. Indeed, the stronger the intermolecular forces, the lower the vapor pressure. For example, the vapor pressure of water (H_2O) is lower than that of hydrogen sulfide (H_2S) or diethyl ether [(C_2H_5)$_2$O] because water molecules are held to one another by hydrogen bonds, that are stronger than the Keesom forces (dipole–dipole interactions) between H_2S molecules or the London dispersion forces between [(C_2H_5)$_2$O] molecules. The substances that have a high vapor pressure at room temperature are called volatiles.

When the temperature increases, the kinetic energy of the liquid molecules increases. Thus, the molecules of the liquid become more excited and escape the surface of the liquid easier, thereby increasing the vapor pressure. Conversely, when the temperature decreases, the kinetic energy of the liquid molecules decreases and the number of molecules in the vapor phase decreases, thereby decreasing the vapor pressure of the liquid. The Clausius–Clapeyron equation gives the quantitative relation between the vapor pressures of a substance at two different temperatures:

$$\ln\frac{P_2}{P_1} = -\frac{\Delta H_{vap}}{R}\left(\frac{1}{T_2} - \frac{1}{T_1}\right)$$

where P_1 and P_2 are the vapor pressures at the temperatures T_1 and T_2, respectively. ΔH_{vap} is the enthalpy of vaporization (expressed in J mol^{-1}) and R is the gas constant (R = 8.3145 J mol^{-1} K^{-1}). Using the Clausius–Clapeyron equation, we can calculate the vapor pressure of a liquid at another temperature, if the vapor pressure is known at some temperature, and if the enthalpy of vaporization ΔH_{vap} is also known.

Practice 9.2 The vapor pressure of water is 1 atm at 373 K and the enthalpy of vaporization $\Delta H_{vap} = 40.7$ kJ mol^{-1}. Calculate the vapor pressure of water at 393 K.

Answer:
We are looking for a vapor pressure change due to the increase in temperature. Taking $P_1 = 1$ atm, $T_1 = 373$ K and $T_2 = 393$ K, we will use the Clausius–Clapeyron equation to determine the vapor pressure of water P_2 at 393 K:

$$\ln\frac{P_2}{P_1} = -\frac{\Delta H_{vap}}{R}\left(\frac{1}{T_2} - \frac{1}{T_1}\right)$$

Rearranging gives $P_2 = P_1 \times e^{-\frac{\Delta H_{vap}}{R}\left(\frac{1}{T_2} - \frac{1}{T_1}\right)}$

Resolving gives $P_2 = 1\text{ atm} \times e^{-\frac{40700\text{ J mol}^{-1}}{8.3145\text{ J mol}^{-1}\text{ K}^{-1}} \times \left(\frac{1}{393\text{ K}} - \frac{1}{373\text{ K}}\right)} = 1.95$ atm.

The increase of temperature from 373 K to 393 K leads to the increase of the vapor pressure from 1 atm to 1.95 atm.

The presence of a solute in the liquid decreases the vapor pressure because of the displacement of the molecules of the solvent by the molecules of the solute. Indeed, some of the solvent molecules at the surface of the liquid are replaced by the solute molecules, thereby decreasing the number of molecules of the solvent at the liquid surface and thereby in the vapor phase. According to Raoult's law, for an ideal solution (where the intermolecular forces between unlike molecules are equal to those between similar molecules), the vapor pressure of a solvent $\left(P_{solvent}\right)$ above a solution is equal to the vapor pressure of the pure solvent $\left(P^0_{solvent}\right)$ at the same temperature multiplied by the mole fraction of the solvent $\left(X_{solvent}\right)$ present:

$$P_{solvent} = X_{solvent} \times P^0_{solvent}$$

The mole fraction of the solvent $\left(X_{solvent}\right) = \dfrac{n_{solvent}}{n_{total}} = \dfrac{n_{solvent}}{n_{solvent} + n_{solute}}$, where n is the number of moles in the liquid phase.

In the same manner, the vapor pressure of the solute $\left(P_{solute}\right)$ above the solution is equal to the vapor pressure of the pure solute $\left(P^0_{solute}\right)$ at the same temperature multiplied by the mole fraction of the solute $\left(X_{solute}\right)$ present:

$$P_{solute} = X_{solute} \times P^0_{solute}$$

The mole fraction of the solute $\left(X_{solute}\right) = \dfrac{n_{solute}}{n_{total}} = \dfrac{n_{solute}}{n_{solvent} + n_{solute}}$, where n is the number of moles.

The vapor pressure of the solution is equal to the sum of the vapor pressures of the solvent and the solute:

$$P_{solution} = P_{solvent} + P_{solute} = X_{solvent} \times P^0_{solvent} + X_{solute} \times P^0_{solute}$$

And since $\left(X_{solvent}\right) + \left(X_{solute}\right) = 1$, we have:

$$P_{solution} = X_{solvent} \times P^0_{solvent} + \left(1 - X_{solvent}\right) \times P^0_{solute}$$

$$= \left(1 - X_{solute}\right) \times P^0_{solvent} + X_{solute} \times P^0_{solute}$$

Rearranging gives

$$P_{solution} = X_{solvent}\left(P^0_{solvent} - P^0_{solute}\right) + P^0_{solute}$$

$$= X_{solute}\left(P^0_{solute} - P^0_{solvent}\right) + P^0_{solvent}$$

Following the above-mentioned equations, the partial pressures of the solvent and solute and hence the vapor pressure of the solution are linear to the mole fraction (Figure 9.15):

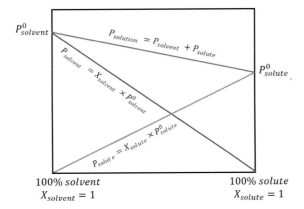

FIGURE 9.15   Vapor pressure of an ideal solution that obeys Raoult's law: the partial pressures of the solvent and solute and the vapor pressure of the solution are linear to the mole fraction (in cases where $P^0_{solvent} > P^0_{solute}$).

In a real solution (where adhesive and cohesive forces are not uniform between the solvent and the solute molecules), a positive or a negative deviation from Raoult's law arises (Figure 9.16). Indeed, when the intermolecular forces between the molecules of the solute, or between the molecules of the solvent (cohesive forces between like molecules), are stronger than the intermolecular

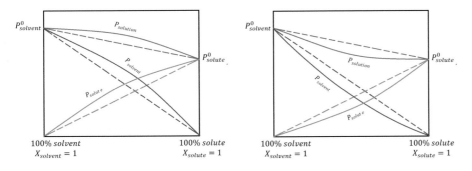

FIGURE 9.16   Deviations from Raoult's law. The dashed lines represent Raoult's law for an ideal solution. The curved lines show positive deviations (on the left) and negative deviations (on the right) from the ideal behavior.

forces between the molecules of the solute and the molecules of the solvent (adhesive forces between unlike molecules), both solvent and solute escape the solution more easily. Therefore, the vapor pressure of the solution is higher than expected from Raoult's law, showing a positive deviation. Conversely, when cohesive forces are weaker than adhesive forces, each component is engaged in the solution by intermolecular attractive forces, which are stronger than in the pure liquid, so that its partial vapor pressure is lower and a negative deviation from Raoult's law arises.

Practice 9.3 The vapor pressure of pure water at 333 K is 149.4 mmHg. Calculate the vapor pressure of an ideal solution containing 5 mol % of sugar at 333 K. Note that sugar is a nonvolatile compound.

Answer:
We are looking for the vapor pressure of an ideal solution containing 5 mol % of sugar at 333 K, so we will use Raoult's law, that states

$$P_{solution} = P_{solvent} + P_{solute} = X_{solvent} \times P^0_{solvent} + X_{solute} \times P^0_{solute}$$

where the solvent is water (H_2O) and the solute is sugar.

$$\text{Taking } X_{solvent} = \frac{(100-5)}{100} = 0.95 \text{ and } P^0_{solvent} = 149.4 \text{ mmHg}$$

Since sugar is a nonvolatile compound, it will not escape the solution and consequently, it will not exert any forces on the solution surface, so that $P_{solute} = 0$.

We have:

$$P_{solution} = P_{solvent} = X_{solvent} \times P^0_{solvent}$$

Solving gives

$$P_{solution} = 0.95 \times 149.4 \text{ mmHg} = 141.9 \text{ mmHg}$$

The presence of sugar in water decreases the vapor pressure of the solution because of the displacement of the molecules of water by the molecules of sugar at the surface of the solution, thereby decreasing the number of molecules of water in the solution surface and thereby in the vapor phase.

9.3.2 BOILING POINT

The temperature at which the vapor pressure equals the external pressure (atmospheric pressure) is denoted as the boiling point. This is the point where bubbles of vapor form inside the bulk of the liquid. The boiling point of a liquid at 1 atm pressure is called the normal boiling point.

The boiling point depends on the intermolecular forces that hold together the molecules of the substance and the external pressure. Indeed, the stronger the intermolecular forces, the higher is the boiling point. For example, the boiling point of water (H_2O) is higher than that of hydrogen sulfide (H_2S) or diethyl ether [$(C_2H_5)_2O$] because water molecules are bound to one another by hydrogen bonds, which are stronger than the Keesom forces (dipole–dipole interactions) between H_2S molecules or the London dispersion forces between [$(C_2H_5)_2O$] molecules. Therefore, the substance having the highest boiling point has the lowest vapor pressure and conversely, the substance having the lowest boiling point has the highest vapor pressure.

When the external pressure increases, the boiling point increases too. Indeed, since the boiling point is the temperature at which the external (atmospheric) pressure equals the vapor pressure, when the atmospheric pressure is high, the vapor pressure of the liquid needs to be high so that the heated substance reaches the boiling point. Thus, an elevated temperature is required to make the vapor pressure equal to the external pressure. Therefore, the boiling point increases. Conversely, at high altitudes, where the external pressure is low because of the low density of the air, the vapor pressure needs to be lower to reach the boiling point. Therefore, low heat is needed to make the vapor pressure equal to the external pressure, which means a decrease in the boiling point (Figure 9.17).

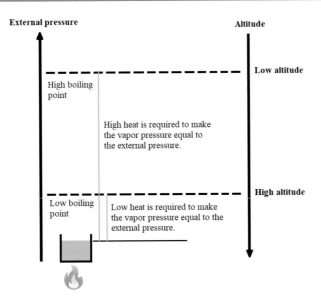

FIGURE 9.17   Dependence of the boiling point on the external pressure

Practice 9.4 Choose the molecule with the lowest boiling point and the highest vapor pressure from each of the following pairs. Explain your answer.

1. **C_3H_8 and C_6H_{14}**
2. **SO_2 and CO_2**

Answer:

1. Hexane (C_6H_{14}) is a longer-chain hydrocarbon than propane (C_3H_8), so hexane is held by higher attractive molecular forces *via* London dispersion forces than propane since there are more points of attachment in C_6H_{14} than in C_3H_8. The boiling point increases and the vapor pressure decreases as attractive forces increase. Therefore, hexane has a lower vapor pressure and a higher boiling point than propane.
2. SO_2 is a polar molecule and the interactions between SO_2 molecules are Keesom forces. However, CO_2 is a non-polar molecule and CO_2 molecules are bound to one another by London dispersion forces. Therefore, SO_2 molecules are held by stronger intermolecular forces than CO_2 molecules since Keesom forces are stronger than London dispersion forces. The boiling point increases and the vapor pressure decreases as attractive forces increase. Therefore, SO_2 has a lower vapor pressure and a higher boiling point than CO_2.

9.3.3 VISCOSITY

The viscosity of a fluid (a liquid or a gas) is a measure of its resistance to flow. High viscosity of a fluid means that the fluid hardly flows. For example, both honey and syrup have a higher viscosity than water. This means that honey and syrup resist deformation more than water. The SI unit of viscosity is Pascal second (Pa s or Nm^{-2} s). Viscosity depends on both the intermolecular forces that hold together the molecules of the fluid and the temperature. The stronger the intermolecular forces, the greater the viscosity. For example, water (H_2O) viscosity is higher than that of hydrogen sulfide (H_2S) or diethyl ether [$(C_2H_5)_2O$] because water molecules are bound to one another by hydrogen bonds, which are stronger than the Keesom forces (dipole–dipole interactions) in H_2S or the London dispersion forces in $(C_2H_5)_2O$. As the temperature increases, the molecules in a liquid move faster, thereby decreasing the average intermolecular forces. As a result, the viscosity of the liquid decreases. However, the viscosity of gases increases when the temperature increases. Indeed, the frequency of collisions between the molecules of the gas increases with increasing temperature, which decreases the ability of the molecules to engage in their coordinated movement.

9.3.4 SURFACE TENSION

Forces between unlike molecules in a liquid are called adhesive forces. For example, the forces between water molecules and the walls of a glass tube are adhesive forces. Forces between like molecules in a liquid are called cohesive forces. For example, in a droplet of water, the molecules are bound to one another by cohesive forces. The cohesive forces at the surface of a liquid are responsible for its surface tension. Indeed, inside a liquid, the cohesive forces between molecules are shared with all neighboring molecules. However, the molecules on the surface do not have neighboring molecules above. They cohere more strongly to their nearest neighbors on the surface and show stronger attractive forces than the molecules within the liquid. The increase of the intermolecular attractive forces at the surface of the liquid is called surface tension (Figure 9.18). In terms of energy, the surface tension is the amount of energy required to increase the surface of a liquid by a unit area. The SI unit of surface tension is Newton per meter (N m^{-1}). Also, it can be measured as surface energy, using J m^{-2} as the unit. The surface tension depends on the intermolecular forces and temperature. The stronger the intermolecular forces, the greater the surface tension. For example, the surface tension of water is higher than that of ethyl alcohol, but both have a lower surface tension than mercury which shows stronger attractive forces, which are metallic bonds. An increase in temperature lowers the attractive forces between the molecules of the liquid, thereby decreasing the surface tension.

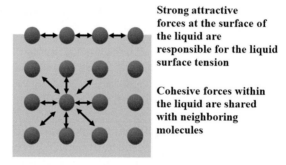

Strong attractive forces at the surface of the liquid are responsible for the liquid surface tension

Cohesive forces within the liquid are shared with neighboring molecules

FIGURE 9.18   The increase in the intermolecular attractive forces at the surface of the liquid is called surface tension.

CHECK YOUR READING

What is the difference between intramolecular forces and intermolecular forces?
What are the types of intermolecular forces? Give examples of each type.
Give the correct order of intermolecular forces from the strongest to the weakest.
Define the following properties of matter: vapor pressure, boiling point, viscosity and surface tension. Which factors affect these properties of matter and how?

SUMMARY OF CHAPTER 9

INTRAMOLECULAR FORCES

The atoms of a molecule are held together by intramolecular interactions which include covalent bonds, ionic bonds, and metallic bonds. The strength of intramolecular forces from the strongest to the weakest is as follows: ionic bond > covalent bond > metallic bond

INTERMOLECULAR FORCES

Intermolecular forces hold molecules to other molecules by intermolecular interactions which include ion–dipole forces, ion-induced dipole forces, van der Waals forces (Keesom forces, Debye forces and London dispersion forces) and hydrogen bonds (Table 9.2). The intramolecular interactions are stronger than the intermolecular forces.

Non-polar molecules can be formed between two atoms of the same non-metal that share electrons equally between them, e.g., H_2, O_2, N_2 and Cl_2. Also, polyatomic molecule, where the sum of the dipole moment of the different bonds in the molecule is zero, such as carbon dioxide (CO_2) and hydrocarbons (C_nH_{2n}), are non-polar molecules.

In polar molecules, such as H_2O, HCl, HBr and HF, the electrons are unequally shared between the atoms forming the molecule and the atoms are held together by a polar covalent bond.

In the presence of an electric field, the electron cloud of an atom or a molecule can be distorted. The ease by which this distortion occurs is defined as the polarizability. The distortion of the electron cloud causes the originally non-polar molecule or atom to acquire a dipole moment and an induced dipole is obtained.

TABLE 9.2 Intermolecular forces

Intermolecular Forces		Between	Examples
van der Waals	Keesom forces	Polar molecules	HCl, HBr and HI, SO_2 and O_3
	Debye forces	Polar molecules and non-polar molecules / atom	H_2O and O_2 Ar and HCl
	London dispersion forces	Non-polar molecules	I_2, Br_2, Cl_2, O_2, F_2, H_2, CO_2 and C_nH_{2n}
Hydrogen bond		H atom on a polar molecule and a N, O or F atom on another polar molecule	H_2O, HF, NH_3 and CH_3COOH
Ion–dipole interaction		An ion and a polar molecule	Na^+ (or Cl^-) and H_2O
Ion-induced dipole interaction		An ion and a non-polar molecule	NO_3^- and I_2 Fe^{2+} and O_2

Order from the strongest to the weakest intermolecular forces:
Ion–dipole interactions > Ion-induced dipole interactions > Hydrogen bonds > van der Waals forces
For the van der Waals Forces:
Keesom forces > Debye forces > London dispersion forces

PROPERTIES OF MATTER

VAPOR PRESSURE

The pressure exerted by the vapor, at a given temperature, on the surface of the liquid at equilibrium is called the vapor pressure.

The vapor pressure of a liquid depends on the intermolecular forces established in the liquid, the temperature, and the presence/absence of a solute in the liquid:

o The stronger the intermolecular forces, the lower the vapor pressure. For example, the vapor pressure of water (H_2O) is lower than that of hydrogen sulfide (H_2S) because water molecules are held to one another by a hydrogen bond, that is stronger than the Keesom force (dipole–dipole interaction) between H_2S molecules.

o When the temperature increases, the vapor pressure increases. The substances that have a high vapor pressure at room temperature are called volatile substances. The Clausius–Clapeyron equation allows the calculation of the vapor pressure of a liquid at another temperature, if the vapor pressure is known at some temperature, and if the enthalpy of vaporization, ΔH_{vap}, is known:

$$\ln \frac{P_2}{P_1} = -\frac{\Delta H_{vap}}{R}\left(\frac{1}{T_2} - \frac{1}{T_1}\right)$$

o The presence of a solute in the liquid decreases the vapor pressure because some of the solvent molecules at the surface of the liquid are replaced by solute molecules, thereby decreasing the vapor phase. For an ideal solution, Raoult's law is given by:

$$P_{solution} = P_{solvent} + P_{solute} = X_{solvent} \times P^0_{solvent} + X_{solute} \times P^0_{solute}$$

$$P_{solution} = X_{solvent} \times P^0_{solvent} + \left(1 - X_{solvent}\right) \times P^0_{solute}$$

$$= \left(1 - X_{solute}\right) \times P^0_{solvent} + X_{solute} \times P^0_{solute}$$

BOILING POINT

The temperature at which the vapor pressure equals the external pressure (atmospheric pressure) is denoted as the boiling point.

The boiling point depends on the intermolecular forces and the external pressure:

- The stronger the intermolecular forces, the higher the boiling point. For example, the boiling point of water (H_2O) is higher than that of diethyl ether [$(C_2H_5)_2O$] because water molecules are held to one another by a hydrogen bond, that is stronger than the London dispersion forces between [$(C_2H_5)_2O$] molecules.
- When the external pressure increases, the boiling point increases.

The substance having the highest boiling point has the lowest vapor pressure and, conversely, the substance having the lowest boiling point has the highest vapor pressure.

VISCOSITY

The viscosity of a fluid (a liquid or a gas) is a measure of its resistance to flow. A high viscosity of a fluid means that the fluid hardly flows.

Viscosity depends on the intermolecular forces and temperature:

- The stronger the intermolecular forces, the greater the viscosity. For example, water (H_2O) viscosity is higher than that of hydrogen sulfide (H_2S) because water molecules are held to one another by hydrogen bonds, that is stronger than the Keesom forces (dipole–dipole interactions) between H_2S molecules.
- When the temperature increases, the molecules in a liquid move faster, thereby decreasing the viscosity of the liquid.
- The viscosity of gases increases when the temperature increases. The frequency of collisions between the molecules of the gas increases with temperature which decreases the ability of the molecules to engage in their coordinated movement.

SURFACE TENSION

The increase in the intermolecular attractive forces at the surface of the liquid is called surface tension. In terms of energy, the surface tension is the amount of energy required to increase the surface of a liquid by a unit area.

The surface tension depends on the intermolecular forces and temperature:

- The stronger the intermolecular forces, the greater the surface tension. For example, the surface tension of water is lower than that of mercury which shows stronger attractive forces, that are metallic bonds.
- An increase in temperature lowers the attractive forces between the molecules of the liquid, thereby decreasing the surface tension.

PRACTICE ON CHAPTER 9

Q9.1   Choose the correct answer
   1. Intermolecular forces
      a) Hold molecules to other molecules.
      b) Are stronger than intramolecular forces.
      c) Include covalent bonds, ionic bonds and metallic bonds.
      d) Are bonds within molecules.

2. **Intermolecular forces**
    a) Hold atoms to one another in a molecule.
    b) Occur between ions.
    c) Include covalent bonds, ionic bonds and metallic bonds.
    d) Include van der Waals forces, hydrogen bonds, ion–dipole forces, and ion-induced dipole forces.

3. **A non-polar molecule**
    a) Exhibits a dipole moment equal to zero.
    b) Is formed between two identical non-metal atoms that equally share electrons.
    c) Is formed between two atoms that unequally share electrons.
    d) Both a) and b) are correct.

4. **A polar molecule**
    a) Exhibits a dipole moment equal to zero.
    b) Is formed between two identical non-metals that equally share electrons.
    c) Is formed between two atoms that unequally share electrons and exhibit a dipole moment different from zero.
    d) Both a) and b) are correct.

5. **Polarizability is defined as**
    a) A share of electrons between identical atoms.
    b) The pressure exerted by the vapor on the liquid at equilibrium
    c) The ease of the distortion of the electronic clouds of the atom or molecule in the presence of an electric field.
    d) The temperature at which a liquid forms a bubble.

Q9.2 **Choose the correct answer**
1. **Keesom forces**
    a) Occur between non-polar molecules.
    b) Are stronger than intramolecular forces.
    c) Occur between a polar molecule and a non-polar molecule.
    d) Occur between polar molecules.

2. **Hydrogen bonds occur between**
    a) Non-polar molecules.
    b) A hydrogen atom in one polar molecule and a strongly electronegative atom, such as nitrogen (N), oxygen (O) or fluorine (F), on another polar molecule.
    c) A polar molecule and a non-polar molecule.
    d) An induced dipole and an ion.

3. **Ion–dipole forces occur between**
    a) Non-polar molecules.
    b) An induced dipole and an ion.
    c) A polar molecule and a non-polar molecule.
    d) A permanent dipole and an ion.

4. **Which is the correct order of intermolecular forces from the strongest to the weakest**
    a) Hydrogen bonds > van der Waals forces > Ion–dipole forces > Ion-induced dipole forces.
    b) van der Waals forces > Hydrogen bonds > Ion–dipole forces > Ion-induced dipole forces.
    c) Ion–dipole forces > Ion-induced dipole forces > Hydrogen bonds > van der Waals forces.
    d) Hydrogen bonds > Ion–dipole forces > Ion-induced dipole forces > van der Waals forces.

5. **For van der Waals forces, the correct order, from the strongest to the weakest intermolecular forces, is**
    a) Keesom forces > Debye forces > London dispersion forces.
    b) Keesom forces > London dispersion forces > Debye forces.
    c) Debye forces > Keesom forces > London dispersion forces.
    d) London dispersion forces > Debye forces > Keesom forces.

6. **Keesom forces increase when**
    a) The dipole moment of the molecule increases.
    b) The temperature increases.
    c) The distance between the molecules increases.
    d) All the above are correct.

7. **Debye forces become stronger when**
    a) The dipole moment or the polarizability of the molecule decreases.
    b) The temperature increases.
    c) The distance between the molecules increases.
    d) The dipole moment or the polarizability increases.

8. **London dispersion forces become stronger when**
    a) The size of the molecule or the atom increases.
    b) The temperature increases.
    c) The distance between the molecules increases.
    d) The dipole moment or the polarizability decreases.

Q9.3   Choose the correct answer

1. **Which of the following exhibit Keesom forces?**
    a) HCl, HBr and HI.
    b) H_2O, NH_3 and HF.
    c) O_2, N_2 and F_2.
    d) CO_2, CH_4 and C_2H_6.

2. **Which of the following exhibit hydrogen bonds?**
    a) HCl, HBr and HI.
    b) H_2O, NH_3 and CH_3COOH.
    c) O_2, N_2 and F_2.
    d) CO_2, CH_4 and C_2H_6.

3. **Debye forces occur between**
    a) HCl and HBr.
    b) H_2O and NH_3.
    c) O_2 and H_2O.
    d) CO_2 and C_2H_6.

4. **Ion–dipole forces occur between**
    a) Na^+ and Cl^-.
    b) H_2O and Cl^-.
    c) O_2 and Na^+.
    d) CO_2 and NaCl.

5. **Ion-induced dipole forces occur between**
    a) Na^+ and Cl^-.
    b) H_2O and Cl^-.
    c) O_2 and Fe^{3+}.
    d) CO_2 and O_2.

6. **O_2, Cl_2, CH_4 and CO_2 exhibit**
    a) Keesom forces
    b) Debye forces.
    c) London dispersion forces.
    d) Hydrogen bonds.

7. **The forces between HBr molecules are**
    a) Ionic bonds
    b) Hydrogen bonds
    c) Keesom forces
    d) London dispersion forces

8. **which one of the following does not exhibit hydrogen bonds?**
    a) HF
    b) H_2O
    c) NH_3
    d) H_2Se

9. **Which of the following shows the strongest intermolecular forces?**
   a) HBr
   b) H_2O
   c) O_3
   d) CO_2

10. **Which of the following shows the strongest London dispersion forces?**
    a) Cl_2
    b) Br_2
    c) I_2
    d) F_2

11. **Which of the following shows the weakest London dispersion forces?**
    a) C_6H_{14}
    b) C_3H_8
    c) C_5H_{12}
    d) C_4H_{10}

Q9.4   Choose the correct answer

1. **Vapor pressure is defined as**
   a) The resistance of a fluid to flow.
   b) The pressure exerted by the vapor on the liquid at equilibrium
   c) The ease of distortion of the electronic clouds of the atom or molecule in the presence of an electric field.
   d) The temperature at which the external pressure equals the vapor pressure.

2. **The conditions under which two opposing processes are occurring simultaneously at equal rates in a closed system is called**
   a) The vapor pressure.
   b) Dynamic equilibrium.
   c) The boiling point.
   d) The surface tension.

3. **The amount of energy required to increase the surface of a liquid by a unit area is**
   a) The viscosity of a fluid.
   b) The vapor pressure.
   c) The boiling point.
   d) The surface tension.

4. **The vapor pressure of a liquid depends on**
   a) The intermolecular forces, the temperature and the presence of a solute.
   b) The external pressure.
   c) The intramolecular forces.
   d) The adhesive forces.

5. **Which one of the following properties of matter depends on the external pressure?**
   a) The vapor pressure.
   b) The boiling point.
   c) The viscosity.
   d) The surface tension.

6. **The vapor pressure of a liquid increases when**
   a) The intermolecular forces become stronger.
   b) The temperature increases.
   c) The liquid does not contain a solute.
   d) All the above are correct.

7. **In non-polar molecules, when the size of the molecule increases and the London forces become stronger, what happens to the liquid vapor pressure?**
   a) It increases
   b) It decreases
   c) It remains the same
   d) It equals 1 atm

8. **The boiling point of a liquid increases when**
   a) The intermolecular forces become stronger, and the external pressure increases.
   b) At low altitudes.
   c) The liquid does not contain a solute.
   d) Both a) and b) are correct.
9. **The viscosity of a liquid increases when**
   a) The intermolecular forces become weaker.
   b) The temperature increases.
   c) The temperature decreases, or the intermolecular forces become stronger.
   d) The external pressure increases.
10. **When the temperature of a liquid decreases, its viscosity**
    a) Increases
    b) Decreases
    c) Remains the same
    d) All the above are correct
11. **The surface tension of a liquid increases when**
    a) The temperature increases, or the intermolecular forces become weaker.
    b) The surface area of the liquid increases.
    c) The temperature decreases, or the intermolecular forces become stronger.
    d) The surface area of the liquid decreases.
12. **When the temperature increases, the surface tension of a liquid**
    a) Increases.
    b) Decreases.
    c) Remains the same.
    d) Increases or decreases, depending on the intermolecular forces of the liquid.
13. **Which of the following has the highest vapor pressure?**
    a) O_2
    b) H_2O
    c) CO_2
    d) HF
14. **Which of the following has the lowest vapor pressure?**
    a) C_6H_{14}
    b) C_3H_8
    c) C_5H_{12}
    d) C_4H_{10}
15. **Which of the following has the highest boiling point?**
    a) C_6H_{14}
    b) C_3H_8
    c) C_5H_{12}
    d) C_4H_{10}
16. **Which of the following has the lowest boiling point?**
    a) O_2
    b) H_2O
    c) CO_2
    d) H_2S
17. **Which is the correct order of boiling point, from the lowest to the highest?**
    a) $F_2 < Cl_2 < Br_2 < I_2$
    b) $I_2 < Br_2 < Cl_2 < F_2$
    c) $F_2 < Br_2 < Cl_2 < I_2$
    d) $I_2 < Cl_2 < Br_2 < F_2$
18. **Which of the following liquids has the lowest viscosity?**
    a) HF
    b) H_2O
    c) $(C_2H_5)_2O$
    d) H_2S

19. **Which of the following has the highest viscosity?**
    a) C_6H_{14}
    b) C_3H_8
    c) C_5H_{12}
    d) C_4H_{10}
20. **Which of the following substances has the lowest surface tension?**
    a) HF
    b) H_2O
    c) $(C_2H_5)_2O$
    d) H_2S

Calculation

Q9.5 **The vapor pressure of a substance is 1.5 atm at 350 K. At what temperature will this substance have a vapor pressure of 0.3 atm?**

The enthalpy of vaporization $\Delta H_{vap} = 40$ J mol^{-1}.

ANSWERS TO QUESTIONS

Q9.1
1. a
2. d
3. d
4. c
5. c

Q9.2
1. d
2. b
3. d
4. c
5. a
6. a
7. d
8. a

Q9.3
1. a
2. b
3. c
4. b
5. c
6. c
7. c
8. d
9. b
10. c
11. b

Q9.4
1. b
2. b
3. d
4. a
5. b
6. b
7. b

8. d
9. c
10. a
11. c
12. b
13. a
14. a
15. a
16. a
17. a
18. c
19. a
20. c

Q9.5 $T_2 = 333\ K$

KEY EXPLANATIONS

Q9.1

1. Intermolecular forces are forces between the molecules, whereas intramolecular forces are forces within a molecule (bonds) which hold the atoms of a molecule together. Intermolecular forces are weaker than intramolecular forces.

2. Intermolecular forces include van der Waals forces, hydrogen bonds, ion–dipole forces, and ion-induced dipole forces. The van der Waals forces include Keesom forces, Debye forces and London dispersion forces. The intramolecular forces include covalent bonds, ionic bonds and metallic bonds.

3. and 5. A non-polar molecule is formed between two molecules of the same non-metal that share electrons equally between them, e.g., Br_2, O_2, F_2 or Cl_2, or when the vectoral sum of the dipole moment of the molecule equals zero, such as in the CO_2 molecule. In the presence of an electric field, the electron clouds of the non-polar molecules can be distorted, and the originally non-polar molecule or atom acquires a dipole moment and an induced dipole is obtained. The ease of this distortion is defined as the polarizability of the atom or molecule.

4. In polar molecules, such as H_2O, HCl, HBr or HF, the electrons are unequally shared between the atoms forming the molecule and the atoms are held together by a polar covalent bond. The symmetry of the electronic cloud disappears, and a permanent dipole is obtained ($\mu \neq 0$).

Q9.2

1., 2. and 3. Keesom forces occur between polar molecules. Debye forces occur between a polar molecule and a non-polar molecule, while London dispersion forces occur between non-polar molecules. The ion–dipole forces occur between an ion and a polar molecule (that has a permanent dipole) and the ion-induced dipole forces occur between an ion and a nonpolar molecule that acquires an induced dipole when in the presence of the ion. The hydrogen bond occurs between a hydrogen atom in one polar molecule and a strongly electronegative atom, such as nitrogen (N), oxygen (O) or fluorine (F), on another polar molecule.

4. and 5. The strength of the intermolecular forces, from the strongest to the weakest, is as follows:

Ion–dipole interaction > Ion-induced dipole interaction > Hydrogen bond > van der Waals forces

And the strength of van der Waals forces, from the strongest to the weakest, is as follows:

Keesom forces > Debye forces > London dispersion forces

6., 7. and 8.   See Table 9.3

TABLE 9.3   Factors affecting van der Walls forces

Van der Waals Forces	Depend on	How?
Keesom forces	Dipole moment Temperature Distance between molecules	**Keesom forces increase when** • The dipole moment increases. • The temperature or the distance between the molecules decreases.
Debye forces	Dipole moment Polarizability Temperature Distance between molecules	**Deby forces increase when** • The dipole moment or the polarizability increases. • The temperature or the distance between the molecules decreases.
London dispersion forces	Size of atoms or molecules	**London dispersion forces increase when** • The size of the atom or molecule (and so the polarizability) increases. • The number of carbon atoms in the chain of hydrocarbons increases.

Q9.3

1. and 2.   Keesom forces arise between polar molecules, such as HCl, HBr and HI. H_2O, NH_3, and HF show hydrogen bonds where the hydrogen bond occurs between hydrogen atom in one polar molecule and a strongly electronegative atom, such as nitrogen (N), oxygen (O) or fluorine (F), on another polar molecule. O_2, N_2, F_2, CO_2 and hydrocarbons such as CH_4 and C_2H_6 are nonpolar molecules, exhibiting London dispersion forces.

3.   Debye forces occur between a polar molecule such as H_2O, HCl and HBr and a nonpolar molecule such as O_2, F_2 and N_2. Therefore, the intermolecular forces between O_2 and H_2O are Debye forces.

4. and 5.   The ion-dipole forces occur between an ion such as Na^+ and Cl^- and a polar molecule (that have a permanent dipole) such as H_2O. The ion-induced dipole forces occur between an ion and a nonpolar molecule such as O_2 that acquires an induced dipole when in presence of the ion.

6. O_2, Cl_2, CH_4 and CO_2 are non-polar molecules. Therefore, they all exhibit London dispersion forces.

7.   HBr is a polar molecule. The intermolecular forces between HBr molecules are Keesom forces.

8.   Hydrogen selenide (H_2Se) is a polar molecule. The intermolecular forces between H_2Se molecules are Keesom forces. HF, H_2O and NH_3 show hydrogen bonds.

9.   HBr and O_3 show Keesom forces, H_2O shows hydrogen bonds, and CO_2 is a non-polar molecule, showing London dispersion forces. The strength of the intermolecular forces, from the strongest to the weakest, is as follows:

Hydrogen bond > Keesom forces > Debye forces > London dispersion forces

Therefore, H_2O shows the strongest intermolecular forces, namely hydrogen bonds.

10. and 11.   London dispersion forces depend on the size of the atom or molecule. When the size increases, London dispersion forces become stronger. For F, Cl, Br, and I atoms, we have the following order, from the lowest to the highest atomic size: $F < Cl < Br < I$; therefore, the following order is established from the strongest to the weakest London dispersion forces: $I_2 > Br_2 > Cl_2 > F_2$. For that reason, I_2 is a solid, Br_2 is a liquid and F_2 and Cl_2 are gases at room temperature. For the hydrocarbons (C_nH_{2n+2}), as the number of carbon atoms in the chain of hydrocarbon increases, the points of attachments *via* London dispersion forces increase and the London forces become stronger. Therefore, we have the following order from the weakest to the strongest London dispersion forces:

$C_3H_8 < C_4H_{10} < C_5H_{12} < C_6H_{14}$.

Q9.4

1. and 2.   The vapor pressure is defined as the pressure exerted by the vapor on the surface of a liquid at dynamic equilibrium. The dynamic equilibrium is defined as the conditions under which two opposing processes are occurring simultaneously at equal rates in a closed system.

3.   The increase of the intermolecular attractive forces at the surface of the liquid is called surface tension. In terms of energy, the surface tension is the amount of energy required to increase the surface of a liquid by a unit area.

4. to 12.   See Table 9.4

TABLE 9.4   Factors affecting properties of matter

Property	Depends on	How?
Vapor pressure	• Intermolecular forces • Temperature • Presence of a solute	**Vapor pressure decreases when** • The intermolecular forces become stronger. • The temperature decreases. • A solute is added to the liquid
Boiling point	• Intermolecular forces • External (atmospheric) pressure	**Boiling point increases when** • The intermolecular forces become stronger. • The external pressure increases • At low altitude, where the external pressure is high.
	The substance having the highest boiling point has the lowest vapor pressure.	
Viscosity	• Intermolecular forces • Temperature	**Viscosity increases when** • The intermolecular forces become stronger. • The temperature decreases.
Surface tension	• Intermolecular forces • Temperature	**Surface tension increases when** • The intermolecular forces become stronger. • The temperature decreases for liquids and the temperature increases for gases.

13.   The substance that has the strongest intermolecular forces should have the lowest vapor pressure; conversely, the substance that have the weakest intermolecular forces should have the highest vapor pressure. H_2O and HF exhibit hydrogen bonds, whereas O_2 and CO_2 exhibit weaker intermolecular forces that are London dispersion forces. Therefore, O_2 and CO_2 have higher vapor pressures than H_2O and HF. London dispersion forces become stronger when the size of the molecule increases. Therefore, O_2, which is a lower size than CO_2, exhibits weaker London dispersion forces and then a higher vapor pressure.

14., 15. and 19.   Hydrocarbon compounds exhibit London dispersion forces which become stronger as the number of carbon atoms in the chain of hydrocarbons increases. Therefore, of the following hydrocarbon compounds C_6H_{14}, C_3H_8, C_5H_{12} and C_4H_{10}, the strongest London dispersion forces are present in hexane C_6H_{14} which has the longest carbon chain and consequently has the lowest vapor pressure, the highest boiling point and the highest viscosity.

16. H_2O exhibits hydrogen bonds, the polar H_2S molecules exhibit Keesom forces and the two non-polar molecules O_2 and CO_2 exhibit weaker intermolecular forces that are London dispersion forces. Therefore, O_2 and CO_2 have higher vapor pressures than H_2O and HF. London dispersion forces become stronger when the size of the molecule increases. Therefore, CO_2, which has a larger size than O_2, exhibits stronger London dispersion forces and thus a lower boiling point.

17. F_2, Cl_2, Br_2 and I_2 are non-polar molecules. They exhibit London dispersion forces. London dispersion forces become stronger when the size of the molecule increases. From the lowest size to the highest size, we have the following order: $F < Cl < Br < I$. Therefore, the correct order from the lowest to the highest boiling point is $F_2 < Cl_2 < Br_2 < I_2$.

18. and 20. HF and H_2O exhibit hydrogen bonds, the polar H_2S molecules exhibit Keesom forces and the non-polar diethyl ether $(C_2H_5)_2O$ molecules exhibit weaker intermolecular forces, namely London dispersion forces. Since the viscosity and the surface tension become higher as the intermolecular forces become stronger, $[(C_2H_5)_2O$ has the lowest viscosity and the lowest surface tension.

Q9.5

We are looking for a temperature change when the vapor pressure decreases. Taking $P_1 = 1.5$ atm and $T_1 = 350$ K, we will use the Clausius–Clapeyron equation to determine the temperature T_2 at which the vapor pressure of the substance $P_2 = 0.3$ atm:

$$\ln \frac{P_2}{P_1} = -\frac{\Delta H_{vap}}{R}\left(\frac{1}{T_2} - \frac{1}{T_1}\right)$$

Rearranging gives $\dfrac{1}{T_2} = -\dfrac{R}{\Delta H_{vap}} \times \ln \dfrac{P_2}{P_1} + \dfrac{1}{T_1}$

and $T_2 = \left(\dfrac{1}{T_1} - \dfrac{R}{\Delta H_{vap}} \times \ln \dfrac{P_2}{P_1}\right)^{-1}$

Resolving gives $T_2 = \left(\dfrac{1}{350\,K} - \dfrac{8.3145\,J\,mol^{-1}\,K^{-1}}{40000\,J\,mol^{-1}} \times \ln \dfrac{0.3\,atm}{1.5\,atm_1}\right)^{-1} = 333\,K$

Note that $\dfrac{P_2}{P_1} = \dfrac{1.5\,atm}{0.3\,atm} = 5$ and $\dfrac{T_2}{T_1} = \dfrac{333\,K}{350\,K} = 0.95$

The decrease in the vapor pressure from 1.5 atm to 0.3 atm is a five-fold decrease, but the decrease in temperature from 350 to 333 K is by a factor of 0.95. The change in vapor pressure with temperature is clearly not a linear process. In other words, the vapor pressure is not directly proportional to the temperature.

Gases

10

10.1 OBJECTIVES

At the end of the present chapter, the student will be able to:

1. Interpret the temperature and the pressure of gases.
2. Recognize the different gas laws and derive the ideal gas law.
3. Calculate the gas density from the general law of the ideal gas.
4. Define the partial pressure of a gas in a mixture of gases and the relationship between the partial pressure and the total pressure.
5. Define diffusion and effusion of gases.
6. Differentiate between an ideal gas and a real gas.

10.2 KINETIC MOLECULAR THEORY OF GASES

The kinetic molecular theory of gases explains the laws that describe the behavior of gases and provides a molecular explanation for observations that led to the development of the ideal gas law. The kinetic molecular theory of gases is based on the following postulates:

(i) The volume of gas molecules is ignored with respect to the volume of the recipient containing the gas (no volume) because the volume of an individual gas molecule is significantly smaller than the volume of the recipient container. The molecules of an ideal gas have no volume, but they have a mass.
(ii) The interaction between the gas molecules is so weak that it can be ignored (no interactions).
(iii) The gas molecules are in a constant random motion, and they collide frequently with one another and with the recipient walls.
(iv) These collisions are elastic which means that they do not affect the average kinetic energy of the gas molecules.
(v) The average kinetic energy of the molecules of any gas depends only on the temperature. At a given temperature, all the molecules have the same average kinetic energy and, when the temperature increases, the average speed of the molecules increases and hence the average kinetic energy increases.

For real gases, however, there are attractive forces between the gaseous molecules. Also, the volume of gas molecules should not be ignored with respect to the volume of the recipient. In the case of a real gas, the theory must be modified to account for the behavior of real gases. In the present chapter, we will focus on how the kinetic molecular theory of gases relates to the properties of ideal gases.

The behavior of gases can be described by four measurable quantities, namely the pressure P, the volume V, the number of moles n and the absolute temperature T (expressed in K; $T(K) = T(°C) + 273$). The pressure of a gas results from the collision of the gaseous molecules with

DOI: 10.1201/9781003257059-10

the recipient walls. The pressure is equal to the force exerted by the molecules that collide with a unit area of the gas recipient surface. The SI unit of pressure is Pascal (Pa) but the units bar, atmosphere (atm) and millimeters of mercury (mmHg) are commonly used:

$$1 \, atm = 1.013 \times 10^5 \, Pa = 760 \, mmHg$$

10.3 BOYLE'S LAW

Boyle's law states that, at constant temperature and gas amount (number of moles n), the volume (V) of the ideal gas is inversely proportional to its pressure (P) (i.e., PV = constant). This means that when the volume of an ideal gas increases by a certain factor, its pressure decreases by the same factor; conversely, when the volume decreases by a certain factor, the gas pressure increases by the same factor. At a constant temperature, the kinetic energy of the gaseous molecules remains unchanged. If a given gas occupies a larger volume, the average distance between the molecules increases and therefore, the gaseous molecules collide with their recipient surfaces (and with one another) less often, leading to a decrease in pressure (Figure 10.1). Conversely, when the gas is subjected to a greater pressure, this forces the molecules closer together and increases the density, until the collective impact of the collisions of the molecules with the recipient surfaces just balances the applied pressure.

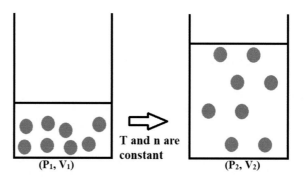

FIGURE 10.1   When the volume of a gas sample increases at constant temperature and constant gas amount, the gaseous molecules collide with the walls of the recipient surfaces less often, leading to decreases in the forces exerted per unit area and hence in pressure ($P_1 > P_2$ and $P_1V_1 = P_2V_2$).

Practice 10.1 A sample of chlorine gas occupies a volume of 945 mL at a pressure of 750 mmHg. What is the pressure of the gas if the volume is reduced to 315 mL at a constant temperature?

Answer:
We are looking for a pressure change due to a volume change at constant temperature and gas amount, so we will use Boyle's law, that is PV = B, and B is a constant. Taking $P_1 = 750$ mmHg and $V_1 = 945$ mL as the initial values, and $V_2 = 315$ mL as the volume where the pressure P_2 is unknown, we have:

$$B = P_1 \times V_1 = P_2 \times V_2$$

Rearranging gives $P_2 = \dfrac{P_1 \times V_1}{V_2}$

Solving gives $P_2 = \dfrac{750 \, mmHg \times 945 \, mL}{315 \, mL} = 2250 \, mmHg$

It is interesting to note that the volume is reduced by a factor of three and the pressure increases by the same factor, three.

10.4 CHARLES'S LAW

Charles's law states that, at constant pressure and gas amount (number of moles n), the volume of the ideal gas is directly proportional to its absolute temperature (T) (i.e., V/T = constant). This means that, when the volume of an ideal gas increases by a certain factor, its absolute temperature increases by the same factor and conversely, when the volume decreases by a certain factor, the absolute temperature of the gas decreases by the same factor. At a constant pressure, when the volume of a gas increases, the gaseous molecules collide with the walls of their recipient (and with one another) less often. This forces the molecules to move faster to increase the number of collisions, until the collective impact of the collisions of the molecules with the recipient walls precisely balances the applied pressure. As a result of the faster movement of gas molecules, the average kinetic energy of the gaseous molecules increases and, since the absolute temperature of the gas is proportional to the kinetic energy of the gaseous molecules, the absolute temperature increases, too.

Practice 10.2 A sample of carbon monoxide gas (CO) occupies 3.2 L at 125°C. At what temperature will the gas occupy a volume of 1.6 L if the pressure remains constant?

Answer:
We are looking for a temperature change due to a volume change at constant pressure and gas amount, so we will use Charles's law that is $V/T = C$, and C is a constant. Charles's law describes how a gas expands as the temperature increases. Conversely, a decrease in temperature leads to a decrease in the gas volume. Taking $T_1 = 125°C$ and $V_1 = 3.2$ L as the initial values, and $V_2 = 1.6$ L as the volume where the temperature T_2 is unknown, we have:

$$C = \frac{V_1}{T_1} = \frac{V_2}{T_2}$$

Rearranging gives $T_2 = \dfrac{V_2 \times T_1}{V_1}$

The temperature should be expressed in K, so $T_1 = 125°C = (125 + 273)$ K $= 398$ K.

Solving gives $T_2 = \dfrac{1.6\,L \times 398\,K}{3.2\,L} = 199\,K$

It is interesting to note that the volume is reduced by a factor of two and that the temperature decreases by the same factor, two.

10.5 AVOGADRO'S LAW

Avogadro's law states that, at a constant temperature and pressure, the volume of the ideal gas is directly proportional to gas amount (number of moles n), that is n/V = constant. This means that, when the gas amount increases by a certain factor, the gas volume increases by the same factor; conversely, when the gas amount decreases by a certain factor, the volume of the gas decreases by the same factor. At a constant temperature and pressure, when the gas amount increases, the gaseous molecules collide with the walls of their recipient (and with one another) more often. The increase in the number of collisions increases the pressure, which forces the gaseous molecules to occupy a larger volume to increase the wall surfaces and hence to decrease the number of collisions, until the collective impact of the collisions of the molecules with the recipient walls precisely balances the applied pressure.

Practice 10.3 A balloon contains 1.1 mol of helium (He) occupying 26.2 L. What volume will the gas occupy when 1.21 mol of He was added to the balloon if the pressure and temperature remain constant?

Answer:
We are looking for a volume change due to a gas amount change in a balloon at constant temperature and pressure, so we will use Avogadro's law, that is $n/V = A$, and A is a constant. Taking

$V_1 = 26.2$ L and $n_1 = 1.1$ mol as the initial values, and n_2 as the number of moles where the volume V_2 is unknown, we have:

$$A = \frac{n_1}{V_1} = \frac{n_2}{V_2}$$

Rearranging gives $V_2 = \dfrac{n_2 \times V_1}{n_1}$

1.21 mol of He was added to the balloon which previously contains 1.1 mol of He. So,

$$n_2 = n_1 + 1.21 = 1.1 + 1.21 = 2.31 \text{ mol}.$$

Solving gives $V_2 = \dfrac{2.31\,\text{mol} \times 26.2\,\text{L}}{1.1\,\text{mol}} = 55$ L

10.6 GAY-LUSSAC'S LAW

Gay-Lussac's law states that, at a constant volume and gas amount (number of moles n), the pressure of the ideal gas is directly proportional to its absolute temperature (P/T = constant). This means that, when the pressure of an ideal gas increases by a certain factor, its absolute temperature increases by the same factor and, conversely, when the pressure decreases by a certain factor, the absolute temperature of the gas decreases by the same factor. At a constant volume, when the temperature of a gas increases, the molecules move more rapidly and the number of collisions increases, which increases the gas pressure.

Practice 10.4 The temperature of a sample of O_2 gas is 27°C at 1.1 atm. Calculate the pressure of the gas if it is heated to 70°C at a constant volume.

Answer:
We are looking for a pressure change due to a temperature change at constant volume and gas amount, so we will use Gay-Lussac's law, that is P/T = G, and G is a constant. Taking $T_1 = 27$°C and $P_1 = 1.1$ atm as the initial values, and $T_2 = 70$°C as the temperature where the pressure P_2 is unknown, we have:

$$G = \frac{P_1}{T_1} = \frac{P_2}{T_2}$$

Rearranging gives $P_2 = \dfrac{P_1 \times T_2}{T_1}$

The temperature should be expressed in K, so $T_1 = 27$°C $= (27 + 273)$ K $= 300$ K and $T_2 = 70$°C $= (70 + 273)$ K $= 343$ K.

Solving gives $P_2 = \dfrac{1.1\,\text{atm} \times 343\,\text{K}}{300\,\text{K}} = 1.25\,\text{atm}$

10.7 IDEAL GAS LAW

The ideal gas law relates four measurable quantities, namely the pressure P, the volume V, the number of moles n and the temperature T, of an ideal gas:

$$PV = nRT$$

where R is the gas constant (R = 0.082 L atm K^{-1}mol^{-1}). The ideal gas law works well at relatively high temperatures and relatively low pressures. At high pressures or low temperatures, the gas deviates from the ideal behavior and becomes real. Under these conditions, the ideal gas law should not be used since significant errors can arise.

By introducing the Boltzmann constant k_b as the gas constant per molecule ($k_B = R/N_A$, N_A is the Avogadro's number), the ideal gas law can be written in an alternative form:

$$PV = Nk_BT$$

Where N is the number of the molecules of the gas and k_B is the Boltzmann constant ($k_B = 1.38 \times 10^{-23}$ JK^{-1}). Note that the Boltzmann constant relates the average kinetic energy K_E of the gas particles with the temperature of the gas $\left(K_E = \frac{3}{2}k_bT \right)$.

GET SMART

WHEN DOES A GAS DEVIATE FROM THE IDEAL BEHAVIOR?

At high pressures, the gaseous molecules become crowded together and gas molecules exert attractive forces upon one another due to the short distances between them. In addition, the physical sizes of the gaseous molecules cannot be ignored, and the gas molecules acquire a finite volume compared to the volume of the recipient. At low temperatures, the kinetic energy of the gas molecules decreases and, consequently, the molecules can no longer overcome the attractive forces upon one another. The ideal gas law does not describe the gas behavior either at high pressures or at low temperatures since, at these conditions, the gas either deviates from the ideal behavior and it becomes a real gas, or it condenses to a liquid. The behavior of a real gas can be described by the following equation:

$$\left(P + a\frac{n^2}{V^2} \right) \times (V - nb) = nRT$$

where a and b are constants. The quantity $a\frac{n^2}{V^2}$ takes into account the interaction between the gaseous molecules, which affects the pressure of the gas, and the quantity nb takes into account the volume of the molecules, which affects the volume of the gas.

Practice 10.5 Calculate the pressure of an ideal gas if 0.215 mol of this gas occupies 338 mL at 32°C.

The gas constant R = 0.082 L atm K⁻¹mol⁻¹.

Answer:
We are looking for the pressure of an ideal gas at constant volume, temperature, and gas amount, so we will use the ideal gas law (PV = nRT, where R is the gas constant). We have:

$$PV = nRT$$

Rearranging gives $P = \dfrac{nRT}{V}$

The temperature should be expressed in K, so T = 32°C = (32 + 273) K = 305 K, and the volume in L, so V = 338 mL = 338×10^{-3} L

Solving gives $P = \dfrac{0.215\,\text{mol} \times 0.082\,\text{LatmK}^{-1}\text{mol}^{-1} \times 305\,\text{K}}{338 \times 10^{-3}\,\text{L}} = 15.9\,\text{atm}$

GET SMART

HOW CAN YOU EASILY DEDUCE THE DIFFERENT LAWS FROM THE IDEAL GAS LAW?

Take the ideal gas law PV = nRT, where R is the gas constant and consider:

- T and n are constant. We have *PV* = constant, which means that P is inversely proportional to V and $P_1 \times V_1 = P_2 \times V_2$. This is Boyle's law.

- P and n constant. We have $V = \text{constant} \times T$, which means that V is proportional to T and $\dfrac{V_1}{T_1} = \dfrac{V_2}{T_2}$. This is Charles' law.

- T and P constant. We have $V = \text{constant} \times n$, which means that V is proportional to n and $\dfrac{V_1}{n_1} = \dfrac{V_2}{n_2}$. This is Avogadro's law.

- V and n constant. We have $P = \text{constant} \times T$, which means that P is proportional to T and $\dfrac{P_1}{T_1} = \dfrac{P_2}{T_2}$. This is Guy-Lussac's law.

10.8 STANDARD TEMPERATURE AND PRESSURE

The volume of a gas is highly dependent on temperature and pressure. In executing calculations on gases, different standard conditions can be used but one common standard used is the STP, which means standard temperature and pressure. The STP is defined as a temperature of 273 K (0°C or 32 °F) and the standard pressure of 1 atm. This corresponds to the freezing point of pure water at sea level atmospheric pressure. However, it should be noted that, in 1982, the International Union of Pure and Applied Chemistry (IUPAC) applied a more rigorous standard of STP as a temperature of 273.15 K (0°C, 32 °F) and a pressure of 10^5 Pa (0.98692 atm).

Practice 10.6 Calculate the volume of one mole of an ideal gas at STP.

The gas constant R = 0.082 L atm K^{-1}mol^{-1}.

Answer:
We are looking for the volume of one mole of an ideal gas at constant temperature and pressure, where the temperature equals 273 K and the pressure equals 1 atm (STP). So, we will use the ideal gas law (PV = nRT, and R is the gas constant). We have:

$$PV = nRT$$

Rearranging gives $V = \dfrac{nRT}{P}$

Solving gives $V = \dfrac{1\,\text{mol} \times 0.082\,\text{L atm K}^{-1}\text{mol}^{-1} \times 273\,\text{K}}{1\,\text{atm}} = 22.4\,\text{L}$

At STP, one mole of an ideal gas occupies 22.4 L. That is called the molar volume.

10.9 DENSITY OF GASES

The density (d) is defined as the mass per unit volume of a substance (d = m/V). Since one mole occupies the same volume, whatever the gas, the density of a gas depends on its molar mass. We are looking for the density (d) of a gas, so we will use the ideal gas law (PV = nRT, and R is the gas constant). We have:

$$PV = nRT \text{ or } n = \frac{m}{M}, \text{where m is the mass of the gas and m its molar mass}$$

$$\text{which means that } P = \frac{mRT}{VM} \text{ or } \rho = \frac{m}{V} \text{ so, } P = \frac{\rho RT}{M}$$

Rearranging gives $\rho = \dfrac{PM}{RT}$

It is clear from the above relationship, between the density d of a gas and its molar mass M, that a gas with a larger molar mass has a greater density than a gas with a small molar mass. Also, at a fixed pressure, the density of a gas decreases as the temperature increases and, conversely, the density of a gas increases as the temperature decreases. At a fixed temperature, the density of a gas decreases as the pressure decreases and, conversely, the density of a gas increases as the pressure increases.

Practice 10.7 Calculate the density of carbon dioxide (CO₂) gas at STP.

Atomic masses: C = 12, O = 16 and the gas constant R = 0.082 L atm K⁻¹mol⁻¹.

Answer:
The molar mass of CO_2 is M = 12 + 2×16 = 44 g mol⁻¹.
The STP is defined as T = 273 K and P = 1 atm.

The density of CO_2 at STP is $\rho = \dfrac{PM}{RT} = \dfrac{1\,atm \times 44\,g mol^{-1}}{0.082\,L atm K^{-1} mol^{-1} \times 273\,K} = 1.97\,g L^{-1}$

Practice 10.8 An unknown gas has a density of 0.72 g L⁻¹ at 1.1 atm and 300 K. Calculate its molar mass.

The gas constant R = 0.082 L atm K⁻¹mol⁻¹.

Answer:
The density of the unknown gas is $\rho = \dfrac{PM}{RT}$

Rearranging gives $M = \dfrac{\rho RT}{P}$

Solving gives $M = \dfrac{0.72\,g L^{-1} \times 0.082\,L atm K^{-1} mol^{-1} \times 300\,K}{1.1\,atm} = 16\,g mol^{-1}$

10.10 DIFFUSION AND EFFUSION

Diffusion is defined as the movement of gaseous molecules from an area of high concentration to an area of low concentration due to their constant random motion (Figure 10.2). The rate of diffusion depends on the temperature and viscosity of the medium, and the molar mass of the gas. It occurs when the gaseous molecules disperse throughout a container. Diffusion results into the gradual mixing of gases.

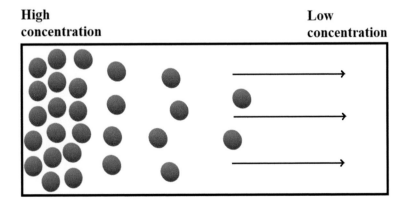

FIGURE 10.2   Diffusion is the movement of molecules of a gas from an area of high concentration to an area of low concentration due to their continuous random motion.

Effusion is defined as the movement of gaseous molecules through a tiny hole smaller than the mean free path of the gas molecules or atoms (Figure 10.3). The mean free path is the average distance traveled by the atoms or molecules of the gas between collisions.

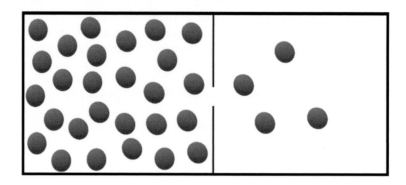

FIGURE 10.3   Effusion is defined as the movement of gaseous molecules through a small hole.

Graham's law of effusion states that the rate of effusion of a gas is inversely related to the square root of its molar mass $\left(\text{Rate of effusion} \, \alpha \, \dfrac{1}{\sqrt{M}} \right)$. This means that heavier gases effuse more slowly. For example, Cl_2 gas effuses more slowly than CH_4 gas, because it has a higher molar mass. The rate of effusion is determined by the number of molecules or atoms of the gas that pass through a small hole in a unit of time, and therefore by the average molecular velocity of the gas molecules or atoms. Graham's law of effusion can be written by the following equation:

$$\frac{Rate\,A}{Rate\,B} = \sqrt{\frac{M_B}{M_A}}$$

where Rate A and Rate B are the rates of effusion of gases A and B, respectively, and M_A and M_B are the molar masses of gases A and B, respectively. It is interesting to note that performing two identical effusion experiments measuring the rates of two different gases, one known and the other one unknown, allows the molar mass of the unknown gas to be determined.

Practice 10.9 Calculate the ratio of the effusion rates of helium (He) and methane (CH$_4$) gases.

Atomic masses: H = 1, He = 4, C = 12

Answer:
We use Graham's law of effusion to determine the ratio of effusion rates of He and CH_4.
    The molar masses of He and CH_4 are respectively, 4 g mol^{-1} and 16 g mol^{-1}. We have:

$$\frac{Rate(He)}{Rate(CH_4)} = \sqrt{\frac{M_{CH_4}}{M_{He}}}$$

Resolving gives $\dfrac{Rate(He)}{Rate(CH_4)} = \sqrt{\dfrac{16}{4}} = 2$

This means that He diffuses two times faster than CH_4.

Practice 10.10 An unknown gas (X) effuses 1.4 times faster than oxygen (O$_2$). What is the molar mass of this unknown gas?

Atomic mass: O = 16

Answer:
We use Graham's law of effusion to determine the molar mass (M_X) of the unknown gas X. We have:

$$\frac{Rate(X)}{Rate(O_2)} = \sqrt{\frac{M_{O_2}}{M_X}}$$

Therefore, we have $\left(\dfrac{Rate(X)}{Rate(O_2)}\right)^2 = \dfrac{M_{O_2}}{M_X}$

Rearranging gives $M_X = \dfrac{M_{O_2}}{\left(\dfrac{Rate(X)}{Rate(O_2)}\right)^2}$

Resolving gives $M_X = \dfrac{32}{(1.4)^2} = 16\,\text{g mol}^{-1}$.

10.11 DALTON'S LAW OF PARTIAL PRESSURES

In a mixture of non-reacting gases, each gas behaves as if it is the only gas present, and its pressure is called the partial pressure. The partial pressure of a gas (i) in a mixture of non-reacting gases is calculated from the ideal gas law $P_i V = n_i RT$, where P_i and n_i are the partial pressure and the number of moles of the gas (i), respectively.

Dalton's law of partial pressures states that the total gas pressure in a container is the sum of the partial pressures of all the gases present. Indeed, the collisions of the gaseous molecules with the container define the pressure of an ideal gas. Each gas collides with the surfaces of the container and exerts its own pressure on the container, which can be added up to determine the total pressure of the mixture of gases in the container:

$$P_{total} = P_1 + P_2 + P_3 + \ldots$$

Since the gases in a mixture are in the same recipient, the volume (V) and temperature (T) for the different gases are the same as well. Now, we are looking for a relationship between the partial pressure P_i of a gas (i) and the total pressure P_{total}. From the ideal gas law, we have:

$$P_{total} V = n_{total} RT$$

where $n_{total} = n_1 + n_2 + n_3 + \ldots$

When applying the ideal gas law for a given gas (i) in a mixture of gases, we have:

$$P_i V = n_i RT$$

where P_i is the partial pressure of the gas (i) and n_i is its number of moles. From the above last two equations, one can obtain the following relationship between P_{total} and P_i:

$$\frac{P_i}{P_{total}} = \frac{n_i}{n_{total}} = X_i$$

Rearranging gives $P_i = X_i P_{total}$

X_i is called the mole ratio of the gas (i). It is used to determine the composition of gases in a mixture. The sum of the mole ratios of each of the gases in a mixture always equals one since they represent the proportion of each gas in the mixture.

It is interesting to note that the partial pressure of a gas (i) represents the amount of pressure that the gas (i) contributes to the total pressure. The contribution of all gases in the mixture gives the total pressure.

Practice 10.11 What pressure in atm is exerted by a mixture of 2.0 g of H_2 and 8.0 g of N_2 at 273 K in a 10.0 L vessel? Calculate the partial pressure of N_2.

Atomic masses: H = 1, N = 14, and the gas constant R = 0.082 L atm $K^{-1}mol^{-1}$.

Answer:
We are looking for the total pressure exerted by a mixture of two gases H_2 and N_2 at T = 300 K and V = 10 L.

The number of moles of H_2 is $n_{H_2} = \dfrac{m_{H_2}}{M_{H_2}} = \dfrac{2g}{2 \times 1 \,\text{gmol}^{-1}} = 1\,\text{mol}$

And the number of moles of N_2 is $n_{N_2} = \dfrac{m_{N_2}}{M_{N_2}} = \dfrac{8g}{2 \times 14\,\text{gmol}^{-1}} = 0.286\,\text{mol}$

From the ideal gas law, we have:

$$P_{total}V = n_{total}RT$$

Rearranging gives $P_{total} = \dfrac{n_{total}RT}{V}$

where $n_{total} = n_{H_2} + n_{N_2} = (1 + 0.286) = 1.286\,\text{mol}$

Resolving gives $P_{total} = \dfrac{1.286\,\text{mol} \times 0.082\,\text{LatmK}^{-1}\text{mol}^{-1} \times 300\,\text{K}}{10\,\text{L}} = 3.164\,\text{atm}$

The partial pressure of N_2 is $P_{N_2} = X_{N_2}P_{total}$

where $X_{N_2} = \dfrac{n_{N_2}}{n_{total}}$

Resolving gives $P_{N_2} = \dfrac{0.286\,\text{mol}}{1.286\,\text{mol}} \times 3.164\,\text{atm} = 0.704\,\text{atm}.$

Therefore, in the mixture of gases N_2 and H_2, N_2 gas contributes 0.704 atm to the total pressure of 3.164 atm. The partial pressure of H_2 can be determined directly using the following equation:

$$P_{H_2} = \dfrac{1\,\text{mol}}{1.286\,\text{mol}} \times 3.164\,\text{atm} = 2.46\,\text{atm}.$$

or by subtracting P_{N_2} from the total pressure:

$$P_{H_2} = P_{total} - P_{N_2}$$

Resolving gives $P_{H_2} = 3.164 - 0.704 = 2.46\,\text{atm}.$

CHECK YOUR READING

What is an ideal gas? What is the difference between an ideal gas and a real gas?
What is the ideal gas law? How can the different laws be deduced from the ideal gas law?
What is STP? How can the density of a gas at STP be calculated?
How can the molar mass of an unknown gas be determined?
Define effusion and diffusion. What is the Graham's law of effusion?
What is the partial pressure of a gas? What is the Dalton's law of partial pressures?

SUMMARY OF CHAPTER 10

For an ideal gas:

(i) The volume of gas molecules is ignored with respect to the volume of the recipient containing the gas (no volume).
(ii) The interaction between the gas molecules is ignored.

(iii) The gas molecules are in constant random motion and they collide frequently with one another and with the recipient walls. These collisions are elastic. The collisions of the gaseous molecules with the container define the pressure of an ideal gas.

(iv) The average kinetic energy of the molecules of any gas depends only on the temperature. At a given temperature, all the molecules have the same average kinetic energy and when the temperature increases, the average speed of the molecules increases and hence the average kinetic energy increases.

For real gases, there are attractive forces between the gaseous molecules, and the volume of the gas molecules is not ignored with respect to the volume of the container.

STP means standard temperature and pressure. STP is defined as a temperature of 273 K (0°C or 32 °F) and the standard pressure of 1 atm. However, it should be noted that in 1982, the International Union of Pure and Applied Chemistry (IUPAC) applied a more rigorous standard of STP as a temperature of 273.15 K (0°C, 32 °F) and an absolute pressure of 10^5 Pa (0.98692 atm).

At STP, one mole of an ideal gas occupies 22.4 L; this is called the molar volume.

The density (ρ) is defined as the mass per unit volume of a substance $\left(\rho = \dfrac{m}{V}\right)$. For an ideal gas:

$$\rho = \frac{PM}{RT}$$

Diffusion is defined as the movement of gaseous molecules from an area of high concentration to an area of low concentration due to their constant random motion.

Effusion is defined as the movement of gaseous molecules through a tiny hole smaller than the mean free path of the gas molecules or atoms.

TABLE 10.1   Gas Laws

Law	Related parameters	Equation
Boyle's law	At constant T and n $V \propto 1/P$	$P_1 \times V_1 = P_2 \times V_2$
Charles' law	At constant P and n, $V \propto T$	$\dfrac{V_1}{T_1} = \dfrac{V_2}{T_2}$
Avogadro's law	At constant T and P, $V \propto n$	$\dfrac{V_1}{n_1} = \dfrac{V_2}{n_2}$
Gay-Lussac's Law	At constant V and n $P \propto T$	$\dfrac{P_1}{T_1} = \dfrac{P_2}{T_2}$
Ideal gas law	It relates four measurable parameters (P, V, n, and T) of an ideal gas.	$PV = nRT$
Real gas law	It takes account of the volume of the gaseous molecules and the interaction between them.	$\left(P + a\dfrac{n^2}{V^2}\right) \times (V - nb) = nRT$
Graham's law of effusion	Rate of effusion $\propto \dfrac{1}{\sqrt{M}}$ M is the molar mass of the gas	$\dfrac{Rate A}{Rate B} = \sqrt{\dfrac{M_B}{M_A}}$
Dalton's law of partial pressures	The total gas pressure in a container is the sum of the partial pressures of each of the gases present.	$P_{total} = P_1 + P_2 + P_3 + \dots$ $P_i = X_i P_{total}$ Where $\dfrac{P_i}{P_{total}} = \dfrac{n_i}{n_{total}} = X_i$

PRACTICE ON CHAPTER 10

Q10.1 Choose the correct answer
   1. **An ideal gas is composed of particles that**
      a) are far away from each other
      b) there are no attractive forces between them

      c) their volume is ignored with respect to the volume of the container

      d) all the above.

2. **A real gas is composed of particles where**
   a) there is attraction between them, and their volume is not ignored
   b) there is no attraction between them
   c) their volume is ignored with respect to the volume of container
   d) none of the above

3. **A gas deviates from the ideal behavior at**
   a) high pressure and temperature.
   b) low pressure and temperature.
   c) high pressure and/or low temperature
   d) low pressure and/or high temperature

4. **The equation PV = nRT represents**
   a) the ideal gas law
   b) the general gas law for real gases.
   c) Boyle's law
   d) Dalton's law of partial pressures

5. **Which of the following equations represents the general gas law for real gases?**
   a) $PV = nRT$
   b) $P_{total} = P_1 + P_2 + P_3 \ldots$
   c) $P_i = X_i \times P_{total}$
   d) $\left(P + a \dfrac{n^2}{V^2} \right) \times \left(V - nb \right) = nRT$

6. **Boyle's law, which relates the pressure and volume at constant amount of a substance and constant temperature, states that the volume of an ideal gas is**
   a) directly proportional to its pressure
   b) inversely proportional to its pressure
   c) inversely proportional to the amount of the substance
   d) proportional to the gas constant R

7. **According to Charles' law, when the volume of an ideal gas increases at constant pressure and gas amount, the temperature**
   a) increases
   b) decreases
   c) remains constant
   d) both increases and decreases depending on the volume

8. **When a balloon is filled with oxygen gas at constant pressure and temperature, its volume increases with the increase in the amount of oxygen. This corresponds to**
   a) Boyle's law
   b) Charles' law
   c) Avogadro's law
   d) Dalton's law

9. **Guy-Lussac's law states that**
   a) the volume is inversely proportional to the pressure at constant temperature and gas amount
   b) the volume is proportional to the temperature at constant pressure and gas amount
   c) the volume is proportional to the gas amount at constant pressure and temperature
   d) the pressure is proportional to the temperature at constant volume and gas amount

10. **The equation $\dfrac{V_1}{T_1} = \dfrac{V_2}{T_2}$ is obtained from**
    a) Boyle's law
    b) Charles's law

c) Avogadro's law

d) Guy-Lussac's law

11. Which equation is obtained from Boyle's law?

a) $\dfrac{V_1}{n_1} = \dfrac{V_2}{n_2}$

b) $P_1 \times V_1 = P_2 \times V_2$

c) $\dfrac{V_1}{T_1} = \dfrac{V_2}{T_2}$

d) $\dfrac{P_1}{T_1} = \dfrac{P_2}{T_2}$

12. The standard temperature and pressure (STP) means that

a) $T = 273$ K and $P = 1$ Pa.

b) $T = 0°F$ and $P = 1$ atm.

c) $T = 0$ K and $P = 1$ atm.

d) $T = 273$ K and $P = 1$ atm.

13. Effusion is defined as

a) the proximity of several measurements to the true (accepted) value.

b) the movement of gaseous molecules through a small hole.

c) the sum of the partial pressures of all the gases present in a container.

d) the movement of gaseous molecules from an area of high concentration to an area of low concentration due to their constant random motion.

14. Which one of the following about the effusion rates of gases O_2 and Cl_2 is correct?
Atomic masses: O = 16, Cl = 35.5.

a) O_2 has a higher effusion rate than Cl_2

b) O_2 has a lower effusion rate than Cl_2

c) O_2 and Cl_2 have equal effusion rates

d) none of these

15. Dalton's law of partial pressure states that

a) the volume of the ideal gas is inversely proportional to its pressure at a constant temperature

b) the total gas pressure in a recipient is the sum of the partial pressures of all the gases present

c) the pressure of the ideal gas is proportional to the temperature at a constant volume

d) the pressure of the ideal gas is proportional to the amount of a substance

16. Which equation describes Dalton's law of partial pressure?

a) $\rho = \dfrac{PM}{RT}$

b) $P_1 \times V_1 = P_2 \times V_2$

c) $\dfrac{V_1}{T_1} = \dfrac{V_2}{T_2}$

d) $P_{total} = P_1 + P_2 + P_3 + \ldots$

17. In the equation $(P_i = X_i \times P_{total})$, which relates the partial pressure of a gas (i) to the total pressure in the container, X_i represents the

a) mole ratio of gas (i)

b) number of moles of gas (i)

c) molar mass of gas (i)

d) density of gas (i)

Calculations

Atomic masses: H = 1, O = 16

The gas constant R = 0.082 L atmK^{-1}mol^{-1}

Q10.2 **A gas occupies a volume of 800 mL at a pressure of 350 mmHg. What is the volume of the gas if the pressure is reduced to 200 mmHg at a constant temperature?**

Q10.3 **A sample of carbon monoxide gas (CO) occupies 4 L at 150°C. Calculate the volume of the gas if the temperature of the gas increases to 300°C at a constant pressure.**

Q10.4 **A balloon contains 3 mol of helium (He), occupying 30 L. Calculate the number of He moles in the blimp if it is expanded to a volume of 45 L at constant pressure and temperature.**

Q10.5 **The temperature of a sample of O_2 gas is 15°C at 1.3 atm. Calculate the temperature of the gas if its pressure reaches 1.7 atm at constant volume.**

Q10.6 **A sample of 9.6 g of O_2 gas is introduced into a balloon at a pressure of 7.4 atm and a temperature of 27°C. What is the volume of this gas?**

Q10.7 **A quantity of He gas occupies a volume of 14.5 L at 88°C and 35.7 atm. What is its volume at STP?**

Q10.8 **Hydrogen peroxide (H_2O_2) dismutation occurs according to the following reaction: $2H_2O_{2\,(l)} \rightarrow O_{2\,(g)} + 2H_2O_{\,(l)}$**
   What mass of hydrogen peroxide must be used to produce 1.0 L of oxygen gas (O_2) at 25°C and 1 atm?

Q10.9 **An unknown gas has a density of 2.73 gL^{-1} at 2.1 atm and 300 K.**
   a) Calculate its molar mass
   b) Calculate its density at STP

Q10.10 **A sample of 0.1 L of O_2 at 273 K and 2 atm is mixed with a sample of 0.20 L of N_2 at 273 K and 1 atm in a 0.40 L recipient container at 273 K. Calculate**
   a) the total pressure in the container
   b) the partial pressures of O_2 and N_2

ANSWERS TO QUESTIONS

Q10.1

1. d
2. a
3. c
4. a
5. d
6. b
7. a
8. c
9. d
10. b
11. b
12. d
13. b
14. a
15. b
16. d
17. a

Calculations

Q10.2   1400 mL

Q10.3   4.95 L

Q10.4   4.5 mol

Q10.5   376.62 K

Q10.6   0.99 L

Q10.7   622.82 L

Q10.8   2.72 g

Q10.9   a) 32 g mol^{-1}

b) 1.42 g L^{-1}

Q10.10   a) 1 atm

b) $P_{O_2} = 0.5$ atm.

$P_{N_2} = 0.5$ atm.

KEY EXPLANATIONS

Q10.1

1. and 2. A gas is formed by molecules which are far from each other. The molecules of an ideal gas have no volume, but they have a mass. According to the kinetic molecular theory of gases, the volume of gas molecules is ignored with respect to the volume of the recipient which contains the gas (which is assumed to have no volume) because the volume of an individual gas molecule is significantly smaller than the volume of its recipient container. However, for a real gas, there are attractive forces between the gaseous molecules and the volume of the gas molecules is not ignored with respect to the volume of the container.

3. The ideal gas law does not describe the gas behavior either at high pressures or at low temperatures since, under these conditions, the gas either deviates from the ideal behavior and it becomes a real gas or it condenses to a liquid. At high pressures, the gaseous molecules become crowded together and gas molecules exert attractive forces upon one another due to the short distances between them. In addition, the physical sizes of the gaseous molecules cannot be ignored, and the gaseous molecules acquire a finite volume compared to the volume of the container. At low temperatures, the kinetic energy of the gaseous molecules decreases and consequently, the molecules can no longer overcome the attractive forces upon one another.

4. and 5. The ideal gas law relates four measurable quantities that are the pressure P, the volume V, the number of moles n and the temperature T of an ideal gas:

$$PV = nRT$$

where R is the gas constant (R = 0.082 L atm K^{-1}mol^{-1}). The ideal gas law works well at relatively high temperatures and relatively low pressures. The ideal gas law should not be used for real gases since significant errors can arise. For real gases, there are attractive forces between the gaseous molecules and the volume of gas molecules is not ignored with respect to the volume of the container. The general gas law for real gases is given by the following equation:

$$\left(P + a\frac{n^2}{V^2}\right) \times (V - nb) = nRT$$

The quantity $a\dfrac{n^2}{V^2}$ takes into account the interaction between the gaseous molecules, which affects the pressure of the gas, while the quantity *nb* takes into account the volume of the molecules, which affects the volume of the gas.

The equation ($P_{total} = P_1 + P_2 + P_3...$) gives the total pressure of a mixture of gases (Dalton's law of partial pressures), and the equation ($X_i \times P_i$), where X_i is the mole ratio of the gas (i), gives the partial pressure of a gas (i) in a mixture of gases.

6. to 11. Please refer to Table 10.1, given in the summary, showing the different gas laws.

12. STP means standard temperature and pressure. STP is defined as a temperature of 273 K (0 C or 32°F) and the standard pressure of 1 atm. However, it should be noted that in 1982, the International Union of Pure and Applied Chemistry (IUPAC) applies a more rigorous standard of STP as a temperature of 273.15 K (0°C, 32 °F) and an absolute pressure of 10^5 Pa (0.98692 atm).

13. Effusion is defined as the movement of gaseous molecules through a small hole. Diffusion is defined as the movement of gaseous molecules from an area of high concentration to an area of low concentration due to their continuous random motion.

    The total pressure of a mixture of gases is the sum of the partial pressures of each of the gases present in a container. This constitutes the Dalton's law of partial pressures.

    The proximity of several measurements to the true (accepted) value is defined as the accuracy.

14. Graham's law of effusion states that the rate of effusion of a gas is inversely related to the square root of its molar mass $\left(\text{Rate of effusion} \propto \dfrac{1}{\sqrt{M}} \right)$. This means that heavier gases effuse more slowly. Therefore, Cl_2 gas effuses more slowly than O_2 gas because it has a higher molar mass. Indeed, the molar mass of O_2 is 32 g mol^{-1} and the molar mass of Cl_2 is 71 g mol^{-1}.

15. to 17. Please refer to Table 10.1, given in the summary, showing the different gas laws.

Calculations

Q10.2

We are looking for a volume change due to a pressure change at constant temperature and gas amount, so we will use Boyle's law, that is $PV = B$, where B is a constant. Note that, since the pressure is reduced, the volume should increase. Taking $P_1 = 350$ mmHg and $V_1 = 800$ mL as the initial values, $P_2 = 200$ mmHg is the pressure where the volume V_2 is unknown, we have:

$$B = P_1 \times V_1 = P_2 \times V_2$$

Rearranging gives $V_2 = \dfrac{P_1 \times V_1}{P_2}$

Solving gives $V_2 = \dfrac{350\,\text{mmHg} \times 800\,\text{mL}}{200\,\text{mmHg}} = 1400\,\text{mL}$

Q10.3

We are looking for a volume change due to a temperature change at constant pressure and gas amount, so we will use Charles' law, that is $V/T = C$, where C is a constant. An increase in temperature leads to an increase in the gas volume. Taking $T_1 = 150°C$ and $V_1 = 4$ L as the initial values, and $T_2 = 300°C$ as the temperature, where the volume V_2 is unknown, we have:

$$C = \dfrac{V_1}{T_1} = \dfrac{V_2}{T_2}$$

Rearranging gives $V_2 = \dfrac{V_1 \times T_2}{T_1}$

The temperature should be expressed in K, so $T_1 = 150°C = (150 + 273)$ K $= 423$ K and $T_2 = 300°C = (300 + 273)$ K $= 573$ K.

Solving gives $V_2 = \dfrac{4\,\text{L} \times 523\,\text{K}}{423\,\text{K}} = 4.95\,\text{L}$

Q10.4

We are looking for a gas amount change due to a volume change in a balloon at constant temperature and pressure, so we will use Avogadro's law that is $n/V = A$, where A is a constant. Taking $V_1 = 30$ L and $n_1 = 3$ mol as the initial values, and $V_2 = 45$ L as the volume where the number of moles n_2 is unknown, we have:

$$D = \dfrac{n_1}{V_1} = \dfrac{n_2}{V_2}$$

Rearranging gives $n_2 = \dfrac{n_1 \times V_2}{V_1}$

Solving gives $n_2 = \dfrac{3\,mol \times 45\,L}{30\,L} = 4.5\,mol$

Note that the number of moles of He in the balloon after expansion can be easily determined. Indeed, the volume increased by 1.5 times and, since the volume is directly proportional to the number of moles at constant pressure and temperature, the number of moles should increase by 1.5 times, i.e., $n_2 = 1.5 \times 3 = 4.5\,mol$.

Q10.5

We are looking for a temperature change due to a pressure change at constant volume and gas amount, so we will use Gay-Lussac's law that is P/T = G, where G is a constant. Taking $T_1 = 15°C$ and $P_1 = 1.3$ atm as the initial values, $P_2 = 1.7$ atm as the pressure where the temperature T_2 is unknown, we have:

$$G = \frac{P_1}{T_1} = \frac{P_2}{T_2}$$

Rearranging gives $T_2 = \dfrac{P_2 \times T_1}{P_1}$

The temperature should be expressed in K, so $T_1 = 15°C = (15 + 273)\,K = 288\,K$.

Solving gives $T_2 = \dfrac{1.7\,atm \times 288\,K}{1.3\,atm} = 376.62\,K$

Q10.6

We are looking for the volume of an ideal gas at constant pressure, temperature, and gas amount, so we will use the ideal gas law (PV = nRT, and R is the gas constant). We have:

$$PV = nRT$$

Rearranging gives $V = \dfrac{nRT}{P}$

The temperature should be expressed in K, so $T = 27°C = (27 + 273)\,K = 300\,K$.

The number of moles of O_2 gas is $n = \dfrac{m}{M} = \dfrac{9.6}{(2 \times 16)} = 0.3\,mol$

Solving gives $V = \dfrac{0.3\,mol \times 0.082\,LatmK^{-1}mol^{-1} \times 300\,K}{7.4\,atm} = 0.99\,L$

Q10.7

First, we are looking for the gas amount of He, at a given pressure, temperature, and volume, so we will use the ideal gas law (PV = nRT), where R is the gas constant. We have:

$$PV = nRT$$

Rearranging gives $n = \dfrac{PV}{RT}$

The temperature should be expressed in K, so $T = 88°C = (88 + 273)\,K = 361\,K$.

Solving gives $n = \dfrac{35.7\,atm \times 14.5\,L}{0.082\,LatmK^{-1}mol^{-1} \times 300\,K} = 21.04\,mol$

Second, we are looking for the volume of this gas at STP. STP means that P = 1 atm and T = 273 K. So, we will use the ideal gas law (PV = nRT, where R is the gas constant). We have:

$$PV = nRT$$

Rearranging gives $V = \dfrac{nRT}{P}$

Solving gives $V = \dfrac{21.04\,mol \times 0.082\,LatmK^{-1}mol^{-1} \times 361\,K}{1\,atm} = 622.82\,L$

Q10.8

First, we are looking for the gas amount of O_2 at given pressure, temperature, and volume, so we will use the ideal gas law ($PV = nRT$, where R is the gas constant). We have:

$$PV = nRT$$

Rearranging gives $n = \dfrac{PV}{RT}$

The temperature should be expressed in K, so $T = 25°C = (25 + 273) \text{ K} = 298 \text{ K}$.

Solving gives $n = \dfrac{1 \text{ atm} \times 1 \text{ L}}{0.082 \text{ LatmK}^{-1}\text{mol}^{-1} \times 298 \text{ K}} = 0.04 \text{ mol}$

Second, we are looking for the mass of H_2O_2 that must be used to produce 0.04 mol of O_2 (which has a volume of 1 L). The mass of H_2O_2 is $m = n \times M$.

So, to determine the mass of H_2O_2, we should first determine the number of moles of H_2O_2. From the balanced equation of H_2O_2 dismutation:

$$2H_2O_{2(l)} \rightarrow O_{2(g)} + 2H_2O_{(l)}$$

We have $n(H_2O_2) = 2 \times n(O_2)$.

Solving gives $n(H_2O_2) = 2 \times 0.04 = 0.08 \text{ mol}$

and the mass of H_2O_2 is $m = n \times M = 0.08 \times (2 \times 1 + 2 \times 16) = 2.72 \text{ g}$.

Q10.9

a)   The density of the unknown gas is $\rho = \dfrac{PM}{RT}$

Rearranging gives $M = \dfrac{dRT}{P}$ where $\rho = 2.73 \text{ g L}^{-1}$, $P = 2.1 \text{ atm}$ and $T = 300 \text{ K}$.

Solving gives $M = \dfrac{2.73 \text{ gL}^{-1} \times 0.082 \text{ LatmK}^{-1}\text{mol}^{-1} \times 300 \text{ K}}{2.1 \text{ atm}} = 32 \text{ gmol}^{-1}$

b)   STP means that $P = 1 \text{ atm}$ and $T = 273 \text{ K}$.

The density of the unknown gas is $\rho = \dfrac{PM}{RT}$

Solving gives $\rho = \dfrac{1 \text{ atm} \times 32 \text{ gmol}^{-1}}{0.082 \text{ LatmK}^{-1}\text{mol}^{-1} \times 273 \text{ K}} = 1.42 \text{ gL}^{-1}$

Q10.10

a)   For O_2, we have $P_1 = 2.0 \text{ atm}$, $T = 273 \text{ K}$ and $V_1 = 0.10 \text{ L}$. The number of moles, n_1, of O_2 is unknown. So, we will use the ideal gas law ($PV = nRT$, where R is the gas constant). We have:

$$P_1V_1 = n_1RT$$

Rearranging gives $n_1 = \dfrac{P_1V_1}{RT}$

Solving gives $n_1 = \dfrac{2.0 \text{ atm} \times 0.10 \text{ L}}{0.082 \text{ LatmK}^{-1}\text{mol}^{-1} \times 273 \text{ K}} = 8.9 \times 10^{-3} \text{ mol}$

For N_2, we have $P_2 = 1.0 \text{ atm}$, $T = 273 \text{ K}$ and $V_2 = 0.20 \text{ L}$. The number of moles n_2 of N_2 is unknown. So, we will use the ideal gas law ($PV = nRT$, where R is the gas constant). We have:

$$P_2V_2 = n_2RT$$

Rearranging gives $n_2 = \dfrac{P_2 V_2}{RT}$

Solving gives $n_2 = \dfrac{1.0\,\text{atm} \times 0.20\,\text{L}}{0.082\,\text{Latm K}^{-1}\text{mol}^{-1} \times 298\,\text{K}} = 8.9 \times 10^{-3}\,\text{mol}$

Now, we are looking for the total pressure exerted by a mixture of $n_1 = 8.9 \times 10^{-3}$ mol of O_2 and $n_2 = 8.9 \times 10^{-3}$ mol of N_2 at $T = 273$ K and $V = 0.40$ L. From the ideal gas law, we have:

$$P_{total} V = n_{total} RT$$

Rearranging gives $P_{total} = \dfrac{n_{total} RT}{V}$

where $n_{total} = n_1 + n_2 = \left(8.9 \times 10^{-3} + 8.9 \times 10^{-3}\right) = 1.8 \times 10^{-2}\,\text{mol}$

Resolving gives $P_{total} = \dfrac{0.018\,\text{mol} \times 0.082\,\text{Latm K}^{-1}\text{mol}^{-1} \times 273\,\text{K}}{0.40\,\text{L}} = 1\,\text{atm}$

b) The partial pressure of O_2 is $P_{O_2} = X_{O_2} P_{total}$

where $X_{O_2} = \dfrac{n_{O_2}}{n_{total}} = \dfrac{n_1}{n_{total}}$

Resolving gives $P_{O_2} = \dfrac{8.9 \times 10^{-3}\,\text{mol}}{1.8 \times 10^{-2}\,\text{mol}} \times 1\,\text{atm} = 0.5\,\text{atm}.$

The partial pressure of N_2 is $P_{N_2} = X_{N_2} P_{total}$

where $X_{N_2} = \dfrac{n_{N_2}}{n_{total}} = \dfrac{n_2}{n_{total}}$

Resolving gives $P_{N_2} = \dfrac{8.9 \times 10^{-3}\,\text{mol}}{1.8 \times 10^{-2}\,\text{mol}} \times 1\,\text{atm} = 0.5\,\text{atm}.$

Note that the mole ratios $X_{O_2} + X_{N_2} = \dfrac{8.9 \times 10^{-3}\,\text{mol}}{1.8 \times 10^{-2}\,\text{mol}} + \dfrac{8.9 \times 10^{-3}\,\text{mol}}{1.8 \times 10^{-2}\,\text{mol}} = 1$

and the sum of the partial pressures $P_{O_2} + P_{N_2} = 0.5 + 0.5 = 1\,\text{atm} = P_{total}$

Easy way to answer question Q10.10:

Note that the temperature does not change after mixing the two gases O_2 and N_2 (T is constant and equals 273 K). Also, note that $V = 4 \times V_1 = 2 \times V_2$

Now, we can use Boyle's law, which states that the pressure of an ideal gas is inversely proportional to the volume at constant temperature and the gas amount:

Since $V = 4 \times V_1$, $P_{O_2} = \dfrac{P_1}{4} = \dfrac{2}{4} = 0.5\,\text{atm}.$

and since $V = 2 \times V_2$, $P_{N_2} = \dfrac{P_2}{2} = \dfrac{1}{2} = 0.5\,\text{atm}$

Using Dalton's law of partial pressure gives us:

$$P_{total} = P_{O_2} + P_{N_2} = 0.5 + 0.5 = 1\,\text{atm}$$

Index

A

Accuracy, measured numbers, 27, 28
Acidic and basic trends, of oxides and hydrides, 166–167
Acids, 167
Actinides, 159, 161, 169
Addition, 29
Adhesive forces, 228
Alkali metals, 160–161, 168, 177
Alkaline earth metals, 161, 168, 176
Allowed orbits, 128
Aluminum oxide, 167
Ammonia (NH_3), 88
    Lewis structure of, 183
    trial structure of, 182
Ammonium cation (NH_4^+), Lewis structure and formal charges of, 208
Amphoteric compound, 167
Angular momentum quantum number, 131
Anions, 49, 55
    cations *vs.*, 47
Arithmetic operation, 26–27, 41
Arrhenius definition, 167
Atomic mass (atomic weight), 57–58, 82
    vs. molecular mass and molar mass, 59
Atomic mass units (amu), 73
Atomic number, 49
Atomic orbitals, 130, 140, 190, 196
Atomic size, 162–164, 169
Atomic structure, 44
Atomic weight, 57–58
Atoms, 13, 45
    of alkali metals, 176
    electronic configuration of, 135–136
    history of, 43
    Lewis dot representations of, 180–182
    number of, 60–62
    properties of, *see* properties of atoms
Attractive forces, 1
Aufbau "building-up" principle, 133–134, 154, 189
Average atomic mass, 55
Average kinetic energy, 251
Avogadro's law, 243–244, 246, 256
Avogadro's number, 48, 50
Axial overlaps, 214, 215

B

Balanced chemical equations, 67–69, 74, 84, 89–90
Bases, 167
Basic geometries and derivatives, 187
Benzene, 214, 215
Beryllium chloride ($BeCl_2$), Lewis structure and molecular geometry of, 210
Blood, 12
BO, *see* Bond order (BO)
Bohr's model, 125–128
Boiling point, 13, 226–227, 230
Boltzmann constant, 245
Bond breaking, 117
Bond enthalpy, 105–107, 110, 117, 121

Bond making, 105
Bond order (BO), 191
Born–Haber cycle, 107–108, 111, 122
Boron trifluoride (BF_3) molecule, planar molecular geometry of, 193
Boyle's law, 242, 245, 259
Bronsted-Lowry definition, 167

C

Calcium carbonate ($CaCO_3$), 12
Calcium fluoride (CaF_2), 184
Calcium sulfate ($CaSO_4$), 12
Calculations in chemistry
    atomic mass (atomic weight), 57–58
    balancing chemical equations and stoichiometry, 67–69
    empirical and molecular formulas, 65–67
    law of definite proportions, 62
    law of multiple proportions, 64
    mass of an element, in compound, 61–64
    molecular mass (formula mass), 58, 59
    number of atoms, molecules, or ions, 60–61
    number of moles, 60
    solutions, 70–73
Calorimeter, 97, 109
Calorimetry, 109
    heat capacity and specific heat, 98–99
    principles of, 97
Carbon dioxide (CO_2)
    Lewis structure of, 184, 198, 210
    molecular geometry of, 210
Carbon, electronic configuration of, 213
Carbon monoxide (CO), molecular orbital diagram for, 191
Cations, 55
    vs. anions, 47
Cellular respiration, 5
Chadwick, J., 43, 49
Charles's law, 243, 246, 256
Chemical bonding (intramolecular forces), 179–180, 196
Chemical change of matter, 5–7
Chemical equation, 68
Chemical formulas, 46
    of ionic compound, 48
Chemical properties
    of matter, 5, 7
    of metalloids, 160
    of metals, 160
    of non-metals, 160
Chemical reaction, 111
Chemistry, 1
    balancing chemical equations and stoichiometry, 67–69
    empirical and molecular formulas, 65–67
    law of definite proportions, 62
    law of multiple proportions, 64
    mass of an element, in compound, 61–64
    molar mass, 59
    molecular mass (formula mass), 58
    number of atoms, molecules, or ions, 60–61
    number of moles, 60
    solutions, 70–73
Chlorine (Cl), Lewis dot diagram of, 180

Clausius–Clapeyron equation, 224, 229
Closed system, 94
Coefficients, 67, 85
Cohesive forces, 228
Compound, 1–2; *see also* Amphoteric compound;
   Hydrocarbon compounds; Ionic compounds
  mass percent of element in, 64
  vs. mixture of elements, 3
  separation of, 3
  units of element, 46
Concentration, 75
Covalent bonds, 179, 196; *see also* Lewis structure
Cross-multiplication method, 89
Cross-over rule, 48, 55
Cyanate (OCN⁻) Lewis structure and formal charges
  of, 209

D

Dalton's law of partial pressures, 43, 249–250, 259
de Broglie equation, 140
de Broglie hypothesis, 129
Debye forces, 219–221, 237
Definite proportions, law of, 62
Democritus, 43
Density, 12, 38
  of gases, 246–247
  of substance, 21
Derived physical quantities, 15, 28
Diamond, tridimensional structure of, 214
Dibromomethane (CH_2Br_2) Lewis structure and formal
   charges of, 208
Diffusion, 247, 251, 256
Difluorine (F_2) Lewis structure of, 183
Dilution, 72–73, 75
Dipole–dipole forces, 219
Dipole moment, 217–218
Distance, 1
Distillation, 3, 13
Distortion, 229
Division operation, 26, 29
Dynamic equilibrium, 223

E

Effective nuclear charge, 162, 169
Effusion, 248–249, 251, 256
Eight valence electrons, 197
Electrolysis of water, 3, 12
Electromagnetic radiation of light, 123
Electron affinity, 165, 169
Electron dot diagrams, 197
Electronegativity, 165–166, 169, 177
Electron groups, 186–187, 199
Electronic configuration, 212
  of atoms and ions, 135–136
  Aufbau ((building-up) principle, 133–134
  of carbon, 213
  Hund's rule, 134
  isoelectronic configuration, 137
  of oxygen, 213
  Pauli's exclusion principle, 135
  short cut for writing, 138
  of transition metals and transition metal cations,
   136–137
  valence shell electrons, 137–138
Electrons, 43, 45, 196
  in orbitals, 136
Electron spin quantum number, 132
Elements, 1, 2, 44, 161, 168

periodic table of, 157–159
  short-cut electronic configuration of, 138–139
Emission spectrum, of hydrogen atom, 125, 140, 150
Empirical ("simplest") formula, 46, 49, 55, 65–67, 87
Endothermic reaction, 109
Energy, 38, 93, 108, 116, 128, 221
Energy conservation law, 94, 116
Energy exchange, 94–95
"Energy is quantized," 149
Enthalpy, 109, 110, 116
  bond enthalpy, 105–106
  definition of, 99
  Hess's law, 104–105
  latent heat, 99–102
  of reaction, 102–103
  standard enthalpy of formation, 103–104
Ethane, 84
Exact numbers, 22, 25, 29
Exothermic process, 107
Exothermic reaction, 109
Extensive physical property, 4
Extensive properties of matter, 4, 7, 12, 110

F

Families, 157, 168
FC, *see* Formal charge (FC)
Fe atom, Lewis dot diagram of, 181–182
Filtration, 3, 13
First ionization energy, 164
First law of thermodynamics, 94
Fletcher, Harvey, 43, 49
Force, 38
Formal charge (FC), 185–186, 198
  Lewis structure and, 207–208
Formaldehyde (CH_2O), Lewis structure and molecular
   geometry of, 210
Formula mass, 58
Freezing, 100
Frequency, 123, 139
Fundamental physical quantities, 15, 28

G

Gases, 1, 255
  Avogadro's law, 243–244
  Boyle's law, 242
  Charles's law, 243
  Dalton's law of partial pressures, 249–250
  density of, 246–247
  diffusion and effusion, 247–249
  Gay–Lussac's law, 244
  ideal gas law, 244–245
  kinetic molecular theory of, 241–242
  laws, 251
  molecules, 250–251
  standard temperature and pressure, 246
Gay-Lussac's law, 244, 246, 257
Graham's law of effusion, 248–249, 256
Graphite, 215
Greatest common divisor, 46
Groups, 157, 168
Gypsum ($CaSO_4 \cdot 2H_2O$), 12, 82

H

Halogens, 161, 168, 175–177
Heat, 93, 108
Heat capacity, 98–99
Heat curve, phase changes of water, 101

Heat decomposition, 3
    of mercuric oxide (HgO), 12
Heat state functions, 96
Heisenberg uncertainty principle, 129–130, 140
Hess's law, 104–105, 107, 110, 111, 117
Heterogeneous mixture, 2, 12, 13
Heteronuclear molecules, 45, 46, 49, 55
Homogeneous mixtures, 2, 12, 70
Homonuclear molecules, 45, 49, 55, 190
Horizontal rows of table, 157
Hund's rule, 134, 189
Hybrid orbitals *vs.* molecular orbitals, 196
Hydrides, 167
Hydrocarbon compounds, 238
Hydrogen, 128
Hydrogen atom, 140
    line spectra of, 125–128
Hydrogen bond, 221–222, 236
Hydrogen cyanide (HCN), Lewis structure and
        molecular geometry of, 210
Hydrogenic species, 128, 150, 154
Hydrogen selenide, 237

I

Ideal gas, 250–251
Ideal gas law, 241, 244–246, 249, 255, 257, 258
Inexact number, 22
Intensive physical property, 4
Intensive properties of matter, 4, 7, 12
Intermolecular forces, 217, 228–229, 236
Intermolecular interactions
        hydrogen bond, 221–222
        ion–dipole interaction, 222
        ion-induced dipole interaction, 223
        van der Waals forces, 219–221
Internal energy, 96, 108–109, 116
International System (SI) unit, 38
International Union of Pure and Applied Chemistry
        (IUPAC), 251, 255
Intramolecular forces, 217, 228
Ion–dipole forces, 236, 237
Ion–dipole interaction, 222
Ionic bond, 179, 197
Ionic compounds, 47–48, 55, 121
        chemical formula of, 48
        Lewis structures of, 184
Ionic hydrides, 167
Ion-induced dipole interaction, 223
Ionization energy, 164–165, 169, 177, 178
Ions, 47, 49, 68
        electronic configuration of, 135–136
        number of, 60–61
Iron (II) nitrate $Fe(NO_3)_2$, 13
Isochore transformations, 95
Isoelectronic configuration, 137
Isotopes, 45, 49, 55
IUPAC, *see* International Union of Pure and Applied
        Chemistry (IUPAC)

K

Keesom forces, 219, 221, 223, 226, 229, 236, 237
Kinetic energy, 96
Kinetic molecular theory of gases, 241–242, 255

L

Lanthanides, 159, 161, 169
Latent heat, 99–102

Lateral overlaps, 214, 215
Lattice energy, 107–108, 111
Law of conservation of mass, 67
Law of definite proportions, 62
Law of multiple proportions, 64
LCAO MO bonding theory, 189, 190, 192
Length, 38
Lewis definition, 167
Lewis dot representations of atoms, 180–182, 197
Lewis structure; *see also* Covalent bonds
        of beryllium chloride ($BeCl_2$), 210
        of carbon dioxide (CO_2), 198, 210
        of difluorochlorate (ClF_2), 211–212
        and formal charges, 185–186, 207–208
        of hydrogen cyanide (HCN), 210
        of iodine pentafluoride (IF_5), 212
        of ionic compounds, 184
        of methanal (CH_2O), 210–211
        of methane (CH_4), 211
        of oxygen (O_2), 189, 198
        of oxygen difluoride (OF_2), 211
        of ozone (O_3), 206, 210–211
        of phosphorus pentachloride (PCl_5), 211–212
        of phosphorus trifluoride (PF_3), 211
        of resonance formulas, 186
        of rules for drawing, 182–184
        of sulfur hexafluoride (SF_6), 212
        of sulfur tetrafluoride (SF_4), 211–212
        of tellurium tetrachloride ($TeCl_4$), 206–207
        of tetrafluoroborate ion (BrF_4^-), 207
        of water (H_2O), 206
        of xenon difluoride (XeF_2), 211–212
        of xenon tetrafluoride (XeF_4), 212
Light
        electromagnetic radiation of, 123
        wave-particle duality of, 124
Limonite, 85
Linear combination of atomic orbitals–molecular
        orbitals (LCAO MO) bonding theory, 189
Line spectra of hydrogen atom, 125–128
Liquids, 1
London dispersion forces, 220, 236, 237

M

Magnesium, 152
Magnetic quantum number, 131–132
Mass number, 44, 49
Mass percent, 38, 74
        of atom, 55
        of electron, 54
        of an element in a compound, 61–62, 64
        of hydrochloric acid, 91
        of neutron, 44
        of proton, 44, 54
        of substance, 69
Matter
        chemical change of, 5
        chemistry and, 1
        physical change of, 5–6
        properties of, 4–5
        states of, 1
        types of, 1–3
Measured numbers, 22
        addition and subtraction, 24–25
        exact numbers, 25
        multiplication and division, 25
        scientific notation, 26
Measurements, 15–18, 27
        accuracy and precision, 27

calculations with, 24–27
  of heat, 116
  physical properties and units, 18–22
  uncertainty in, 22–24
Mercuric oxide HgO, 12
Metallic bond, 180, 197
Metalloids, 159–161, 168
Metals, 159–161, 168
Methane (CH_4) molecule, tetrahedral molecular
    geometry of, 192
Millikan, Robert A., 43, 49
Mixture, 2
  of elements *vs.* compound, 3
  of isotopes, 12
  separating compounds and, 3
Molality, 71–72, 75, 84
Molarity, 71, 75, 84
Molar mass, 59, 82–87
  vs. molecular mass and atomic mass, 59
Molecular forces, 217, 223
  dipole moment and polarizability, 217–218
  intermolecular interactions, 219–223
Molecular formula, 46, 49, 65–67, 87
Molecular geometries, 187–189
  basic geometries and derivatives, 187
  of beryllium chloride ($BeCl_2$), 210
  of carbon dioxide (CO_2), 210
  of difluorochlorate (ClF_2), 211–212
  electron groups and molecule notation, 186–187
  of hydrogen cyanide (HCN), 210
  of iodine pentafluoride (IF_5), 212
  of methanal (CH_2O), 210–211
  of methane (CH_4), 211
  of oxygen difluoride (OF_2), 211
  of ozone (O_3), 210–211
  of phosphorus pentachloride (PCl_5), 211–212
  of phosphorus trifluoride (PF_3), 211
  of sulfur hexafluoride (SF_6), 212
  of sulfur tetrafluoride (SF_4), 211–212
  VSEPR theory, 187–189
  of XeF_4, 212
  of xenon difluoride (XeF_2), 211–212
  of xenon tetrafluoride (XeF_4), 211–212
Molecular mass (formula mass), 58, 73, 82
  vs. atomic mass and molar mass, 59
Molecular orbital theory, 189–192
  hybrid orbitals *vs.*, 196
Molecules, 2, 45, 48, 49, 55
  notation of, 186–187
  number of, 60–61
Moles, 48, 50, 73, 85–87, 89, 110
  number of, 60–62
  percent of, 70–71
  ratio of, 70–71
Monatomic elements, 74
Multiple proportions, law of, 64
Multiplication operation, 26, 29

N

Neutral atom, 44, 45, 47, 49
Neutralization reaction, 5
Neutrons, 43–45
Nitrogen dioxide (NO_2)
  Lewis structure of, 187
  resonance structures of, 210
Nitrogen (N_2), molecular orbital diagram for, 191
Noble gases, 161, 169, 176, 177
Non-metals, 159–161, 168
Non-polar covalent bonds, 179
Non-polar molecules, 217, 229, 236

O

O^{2-} anion, Lewis dot diagram of, 181
O atom, Lewis dot diagram of, 181
Octahedral, SF_6 molecule, 195
Octet rule, 182, 197
Open system, 94
Orbital hybridization theory, 199
  hybridized dsp^2 orbitals, 195–196
  hybridized sp orbitals, 193–194
  hybridized sp^2 orbitals, 192–193
  hybridized sp^3 orbitals, 192
  hybridized sp^3d orbitals, 194
  hybridized sp^3d^2 orbitals, 195
  hybridized sp^3d^3 orbitals, 195
Orbitals, electrons in, 136
Oxidation–reduction (Redox) reactions, 5
Oxides, 167
Oxygen (O_2), 12
  atom of, 152
  electronic configuration of, 213
  Lewis structure of, 189, 198
  molecular orbital diagram for, 191
Ozone (O_3)
  Lewis structure of, 206, 210
  molecular geometry of, 210

P

Partial pressures, Dalton's law of, 249–250
Pauli's exclusion principle, 135, 189
Pentagonal bipyramid, IF_7 molecule, 195
Periodic table of elements, 157–159, 167
Periods, 157, 168
Phase changes, 98, 109, 116–117
  enthalpy changes during, 99–102
  of matter, 5
Photoelectric effect, 124–125, 140
Photons, 140, 149
Physical change, 6, 7
  of matter, 5–7
Physical properties and units, 7
  density and specific gravity, 21–22
  of matter, 4, 7
  of metalloids, 160
  of metals, 160
  of non-metals, 160
  temperature of, 20–21
  volume of, 18–19
Physical quantities, 15
Planar molecular geometry of BF_3 molecule, 193
Planck's constant, 124, 129, 140
Polar bonds, 179
Polarizability, 217–218, 229
Polar molecules, 218, 229, 236
Polyatomic molecule, 229
Potassium oxide (K_2O), Lewis structure of, 184
Potential energy, 96
Precipitation reaction, 5
Precision, measured numbers, 27, 28
Pressure, 38
Principal quantum number, 130–131
Principles of calorimetry, 97
Products, 74
Properties of atoms, 169–170
  acidic and basic trends, of oxides and hydrides,
      166–167
  atomic size, 162–164
  effective nuclear charge and shielding effect, 162
  electron affinity, 165
  electronegativity, 165–166

ionization energy, 164–165
Properties of matter, 229
    boiling point, 226–227
    factors affecting, 238
    surface tension, 227–228
    vapor pressure, 223–226
    viscosity, 227
Protons, 43–45, 49
Proust's law, 62
$PtCl_4^{2-}$ ion, square planar, 196

Q

Quantity, 15, 69
Quantum numbers
    angular momentum quantum number, 131
    electron spin quantum number, 132
    magnetic quantum number, 131–132
    principal quantum number, 130–131
Quantum theory
    atomic orbital, 130
    de Broglie hypothesis, 129
    electromagnetic radiation of light, 123
    electronic configuration, 133–139
    Heisenberg uncertainty principle, 129–130
    line spectra of hydrogen atom, 125–128
    photoelectric effect, 124–125
    quantum numbers, 130–133
    wave-particle duality of light, 124

R

Raoult's law, 224, 225–226, 229
Ratio, of physical and extensive property, 5
Reactants, 74
Reaction, enthalpy of, 102–103
Relative density of substance, 21
Resonance formulas, 186, 199
Resonance structures, 186
    of nitrogen dioxide (NO_2), 210
    of sulfur trioxide (SO_3), 186
Rounding-off numbers, rules for, 23–24, 28
Rutherford, E., 43, 49
Rydberg constant, 149
Rydberg formula, 127, 128

S

Sand, 13
Scalar quantities, 16, 28
Scientific notation, 26, 29
Seawater, 12
Second ionization energy, 164
SF, *see* Significant figures (SF)
Shielding effect, 162, 169
Short-cut electronic configuration, of element, 138–139
SI derived units, 16, 28
Significant figures (SF), 22–23
    in calculation, 29
    rules for, 28
Simplest formula, 46
Sodium (Na) atom, Lewis dot diagram of, 181
Sodium chloride (NaCl), Lewis structure of, 184
Sodium hydroxide (NaOH), 12
Sodium ion (Na^+) cation, Lewis dot diagram of, 181
Solids, 1
Solubility, 2, 7, 13, 72, 75
Solute, 2, 75
Solutions, 2, 7, 70–73, 75
Solvent, 2, 75

Specific gravity of substance, 21
Specific heat capacity, 98–99
Spectral line series, of hydrogen atom, 127, 128
Speed of light, 123
Spin quantum number, 151
Standard enthalpy
    of formation, 103–104, 117
    of reaction, 110
Standard temperature and pressure (STP), 246, 251
State function, 108, 116
States of matter, 1
Stoichiometry, 68–69, 75
STP, *see* Standard temperature and pressure (STP)
Structural formula, 46, 49, 55
Sublevels, 132–133
Subscripts, 67
Substances, 1–2
Subtraction, 29
Sugar, 12
Sulfate ($SO_4$2-)
    Lewis structure of, 185, 186
    trial structure of, 185
Sulfur dioxide (SO_2)
    formal charges of, 208
    Lewis structure of, 184, 208
Sulfur trioxide (SO_3), resonance structures of, 186
Surface tension, 227–228, 230, 238
Surroundings, 93–94, 108
Symbol, of an element, 44–45, 49
System, 93–94, 108

T

Tellurium tetrachloride ($TeCl_4$), Lewis structure of, 206–207
Temperature, 38
    conversion of, 28
    remains constant, 109
    SI unit of, 20–21
    values, 20
Tetrahedral molecular geometry of CH_4 molecule, 192
Thermochemistry
    bond enthalpy, 105–106
    calorimetry, 97–99
    energy exchange, 94–95
    energy, heat and work, 93
    enthalpy, 99–105
    internal energy, 96–97
    lattice energy and the Born–Haber cycle, 107–108
    system and surroundings, 93–94
Thermodynamics, 2
Third ionization energy, 164
Thomson's plum pudding model, 43
Time, 38
Transition metal cations, electronic configuration of, 136–137
Transition metals, 161, 168, 176, 177
    electronic configuration of, 136–137
Trial structure
    of ammonia (NH_3), 182
    of sulfate ($SO_4$2-), 185
Triangular bipyramid, 194
Tridimensional structure of diamond, 214

U

Uncertainty in measurements, 22–24
Unit conversion, 17–18
Universe, 108

V

Valence shell electron pair repulsion (VSEPR) theory,
187–189
Valence shell electrons, 137–138, 158–159, 176, 192–194,
196
van der Waals forces, 219–221
factors affecting, 237
Vaporization, 100
Vapor pressure, 223–226, 229–230, 238
Vector quantities, 16, 28
Vertical columns of table, 157
Viscosity, 227, 230
Volatiles, 224
Volatile substances, 229
Volume, 38
extensive physical property, 18–19

VSEPR theory, *see* Valence shell electron pair repulsion
(VSEPR) theory

W

Water (H$_2$O), 12, 238–239
Lewis structure of, 206
Wavelength, 123, 139
Wave-particle duality
of light, 124, 140, 149
of matter, 129, 149
Work, 93, 96, 108

Z

Zeeman effect, 128
Zinc (Zn), 11